Automotive Suspension, Steering, and Brakes

Herbert E. Ellinger

Associate Professor
Transportation Technology Department
Western Michigan University

and

Richard B. Hathaway

Instructor
Transportation Technology Department
Western Michigan University

Automotive Suspension, Steering, and Brakes

Prentice-Hall, Inc.
Englewood Cliffs, N.J.

Library of Congress Cataloging in Publication Data

Ellinger, Herbert E.
   Automotive suspension, steering, and brakes.

   Includes index.
   1. Automobiles—Springs and suspension—Maintenance and repair. 2. Automobiles—Steering-gear—Maintenance and repair. 3. Automobiles—Brakes—Maintenance and repair. I. Hathaway, Richard B., joint author.
II. Title.
TL257.E44   629.2'4   78-32045
ISBN 0-13-054288-1

Editorial/production supervision by Steven Bobker
Interior design by Phyllis Springmeyer
Page layout by Martin Behan
Cover design by George Alon Jaediker
Manufacturing buyer: Gordon Osbourne

© 1980 by Prentice-Hall, Inc., Englewood Cliffs, N.J. 07632

All rights reserved. No part of this book
may be reproduced in any form or
by any means without permission in writing
from the publisher.

10 9 8 7 6 5 4 3 2 1

Printed in the United States of America

Prentice-Hall International, Inc., *London*
Prentice-Hall of Australia Pty. Limited, *Sydney*
Prentice-Hall of Canada, Ltd., *Toronto*
Prentice-Hall of India Private Limited, *New Delhi*
Prentice-Hall of Japan, Inc., *Tokyo*
Prentice-Hall of Southeast Asia Pte. Ltd., *Singapore*
Whitehall Books Limited, *Wellington, New Zealand*

Contents

Preface ix

1 Introduction to the Automobile   1

    1-1   Body and Frame   1
    1-2   Engines   2
    1-3   Drive Line   3
    1-4   Running Gear   4
    1-5   Automotive Service   5

2 Routine Maintenance   7

    2-1   Safety in the Service Area   8
    2-2   Lifting and Supporting Vehicles   10
    2-3   Lubricants   13
    2-4   Chassis Lubrication   15
    2-5   Wheel Bearing Maintenance   17
    2-6   Lug Bolt Replacement   22
    2-7   Towing the Vehicle   23

Contents

3 Tires and Wheels   25

    3-1   Tire Requirements   25
    3-2   Tire Construction   27
    3-3   Tire Performance   30
    3-4   Replacement of Tires and Wheels   34
    3-5   Tire and Wheel Inspection   37

4 Tire and Wheel Service   42

    4-1   Tire Service   42
    4-2   Wheel Service   46
    4-3   Balance   47
    4-4   Tire Grinding   51
    4-5   Tire Repair   52

5 Axles, Bearings, and Housings   57

    5-1   Bearings   58
    5-2   Bearing Support   60
    5-3   Seals   63
    5-4   Problem Diagnosis   64
    5-5   Wheel Bearing Service   66
    5-6   Drive Axle Service   68

6 Fundamentals of Brakes   76

    6-1   Braking Requirements   77
    6-2   Automotive Brake Types   78
    6-3   Brake Lining   85
    6-4   Parking Brakes   87

7 Brake Operating Systems   91

    7-1   Hydraulic Principles   91
    7-2   The Brake System   93
    7-3   Master Cylinder   94
    7-4   Wheel Cylinders   98
    7-5   Hydraulic Valves and Their Function   102
    7-6   Brake Lines   107
    7-7   Brake Fluid   109
    7-8   Wheel Slip Brake Control Systems   110

8 Brake Inspection   115

    8-1   Brake System Inspection   115
    8-2   Brake Drum Removal   116
    8-3   Lining Inspection   117
    8-4   Wheel Cylinder Inspection   119
    8-5   Brake Line Inspection   120
    8-6   Brake Drum Inspection   120
    8-7   Disc Brake Inspection   123
    8-8   Summary   125

9 Drum Brake Service   126

    9-1   Brake Drum Resurfacing   126
    9-2   Brake Disassembly   129
    9-3   Wheel Cylinder Service   130
    9-4   Lining Selection   133
    9-5   Brake Shoe Grinding   134
    9-6   Backing Plate Service   136
    9-7   Parking Brake Cable Service   137
    9-8   Shoe Installation   138
    9-9   Brake Shoe Adjustment   140
    9-10   Parking Brake Adjustment   142

10 Disc Brake Service   144

    10-1   Disassembly   144
    10-2   Rotor Resurfacing   147
    10-3   Caliper Service   149
    10-4   Reassembly   151

11 Hydraulic System Service   153

    11-1   Hydraulic Control Valve Testing   153
    11-2   Master Cylinder Service   156
    11-3   Brake Bleeding   160
    11-4   Methods of Bleeding   161
    11-5   Brake Warning Light   163

12 Power Brake Systems   166

    12-1   Theory of Operation   166
    12-2   Vacuum Power Brake Operation   169
    12-3   Hydraulic Power Booster   172
    12-4   Dual Power System   176
    12-5   Power Brake Service   175
    12-6   Power Brake Removal and Replacement   178

13 Suspension Systems   181

    13-1   Suspension Types   182
    13-2   Sprung and Unsprung Weight   183
    13-3   Spring Requirements   184
    13-4   Spring Types   185
    13-5   Suspension Control Devices   188
    13-6   Shock Absorbers   199
    13-7   Steering Pivots   202
    13-8   Level Control Systems   204

14 Suspension Control   209

    14-1 Suspension Characteristics   210
    14-2 Steering Linkage Characteristics   215
    14-3 Ride Height and Handling   216
    14-4 Vehicle Steer   217
    14-5 Ride Quality   220

15 Steering and Wheel Alignment   223

    15-1 Steering System Geometry   224
    15-2 Steering Linkages   231
    15-3 Wheel Alignment   233
    15-4 Problem Diagnosis   242

16 Front Suspension Service   244

    16-1 System Inspection   245
    16-2 Front Suspension Service   248
    16-3 Steering Linkage Service   259

17 Rear Suspension Service   264

    17-1 Shock Absorbers   264
    17-2 Stabilizer Bar   265
    17-3 Track Bar   266
    17-4 Spring Replacement   267
    17-4 Control Arm Service   270
    17-6 Rear Suspension Alignment   272

18 Steering Gears   277

    18-1 Standard Steering Gear   278
    18-2 Power Steering Gear   280
    18-3 Power Steering Pumps   285
    18-4 Power Steering Service   288

    Glossary   293

    Index   299

Preface

In automotive engineering courses, the engine with its operating accessories is generally the part of the automobile that is given the most attention. The drive line and chassis are usually the last to be studied. Because of this, engine performance has been the most important part of automechanics study. However, as a result of federal regulations emissions, economy, and safety have come to the forefront in the 1970's. Of these three, safety is most greatly affected by the chassis. Suspension, steering, and brakes all play a dominant role in safety.

This book is written to help the reader understand how vehicle suspension, steering, and brakes contribute to vehicle safety. It further provides the reader with enough information and guidance to repair and service these three component systems of the vehicle. The requirements of each of the systems are considered first.

Preface

This is followed by a discussion of the designs that are necessary to meet these requirements. Finally the proper service of each of the systems is described.

It is often necessary to disassemble the steering system when working on the suspension. When the front suspension is reassembled after service, the suspension and steering are generally realigned. Likewise it is often necessary to disassemble the brakes when working on either the steering or suspension. Wheels and hubs must be removed to work on the brakes, bearings, and grease seals. This book covers these items in the order following a normal servicing sequence, so the reader will have previously studied all components that must be dealt with while each system is being serviced.

The material is presented in a manner that will provide the reader with a knowledge of useful, common routine service operations in the early chapters. These routine operations are encompassed in the more complex service operations described in the later chapters. Routine service can usually be done with basic tools. More complex service operations require the use of special tools and shop equipment.

Automotive enthusiasts usually overlook the fact that all vehicle control passes through the tire-to-road contact area. This book discusses the loads on the tire and how these loads react through the suspension, steering, and brakes to give the driver positive control of the vehicle. The importance of the brakes can be seen when the reader realizes the brakes are required to provide as much as three times the power produced by the engine, when fully applied. The importance of the tires cannot be overstressed. They are basic to all ride and handling characteristics of the vehicle. Replacing tires with new ones having the desired properties is often the best solution to a ride or handling complaint.

Brakes are a major contributor to vehicle safety and control. Their service usually includes all the routine service maintenance operations. In a discussion of brakes, it is assumed that the technician will do the complete brake job and do it properly. This is the only way the technician can be sure customers will have safe brakes on their vehicles.

The discussion of brakes is followed by a discussion of the interrelationships of the suspension system and the steering system. Alignment is covered next, followed by service of the front and rear suspensions. The book concludes with a discussion of the steering gear, both standard and power types.

Generally, a discussion of specific make and model automotive component disassembly an reassembly has been avoided. Obsolete features are purposely omitted unless they are required to show the development of a current design feature. Experimental and limited production designs have also been avoided, except where high interest level dictates a need for a short discussion. The text deals specifically with the automotive products that a student will most likely work with in the field.

The authors wish to express their thanks to the large number of individuals who have provided material used in this book and those who have constructively criticized the original material. Insofar as possible, their recommendations have been incorporated in this book. The unresticted use of the automotive laboratories, automotive equipment, and training aids at Western Michigan University for preparation of the photographs is especially appreciated.

A NOTE TO THE STUDENT AND INSTRUCTOR ABOUT THE USE OF METRICS IN THIS BOOK

The metric system of measurement is starting to be used in domestic manufacturing. Cars imported into this country use metric standards. Manufacturers of domestic cars are switching over. Both customary and metric units of measurement are used throughout this book. The only exceptions are several complicated examples where the metric equivalents were omitted so that the examples would not be confusing.

Using metric units is no harder than using customary units. It is simply a matter of getting used to using them. When customary units used in this text were converted to metric units they were kept at the same level of precision. If the customary unit was "approximately 1 to 2 inches" the commonsense metric conversion would be "approximately 25 to 50 mm" or simply "approximately 1 to 2 inches (25-50 mm)." Of course, the exact conversion values are 25.4 mm to 50.8 mm but the customary unit used wasn't exact to begin with so measurements to tenths of millimeters are not necessary. Where high precision is called for conversion is made to high precision metric values.

It is important to keep in mind how big the basic length metric unit is; there are roughly 25 mm to 1 inch; thus, 0.100 inch = 2½ mm (2.5 mm) and 0.010 inch = ¼ mm (0.25 mm). If the specifications are given in hundreth of an inch, the specification of tenths of a milimeter is actually more exact.

Don't be alarmed if you try to convert a customary unit to a metric unit and find the value given in the text is somewhat different than your results. In the text the metric values were rounded to a commonsense value, whenever doing so would not affect the operation or description being discussed. If the results you get in making the conversion differ from that shown by more than 5 percent, recheck your calculations. Once you become accustomed to the metric system, you'll probably find it easier to use than customary units. The important thing is to practice using metrics.

ACKNOWLEDGEMENTS

The authors wish to thank the following organizations for their assistance.

American Motors Corporation
Auto Parts Distributors
The Bendix Corporation
Chrysler Corporation
The Firestone Tire & Rubber Company
General Motors Corporation
   Cadillac Motor Car Division
   Chevrolet Motor Division
   Buick Motor Division
   Oldsmobile Division
Kelsey-Hays Company
Monroe Auto Equipment Company
TRW Replacement Division
Uniroyal Incorporated

Herbert E. Ellinger
Richard B. Hathaway

Introduction to the Automobile

Vehicle design must start with occupant seating. No matter what the designer wishes, the vehicle must have room for the driver, passengers, and load. The vehicle must have enough engine power to move the load at the desired speed and enough braking power to safely stop. While moving, the vehicle must be controllable, so that it can be driven to follow changes in the roadway and in traffic conditions.

1-1 BODY AND FRAME

The occupants expect a comfortable ride to minimize fatigue. The seats must be firm with adequate support. Driver visability must be excellent in all directions. Noise level and vibration should be as low as possible.

Several different construction methods used by manufacturers in their vehicles are shown in Figure 1-1. Of these methods, separate body and frame construction has been used for the longest time. In this type of construction, the engine, drive line, and running gear are firmly fastened to the frame. The body is then mounted to the frame. Insulators are used between the body and frame to keep the noise and vibration from entering the body.

A second type of construction is the unitized body. In this type of construction the frame is welded into the body as a part of the structure. Body panels add strength to the frame pieces. The running gear and drive line are attached to the

Introduction to the Automobile

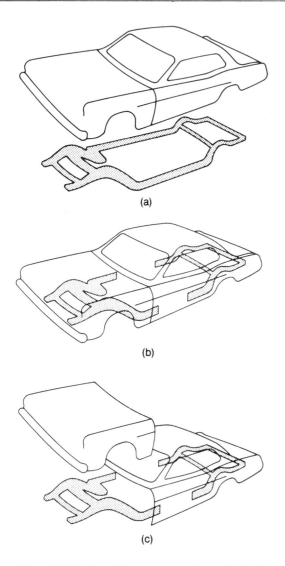

Figure 1-1 Body and frame construction methods: (a) separate body and frame; (b) unitized construction; (c) unitized body with stub front frame.

unitized body with large soft insulators. These are also used to minimize noise and vibration. If the insulators are too soft they will allow excessive movement of the running gear and drive line. In turn, this will adversely affect handling and control. If they are too hard, they cannot adequately insulate the noise and vibration. The manufacturer carefully designs the insulators and selects locations that will produce an automobile that is satisfactory to drive and ride in. Because insulator properties change with age, this will affect the original design characteristics.

A third type of construction combines features from both of the preceding types. It uses a stub frame from the fire wall forward and a unitized body from the fire wall back. The unitized portion is very rigid, while the stub frame provides an opportunity for good insulation.

Manufacturers select their construction method by deciding which type is most economical for them to build, while still providing the noise, vibration, and ride characteristics they want. Large and mid-size automobiles generally use separate body and frame construction. The majority of intermediate and smaller cars use unitized construction.

1-2 ENGINES

Most automobiles use a gasoline-fueled reciprocating engine mounted ahead of the occupants. This location provides the most room for the occupants and load, as well as being a safety factor in a head-on collision. It also allows placement of the cooling radiator at the front of the vehicle using minimum ducting and hoses. Engines have been placed behind the rear axle and between the ocupants and the rear axle. Neither of these arrangements has extensive application in modern automobiles.

Some types of automobiles use a different engine. The diesel engine has been used in taxicab fleets. Because it uses less fuel than the gasoline engine, the diesel is gaining acceptance as an engine for personal automobiles. The diesel engine has been used for a number of years in Europe where the cost of fuel is much greater than in the United States.

The turbocharger has been adapted to domestic passenger cars. An engine used with a turbocharger is small relative to the weight of the vehicle. This gives good gas mileage. The turbocharger comes into operation at full throttle to give the small engine performance equal to that of a larger engine. This gives an automobile with the turbocharger both good cruising economy and good acceleration performance.

A rotating combustion chamber engine has had some popularity in Europe and Japan. These engines are very small and light in weight for the power they produce. They also run very smoothly. Continued development may increase their use in small automobiles where weight and size are critical.

Turbine engines show some promise, especially for commercial vehicles. They are powerful, lightweight, and have low hydrocarbon and carbon monoxide emissions. Turbine engines are ideally suited to replace some diesel engines in over-the-road load-carrying vehicles.

1-3 DRIVE LINE

The drive line carries power to the drive wheels as shown in Figure 1-2. A clutch or torque converter is connected to the engine crankshaft. Either one provides a means to effectively disconnect the engine from the drive line. This is done so the engine can idle while the automobile is stopped. Power to drive the automobile goes through the clutch or torque converter.

A transmission is located directly behind the clutch or torque converter. Its function is to control the ratio between the crankshaft and driveshaft by means of gear selection. It will produce high torque to start the automobile moving and drive it up steep grades. The transmission also provides a reverse gear for backing the automobile. Gear range selection may be either manual or automatic.

In front-engine, rear-wheel drive vehicles, the transmission is located under the front floor of the passenger compartment. A propeller or drive shaft is used to carry engine power to the rear axle. It has a universal joint on each end to allow power flow through the changing drive line angles as the suspension moves.

Figure 1-2 Common engine and drive lines: (a) front engine with rear drive wheels; (b) front engine with front drive wheels; (c) rear engine with rear drive wheels.

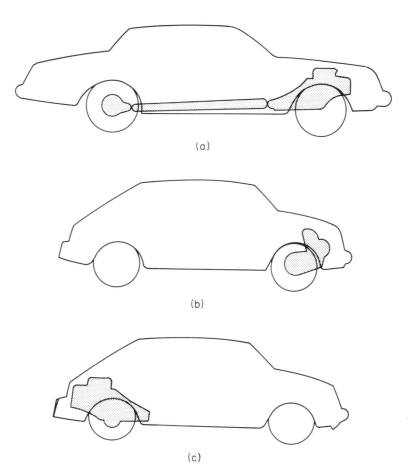

Introduction to the Automobile

A differential in the rear axle splits the incoming power to each drive wheel. It also allows the drive wheels to turn at different speeds as they go over bumps and around corners.

Automobiles with front-wheel drive, midship, or rear engines often combine the transmission and differential. They do not use a propeller shaft. Universal joints are required on each end of short axle shafts between the differential and the drive wheels because these wheels drive during suspension movement.

1-4 RUNNING GEAR

The running gear consists of the suspension components, brakes, wheels, and tires. This book covers the design, operation, and service of the running gear.

The suspension includes the springs, shock absorbers, and linkages or arms. It must be strong enough to support the vehicle body and load. Depending on design, the suspension may resist axle housing twist from the engine power and brake reaction. One of the main functions of the suspension is to keep the tires in contact with the road as much of the time as possible. This is done while supporting the automobile even when travelling over rough roads. It is necessary because the four *tire footprints* are the only place the vehicle touches the road surface. All of the engine power, steering, and braking forces operate through these tire-to-road footprint areas. Control of the vehicle is reduced or lost any time a tire does not contact the road or when skidding begins.

The vehicle body is supported by springs. They can be coil springs, leaf springs, or torsion bars, all shown in Figure 1-3. Coil springs are the most popular. Both coil springs and torsion bars require links and arms to hold the wheel in position. Leaf springs hold the axle in position without any additional devices.

The front wheels of rear-wheel drive vehicles rotate on a spindle which is part of the steering knuckle. The knuckle is attached to the front

Figure 1-3 Suspension spring types (Courtesy of Monroe Auto Equipment Company).

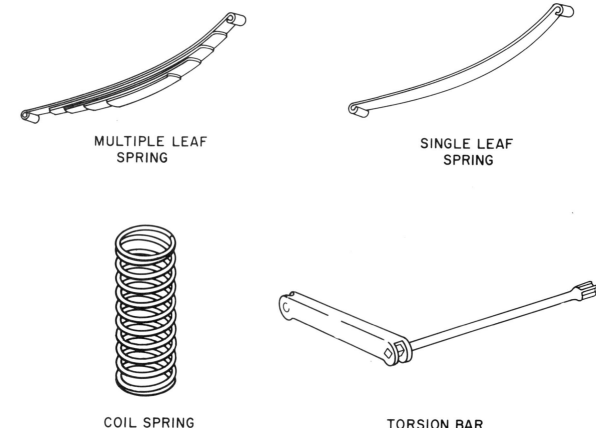

MULTIPLE LEAF SPRING

SINGLE LEAF SPRING

COIL SPRING

TORSION BAR

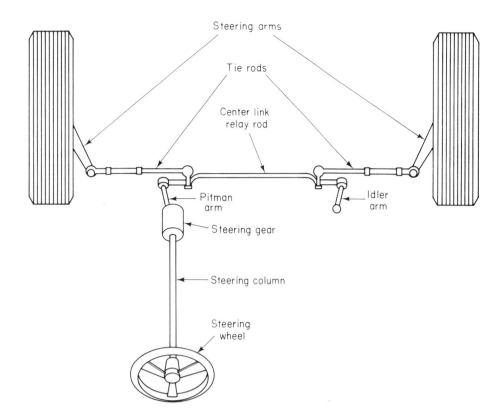

Figure 1-4 Typical steering linkage.

suspension members with ball joints. The ball joints allow for steering as the suspension flexes. A steering wheel controls a steering gear assembly. This, in turn, moves the steering knuckle through the linkage. A typical steering linkage can be seen in Figure 1-4. On some vehicles the steering linkage is in front of the suspension while on others it is behind the suspension.

Brakes are mounted on the steering knuckle in the front suspension and on the axle housing in the rear suspension. The brake drum or disc rotates with the wheel. Hydraulic force from the master cylinder pushes brake shoes against the cast iron braking surface with enough force to slow wheel rotation. This slows the vehicle as it turns motion energy into heat energy.

1-5 AUTOMOTIVE SERVICE

Three things are needed for proper automotive service. First, the technician must have the tools to make the repair. Second, the parts must be available. Third, the technician has to know what is to be done and how to do it. The following chapters are designed to help you master the knowledge and skill necessary to service the suspension, steering, and brake systems.

A vehicle is serviced only under three conditions. First, when preventive maintenance is due. Second, when the owner wants a change in performance, appearance, or comfort items. Third, when there is a specific problem to be corrected. Preventive maintenance includes service that is done at regular intervals. This service is done to extend vehicle life and to prevent problems on the road. Preventive maintenance includes lubrication, brake inspection, tire rotation, etc. It is routine when done according to standard operating procedures.

Changes in economy, performance, appearance, and comfort items usually involve installing a manufactured item such as power brakes, air-leveling shock absorbers, or special tires. In many cases, these items are available in a kit. The kit includes all of the necessary parts and detailed installation instructions. Items may need to be replaced as well as added to give a performance change. This type of service could include a change in ring and pinion gears, springs, or wheels. Here

Introduction to the Automobile

again, the service procedures are specified. The technician knows exactly what the customer wants.

Most service work involves a repair or replacement of the part causing the customer problems. This kind of work should start with a clear understanding of what the customer is concerned about. The customer wants two things: 1) the vehicle repaired so it will operate correctly and 2) a minimum repair cost.

In general, problems can be corrected by straightforward service procedures. These involve checking the system to identify the problem. The system is then tested to pinpoint the cause. Required repairs are made and the system is rechecked for proper operation. In some cases, the cause of the problem is extremely difficult to pinpoint. Under these circumstances an automotive service technician must have a detailed knowledge of part functions and their interrelationships with the other parts. This, along with the ability to use test equipment, allows the service technician to rapidly and accurately correct the problem.

In future chapters we will consider why the vehicle parts are made as they are, what the parts are required to do, and how they accomplish their required tasks. Repair procedures are also discussed.

REVIEW QUESTIONS

1. Where must vehicle design start? [INTRODUCTION]
2. Name three body and frame construction methods. [1-1] Name one current production automobile that uses each.
3. What are the advantages of placing an engine in the front of a vehicle? [1-2]
4. List the components in the running gear. [1-4]
5. Why are the tire footprints important to vehicle control? [1-4]
6. Name three types of springs used in the vehicle suspension. [1-4]
7. Describe the operation of a brake. [1-4]
8. What three things are needed to properly service a vehicle? [1-5]
9. Under what three conditions is a vehicle serviced? [1-5]
10. What is the first thing an automotive service technician must know before starting to service a vehicle? [1-5]

Routine Maintenance

An operator will get maximum service life from a vehicle when the vehicle is given scheduled periodic maintenance. This service will include lubrication of various moving components, fluid level inspections, fluid and oil changes, as well as an overall visual inspection. Properly performed periodic maintenance and inspection will result in reduced total operating cost to the owner during the useful life of the vehicle. This maintenance will extend the service life of each component, while inspection will reveal worn or otherwise faulty parts before they become dangerous or damage other related components.

The automotive service technician of today must not only be completely familiar with the function of each component but must also be familiar with the correct service procedures including the interrelationship of the various operating components. Periodic lubrication is an important part of preventive maintenance. Lubrication requirements include the selection of the proper lubricant for each component. Improper lubricants can shorten the service life of a component as rapidly as no service at all.

Safety is always of primary importance in the work area. Learning proper service procedures will include the development of safe working habits. Properly developed safety habits are not time-consuming, as some believe. If they are used regularly they will save time. These habits will not

Routine Maintenance

only help the automotive service technician, but they will reflect positively on coworkers as well.

Some unsafe working practices may not only be dangerous to you, but they may be a danger to others working around you. Vehicle owners prefer to have repairs done by one who works safely and is careful in performing repairs. An automotive service technician who is careless about personal safety will generally be careless about repair procedures and the protection of the vehicle being serviced. Vehicle protection in the work area involves fender covers and seat protection, as well as the proper use of safety stands.

2-1 SAFETY IN THE SERVICE AREA

Safety takes many forms in a repair facility, although all are aimed at achieving the same goal: that of preventing harm from occurring to the service technician, the equipment, and others in the service area. Using support stands, as shown in Figure 2-1, protects the service technician while performing under vehicle repairs. Connecting exhaust hoses to running vehicles for removing exhaust from the building is a safety precaution that has an effect on everyone in the building.

Personal Protection. Safety begins with the clothing worn while working. Excessively loose clothing should not be worn. It could become caught in rotating parts of the vehicle or machinery and cause personal injury. Good work shoes should be worn to protect the feet from objects on the floor that could puncture thin rubber soles or fabric shoes. Heavy shoes offer protection from spills and from parts that may be dropped on them. The best foot protection is offered by work shoes with steel reinforcement in the toe area.

Eye protection is of primary importance in the service area. *Always wear safety glasses when you are in the work area.* Although you may not be performing a function that would normally cause eye injury, the person in the next work area may cause a small object to fly toward you. Always purchase the best eye protection possible—it will usually be the most comfortable to wear.

Head protection is recommended whenever work is being done under the vehicle or under the hood. A construction or hard hat will give adequate head protection. Even if not worn at all times, a hard hat should be within reach when it is necessary to go under a vehicle. Anyone who has serviced vehicles on hoists for a period of time will be able to describe the need for these hats to prevent lumps and headaches.

Breathing protection is required when working in an environment where dust is present. For example, in the brake service area the dust that becomes airborne when removing the brake drums, cleaning brake components, and fitting brake shoes contains asbestos. Since asbestos is harmful to health, brake dust should not be blown free with an air hose because it will travel throughout the shop and affect everyone. The best method of reducing airborne asbestos dust is to loosen brake dirt with a brush, scraper, or vacuum cleaner.

Gloves are an item of clothing that are priceless when they are needed. A pair of gloves will always be in a well-equipped technician's tool box. For example, gloves are indispensable when working on an old exhaust system or on other parts of the car where high temperature or rough metal exists.

Figure 2-1 Support stand under a vehicle.

Safety Through Proper Equipment. Safety can even reflect back to the type of tool chosen to perform a service operation. The condition of the tool, its use, its cleanliness, and its maintenance all contribute to safety. All of these factors, some involving personal choice and others involving shop training and management, have an influence on the safety of the worker.

The tool selected should always be the tool that is *best suited* for the job, *not* just the tool that is *available*. Time taken to locate the correct tool will usually result in saving time in completing the job safely. For example, choosing a fitting wrench to remove a brake tube rather than an open-end wrench, will usually prevent damage to the tube nut. Many cuts and bruised knuckles have resulted from the poor choice of a tool. For example, using an adjustable wrench rather than taking time to select the proper-sized wrench is a poor decision. Pliers should never be used to turn fasteners that are intended to be turned with a wrench.

The condition of a tool is just as important as the proper selection of a tool. Screwdrivers with broken blades should be replaced immediately. A screwdriver that has a blade that someone has attempted to grind, as pictured in Figure 2-2, will create more problems than it will solve. It may slip, not only damaging the screw, but also injuring the service technician or possibly marring some part of the vehicle. Screwdrivers are not intended to be pry bars; thus, they are rarely broken when used properly.

Cleanliness of the tools also plays an important part in safety. It is desirable to pick up a clean tool, which provides a good grip, prevents slippage, and reduces the chance of injury. In addition to cleaning, the tool itself should be properly maintained.

Training in the correct use of tools and equipment is important. Service technicians have been injured because they failed to get instruction on the use of service equipment. Instruction on the proper use applies to small hand tools as well as complex equipment.

Chisels are dangerous if they are not properly maintained or if they are not used correctly. Chisel holders, similar to the one shown in Figure 2-3, should be used with a hand chisel. Chisels will not cut items that are harder than the chisel itself. This includes hard steel bearings. When a surface is struck with the chisel, a chip may fly off and cause harm. Eye protection is a *must* when using a chisel. Always use chisels that are properly suited to the task at hand. Do not use chisels that are made for sheetmetal, as shown in Figure 2-4, to cut heavy steel or bolts. Pneumatic chisels require additional

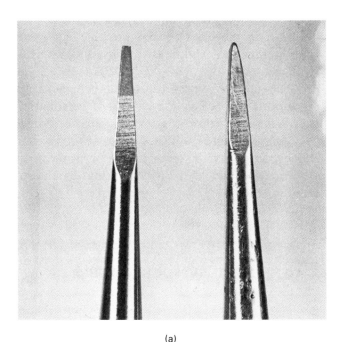
(a)

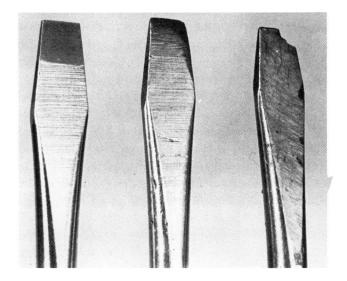
(b)

Figure 2-2 Screwdriver blades: (a) a new blade is on the left and an improperly ground blade is on the right; (b) a new blade is on the left, an improperly ground blade is in the center, and a broken blade is on the right.

Routine Maintenance

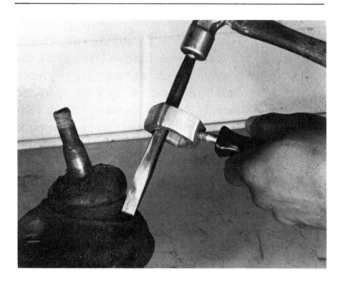

Figure 2-3 Using a chisel holder.

precautions. They must be properly sized for the job and they must be positioned correctly.

Correct use of drills is important to safety in the service area. Always maintain a firm grip on the drill motor to keep it from twisting when the drill bit suddenly breaks through the metal. Whenever possible, use drill handles to increase the holding leverage. Do not use a drill in such a tight location that any movement will pinch your hand. When using a drill press, always clamp the part in a vise or with a C-clamp. Be certain that the vise is securely fastened to the drill press table.

When using a bench grinder, always keep the protective shields in place. Wear eye protection and keep a firm grip on the part being ground. Fingers should be kept clear of the grinding wheel. If the piece is small, clamp it firmly in locking pliers to hold it while grinding. Never grind on the side of the grinding wheel. This can produce unnatural wheel loading that may cause the grinding wheel to fly apart.

The proper use of tools and proper safety precautions will reduce the chance of personal injury. Whenever you use a tool or piece of equipment, ask yourself how it should be used to prevent damage or injury. Then adjust your position or your methods so that safe working conditions result.

2-2 LIFTING AND SUPPORTING VEHICLES

The vehicle must be lifted and supported so the suspension is free when most service work is done on the chassis, steering, and brakes. Whenever the vehicle is raised with a jack for service work it must be supported with *safety stands* before anyone works beneath it. Technicians have been severely injured when a vehicle fell because it was not properly supported.

Lifting devices range from the simple one-post jack that is included with all new vehicles to the heavy-duty, in-floor, twin-post lifts used in the service area. Each is designed for a specific purpose. Jacks should not be used for purposes for which they are not intended.

One-Post Jack. The *one-post bumper jack* provided with new vehicles is designed *only* for the emergency installation of the spare tire. It is unsafe to work under any vehicle that is supported by a bumper jack. In most cases these jacks are designed to lift one corner of the vehicle at a time. They are not designed to lift the entire end of the vehicle and no

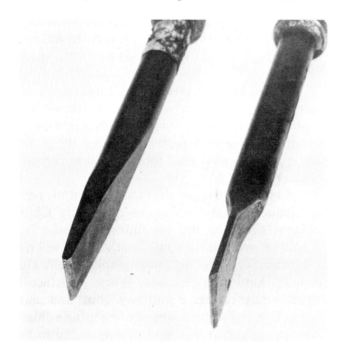

Figure 2-4 Cold chisel on the left and a sheetmetal chisel on the right.

Figure 2-5 One-post emergency bumper jack.

attempt should be made to do so. Jacks that lift at the bumper are only stable when three wheels are on the ground and properly blocked. This type of jack is shown in Figure 2-5.

Hydraulic Jack. The *hydraulic jack,* as shown in Figure 2-6, is used in the service area. They are sometimes carried in the vehicle. The hydraulic jack is placed directly under the part to be lifted. This requires reaching under the vehicle to position and operate the jack. This type of jack is handy due to its small size and its direct lift under the axle or suspension component. The wheel will lift without first lifting the body to the full extended length of the shock absorbers, as a one-post jack does when it lifts from the bumper. The hydraulic jack is stable, but it is still used in conjunction with a safety stand if work is to be performed under the vehicle.

Hydraulic Bumper Jack. A third type of jack, one that is common in service stations, is the pneumatic or hydraulic bumper jack, pictured in Figure 2-7. It is used to lift the entire end of the vehicle until the shock absorbers are fully extended. At this point it will begin to raise the wheels from the floor. The use of this type of jack is not recommended on some vehicles because it produces a high stress that

Figure 2-6 Use of a hydraulic jack.

Figure 2-7 Hydraulic bumper jack.

11

Routine Maintenance

may cause some of the vehicle glass to crack. A hydraulic bumper jack is handy for tire changes. It has a relatively large base to keep the jack from tipping. There is still danger that the vehicle could slip from the jack or that a bumper support could fail. Safety stands are always used in conjunction with this type of jack if work is to be performed under the vehicle. The hydraulic bumper jack is only used on a level floor surface so the vehicle can move toward the jack as it is being raised. This is necessary because the bumper jack does not move on the floor when it is supporting weight.

Hydraulic Floor Jack. Another of the popular service jacks is the hydraulic floor jack shown in Figure 2-8. This type of jack is available for use on all types of vehicles. It comes in different sizes depending on the lifting capacity of the jack. The hydraulic floor jack is on wheels and has a large base to give stable support. The long handle enables it to be properly positioned and operated without having to reach under the vehicle. The wheels on the jack allow it to move toward the vehicle as it is raised. This is necessary because the jack lifting pad moves in an arc upward and toward the jack handle. The hydraulic floor jack can be used to lift a single wheel or the entire front or rear of the vehicle. It can also be used on the side frame rail to lift one corner or an entire side of the vehicle. When lifting from the side, the suspension will fully extend for tire removal. It is important to position the floor jack only at recommended locations that are reinforced lifting points. Even with this heavy jack, safety stands are used to support the vehicle whenever the technician has to go underneath to make repairs.

Safety Stands. Safety stands or jack stands are small supports placed under a raised vehicle as a safety measure. Jack stands, such as shown in Figure 2-1, are available in a number of different designs based on their load-carrying capacity and their extension length. They have large bases and a positive locking adjustment. Safety stands are perfectly adequate to support the vehicle for service after the jack has been removed.

Hoists. In the modern service area, vehicles are most often raised with a *hoist*. This device lifts the vehicle quickly and safely, high enough so the service technician can walk under it. A typical hoist is pictured in Figure 2-9. Different types of hoists lift on the tires, the rear tires and front suspension, rear axle housing and front suspension, or on the frame. There are advantages to each type. The type of service work to be done determines the type of hoist that should be used. Hoists are usually constructed with the lifting posts below the floor, although above-ground hoists are also available.

Below-the-floor hoists have one or two hydraulically operated cylinders. Air is used to pressurize the hydraulic oil. This forces the oil into the hoist cylinders for lifting, in a process called "air over hydraulic operation." Above-ground hoists usually operate electrically. Single-post hoists all have a mechanical safety lock that engages when the hoist is raised to its maximum height. Due to their design, dual-post hoists have no need for a mechanical safety device. Front/back dual-post hoists have one post adjustable so the between post distance will match the wheel base of the vehicle.

Proper positioning of the vehicle on the hoist is of primary importance. This is achieved by positioning the vehicle over the center of the hoist, side to side. When using a frame contact hoist, the front-to-rear balance point or center of gravity is positioned close to the lifting post. Always position the lifting arms of the hoist as far apart as possible

Figure 2-8 Hydraulic floor jack.

Figure 2-9 A twin post hoist supporting an entire automobile.

to offer the greatest stability. The shop manual will indicate the location of the reinforced frame contact points that are to be used with a frame contact hoist. Vehicles placed improperly on a hoist have been known to fall off, doing extensive damage to the vehicle, shop, and equipment!

2-3 LUBRICANTS

Lubrication is the most common preventive maintenance service operation. The lubricant increases the service life of the parts by reducing friction, heat, and wear. In some cases the lubricant prevents contaminants from reaching the lubricated surfaces. There are many different types of lubricants used in vehicles. Each has a specific purpose and provides special characteristics. If they are used in the wrong application or in the wrong quantity they will not provide the required protection.

Common lubricants used in modern vehicles include engine oil, differential lubricant (gear oil), and various greases. Some lubricants are fluids that perform additional functions in the vehicle, such as transferring motion and force in the brakes, power steering, and automatic transmission.

Other special-purpose lubricants are used around the shop. Some examples are: penetrating oils, silicone lubricants, and other specialized products that make disassembly and reassembly of vehicle components easier.

Oils and Fluids. Engine oils are classified using SAE (Society of Automotive Engineers) viscosity number and API (American Petroleum Institute) service classification. The viscosity grade numbers 5W, 10W, 15W, and 20W are rated at 0°F (−18°C) while the grade numbers 20, 30, 40, and 50 are rated at 210°F (99°C). Viscosity is the thickness of the oil at a specific temperature. Viscosity is an engine oil's most important property. Most oils presently being used in passenger vehicles are of a multiviscosity grade, which means they are rated at two different temperatures. An SAE 10W-40 motor oil is an example of a multigrade oil. API service classifications designate the type of use for which the oil is compounded. The SA, SB, SC, SD, and SE service classifications are made for gasoline engines. Modern engines require oil that has an SE service classification. Some old engines can satisfactorily use the other service classifications.

Automatic transmission fluid acts to transfer energy through the transmission as well as to lubricate the transmission parts. Automatic transmission fluid has specified lubrication and frictional properties. These properties are important in relation to the characteristics of the different transmission types. Two types of transmission fluids satisfy the requirements of virtually all automatic transmissions: Type F and Dexron II.

Both are colored red. The primary difference between them is their coefficient of dynamic friction. If they are installed in the wrong transmission or if they are mixed, the shift quality of the transmission will be affected.

Power steering fluid, though similar to automatic transmission fluid, is not the same. It has no coloring added. The two types of fluids should not be used interchangeably. Power steering has different fluid requirements than automatic transmissions. Power steering systems are designed to operate on the fluid specifically compounded for their use.

Brake fluid is a nonpetroleum fluid that is compatible with the rubber and the metal parts of the brake system. Many requirements are placed on brake fluids by DOT (Department of Transportation). These include minimum boiling point, viscosity, and effect of the fluid on rubber components. Brake fluid is hygroscopic, which means it will absorb water. It will pick up moisture from the atmosphere if the container or brake fluid reservoir is left open. Brake fluid (except for silicone types) will damage automobile paint and it is harmful to eyes as well. Brake fluid should always be stored and handled carefully.

Gear oils that are used in differentials and in many manual transmissions are specified by an API-GL (American Petroleum Institute-Gear Lubricants) service classification and an SAE viscosity number. Some manual transmissions use engine oil, while others use automatic transmission fluid. Always consult the appropriate shop manual to determine the type of recommended oil. Do not mix lubricant types. The service classification that meets today's vehicle standards for differentials is the GL-5 classification. The viscosity range that should be used depends upon the anticipated temperatures. In most of the United States either SAE 90 or multigrade SAE 85W-90 gear oil is used. If extreme low temperatures are expected, SAE 75W or multigrade 80W-85W gear oil might be used. Although they are not to be used interchangably, at 210°F (99°C), the viscosity of SAE 90 gear oil is equivalent to SAE 50 engine oil. A limited slip differential requires lubricant with special properties. It has the same GL and SAE numbers plus additives to make it compatible with limited slip differential clutches. Gear oil that is intended for use in limited slip differentials will work satisfactorily in conventional differentials and in manual transmissions, but it is more expensive.

Greases. A lubricating grease is a solid or semisolid consisting of a thickening agent in a liquid lubricant. Lubricating greases are classified by their NLGI (National Lubricating Grease Institute) number, the type of thickener used, and the texture. The NLGI number is a means of comparing the relative hardness or consistency of the grease. Consistency is a lubricating grease's most important property. The type of thickener (aluminum, barium, calcium, sodium, and lithium) and the amount used in the grease determines its general properties. These properties include the melting point (technically called the dropping point), its texture (smooth, fibrous, stringy, or long fiber), and its water resistance. Additives are put into the grease to improve some of its properties. Fillers are also added to grease to help improve their load-carrying capability. Some of the more commonly used fillers are graphite and molybdenum disulfide. The most common grease used in automotive service facilities is lithium. The so-called "moly" grease, is usually lithium grease with molybdenum disulfide used as a filler. Lithium grease has a high dropping point temperature. This helps make it desirable as a wheel bearing lubricant. It also has excellent water resistance, making it suitable for numerous other chassis applications. As a result of its ability to provide satisfactory lubrication for many applications, it is called a *multipurpose grease*. In automotive applications, NLGI #2 multipurpose grease is recommended for wheel bearings and for suspension components. Often the grease that is pumped through the overhead grease systems in the lubrication bay is not NLGI #2. This is most critical for wheel bearing applications where only NLGI #2 grease is recommended. If the NLGI number or the filler type of a grease is not known, or if it is not specifically labeled for wheel bearing use, it should not be used to repack wheel bearings. Grease that is packaged specifically for wheel bearing lubrication may be barium, sodium, or lithium grease. These greases do not intermix; therefore grease should never be added to the used grease remaining in wheel bearings. All of the used grease must be thoroughly cleaned from the bearings before they are repacked with fresh wheel bearing grease.

2-4 CHASSIS LUBRICATION

The chassis lubrication period is based on both a time and a mileage interval. These intervals vary, depending on the make of vehicle and the operating conditions; from one year or 12,000 miles (19,312 km) to three years or 35,000 miles (56,327 km), whichever comes first. When operating under severe conditions, such as dusty environments, stop-and-go driving, or fleet service operations, the vehicle will require lubrication at more frequent intervals. Some vehicle suspension components may be sealed at the factory with adequate lubrication for the expected life of the part. No means is provided for the technician to add lubricant on such parts.

Exact lubrication requirements of the chassis cannot be predicted by the manufacturer. These requirements can only be determined by careful inspection of each component requiring periodic lubrication. The boot or seal around a ball joint or tie rod end will collapse when it is in need of lubrication. The boot having sufficient lubricant will appear slightly swollen or inflated, provided the boot seal is in good condition. Examples of these conditions are shown in Figure 2-10. The ball joint or tie rod end must have enough lubricant to hold the seal slightly swollen to provide an effective seal against moisture and contamination.

Lubrication begins with raising and supporting the vehicle, either with a hoist or on safety stands. While the vehicle is raised for lubrication, the technician has an excellent opportunity to perform other service operations. These include oil changes, a complete inspection of all underbody components, and a fluid level check. A systematic approach to lubrication and inspection should be used. It will speed up the process as well as eliminate any chance of overlooking a lubrication or an inspection point.

(a)

(b)

Figure 2-10 Greasing a tie rod end: (a) collapsed boot seal; (b) inflated boot seal as new grease causes a slight swelling.

Routine Maintenance

Lubrication Procedure. The first step in the lubrication procedure is to consult the lubrication handbook to determine the type of lubricant required and to identify the location and number of lubrication fittings on the vehicle. At each point in the suspension where relative motion occurs, there will be one of the following: a rubber bushing, a sealed joint, or a joint requiring lubrication. Lubrication fittings are usually found at each tie rod end, on both ends of the center steering link, and at the upper and lower ball joints. Other lubrication locations are sometimes found at the universal joints and on the clutch linkage. Each lubrication fitting is wiped clean of accumulated dirt with a shop towel before it is lubricated. Carefully inspect the seal at each suspension joint. Some vehicles do not come from the manufacturer with lubrication fittings installed. They have a small plug located in place of the fitting, as shown in Figure 2-11. The plug is removed and a lubricating fitting temporarily installed to lubricate the joint. The plug is reinstalled after lubrication or the lubricating fitting can be left in place (Figure 2-12). Each fitting is lubricated until the seal starts to balloon or grease begins to seep from around the base of the boot. Always wipe excess grease from

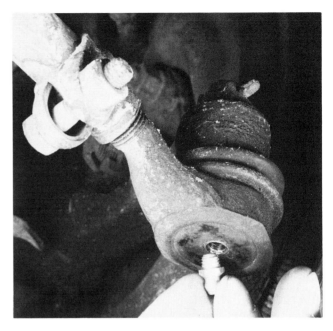

Figure 2-12 Installing a grease fitting in place of the plug.

the area. This procedure is continued until all parts having lubrication fittings have been lubricated. Each time a joint is lubricated the technician should watch for excessive movement of the parts. Movement indicates a worn component.

Maintenance Inspection. If an engine oil change is part of the periodic service, an inspection can be done while the oil is draining. The differential and manual transmission oil levels should be checked. This is accomplished by removing the oil-fill plug located approximately half way up the housing on each unit. The level should be within one-half inch (13 mm) of the filler hole. Refill if necessary. A visual inspection of the exhaust system and rear suspension components can take place while the oil levels are being inspected. Any oil leakage, or any loose or broken components should be noted. The tires are inspected for wear, damage, and inflation pressures. Any abnormal wear should be noted and the cause for the wear determined. The front suspension bushings are inspected visually for deterioration or for a condition where the rubber has relaxed so the pivot is allowed to move off center as shown in Figure 2-13. The engine mounts are inspected for signs of fatigue or failure. If an oil change has been performed, the drain plug and filter are installed.

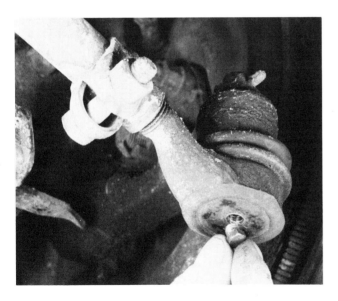

Figure 2-11 Removing a plug from a tie rod end.

The vehicle is lowered and the lubrication completed under the hood and on the body. Oil should be put in the engine and the engine started to establish oil pressure and check for leaks. The coolant and battery fluid levels are inspected and adjusted as required. All hoses are inspected for signs of deterioration. The upper suspension bushing condition is usually inspected from under the hood.

The accessory drive belts are inspected for looseness, wear, and fatigue. The brake fluid, automatic transmission fluid, and power steering fluid levels are checked and adjusted as necessary. White grease is applied sparingly to all hood, deck, and door latches. A few drops of engine oil is put on the hinge pins. Lubricant should never be applied to any carburetor linkage, to any rubber bushings, to rear springs, to any sealed bearings, or to any drive belts.

A lubrication sticker is marked and placed on the door jamb of the vehicle for future reference. Other inspections that are advisable for the safety of the operator are lights and windshield wiper/washer inspection. Although suspension lubrication has been mentioned in terms of periodic maintenance, it is customary to perform a lubrication job whenever front suspension components have been replaced.

Figure 2-13 A deteriorated suspension bushing.

2-5 WHEEL BEARING MAINTENANCE

Wheel bearings are used on the nondriving wheels. The condition of the lubricant in the wheel bearings should be checked every time the brakes are inspected. The wheel bearings should be repacked whenever there is an inadequate supply of grease or when the brakes are serviced. Lubricant should *never* be added to wheel bearings.

Disassembly. The vehicle must be raised and supported to allow removal of the wheels. When the wheels are removed, one lug bolt and one lug hole should be marked to identify their existing position, as pictured in Figure 2-14. This will allow the wheel to be replaced without disturbing the wheel balance.

If the vehicle is equipped with disc brakes, the brake caliper must be removed. On a floating caliper a C-clamp is placed between the outer pad and the cylinder to force the piston inward as pictured in Figure 2-15. The caliper can then be unbolted from the steering knuckle (Figure 2-16) and supported (Figure 2-17) to prevent damage to the brake hose. If the vehicle is equipped with drum brakes, the brake shoes may have to be retracted to allow the drum to be removed. Retraction is done by inserting a wire or small screwdriver into the adjusting slot to hold the automatic adjuster away from the star wheel as it is turned in the proper direction. This is illustrated in Figure 2-18.

To remove the drums or rotor, first take off the dust cover on the wheel hub. Figure 2-19 shows how it can be removed with a special tool. This will allow access to the cotter pin in the front-wheel spindle nut. Removal of the cotter pin and spindle nut releases the bearing washer and outer bearing. The bearing and washer can be lifted from the spindle as shown in Figure 2-20, and placed where they will not be mixed with parts from the other wheel. The dust cap is a good container for these parts. The spindle nut can be threaded back onto the spindle with the hub still in place. The inner bearing

Figure 2-14 A wheel and lug position marked before wheel removal.

Figure 2-17 Supporting the caliper so it does not hang on the brake hose.

Figure 2-15 A C-clamp used to move the caliper piston inward to provide clearance between the shoes and the rotor.

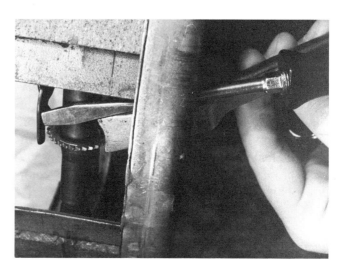

Figure 2-18 The method used to release an automatic adjuster to retract the brake shoes.

Figure 2-16 Lifting a caliper from the rotor.

Figure 2-19 Removing a hub dust cover.

2-5 Wheel Bearing Maintenance

Figure 2-20 Removing an outer hub taper roller bearing.

Figure 2-21 Removing an inner hub taper roller bearing and grease seal.

Figure 2-22 Packing a wheel bearing using a bearing packer.

These conditions are discussed in Section 5-5. After a thorough cleaning in petroleum solvent, the bearing should be blown dry with clean compressed air to remove any contaminants that were not flushed out during cleaning. Do not "spin dry" bearings with air pressure. This will overspeed and score the bearing. Overspeeding may even cause the bearing to fly apart and injure someone in the service area. All old grease must be cleaned from the hub cavity, bearing cups, and the spindle. Not all types of wheel bearing lubricants mix together. Often they will catch on the nut as the front drum or disc and hub are pulled over it. In this way, the inner bearing and the wheel seal are easily removed. This procedure can be seen in Figure 2-21. If the seal and bearing cannot be easily removed in this manner, the hub and drum should be removed and the rear bearing driven out with a large wooden drift and a hammer. Always keep the bearings for the wheels separate so they can be returned to their original location.

Packing Wheel Bearings. As the wheel bearing cones are being cleaned, they should be inspected for any abnormal conditions such as galling, spalling, and brinnelling, or any other signs of distress.

will separate from each other in service. If this happens, bearing failure will result. The bearing cups should be inspected to make sure they are in good condition and secure in the hub. If all parts of the bearings are in good condition, they are packed with a high temperature NLGI #2 wheel bearing grease. Packing is a critical process and therefore it should be done carefully. A bearing packer (Figure 2-22) can be used or the bearing cones can be packed by hand (Figure 2-23). Packing by either method will force the grease around each roller, completely filling the bearing cone cavity. The hub cavity should have a small amount of grease to act as a reservoir of grease as shown in Figure 2-24. The spindle is coated with a thin film of grease to pro-

Routine Maintenance

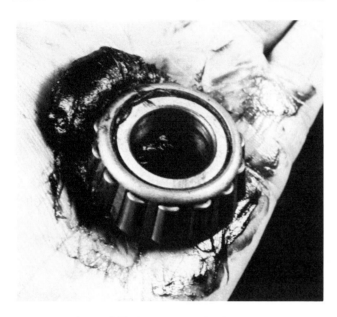

Figure 2-23 Packing a bearing by hand.

Figure 2-24 Reserve grease in the hub cavity.

Figure 2-25 Installing a new grease seal in a hub.

Figure 2-26 Torquing the spindle nut to seat the bearing before final adjustment

vide rust protection and lubrication. Lubrication is needed because the inner bearing race rotates slowly on the spindle during operation. The inner wheel bearing and new grease seal are installed in the hub (Figure 2-25). The hub and drum are placed over the spindle and the outer bearing, washer, and spindle nut are installed.

Wheel Bearing Adjustment. Bearing adjustment for all tapered roller front-wheel bearings is accomplished in approximately the same manner. While the drum is being rotated, the spindle nut is tightened to 20 to 25 ft-lbs (27 to 34 N•m) to seat the bearing (Figure 2-26). The nut is then loosened and retightened *finger tight*. When a nut lock is used on a spindle nut it is properly positioned and the cotter pin installed. When a castle-type adjuster nut is used, the cotter pin slot in the adjuster nut may be backed off just enough to install the cotter pin. This should provide a *preload free* bearing adjustment having *no more than* 0.003 inch (0.08 mm) free-play movement at the hub. If this cannot be obtained without preloading the bearing, either the bearing cup is loose or excessive clearance is present between the bearing cone and the spindle. Either condition requires further inspection. After proper adjustment is obtained, the lock nut or castle nut is aligned with the hole and the cotter pin is installed, as shown in Figure 2-27. This will provide a secure lock for the spindle nut. The dust caps are installed and the brakes are reassembled and adjusted. Drum brakes will require manual shoe adjustment as described in Section 9-9. Disc brake-equipped vehicles will require resetting the pads by

2-5 Wheel Bearing Maintenance

warp. This will result in brake problems or it might cause the wheel to come loose. Wheel lug torque is critical in preventing loosening of the wheel. The raised portion of the wheel around the lug, as shown in Figure 2-29, actually deflects slightly when it is properly tightened. This holds the lug tension and acts to lock the lug nut to prevent it from loosening. Overtightening can result in deflection of the braking surface or failure of the lug bolt.

Figure 2-27 A properly installed cotter pin used to securely lock a spindle nut.

depressing the brake pedal. Never depress the brake pedal until all brakes have been completely assembled. The brake fluid reservoir should be topped off to complete the assembly.

The wheels are reinstalled and lug nuts tightened. It is important to use the proper tightening sequence and to tighten them to specifications. The best way to do this is to use a torque wrench. Lug nut tightening sequences are shown in Figure 2-28. Incorrect tightening can cause the braking surface to

(b)

(a)

Figure 2-28 Normal lug tightening sequences: (a) five-lug pattern; (b) four-lug pattern.

Routine Maintenance

Figure 2-29 Wheels are slightly dished somewhat like a Belleville spring to help hold the lug tension.

2-6 LUG BOLT REPLACEMENT

Occasionally a lug bolt will break when attempting to loosen a wheel. This can be the result of previous overtightening, or the bolt may have become rusted in service. If the lug bolt fails, it must be replaced to make the vehicle safe to drive.

Rear-Wheel Lug. Replacement is rather straightforward if a lug bolt breaks on a rear-wheel drive axle flange. The remaining lug nuts and the wheel are removed. The brake drum lock screw or sheetmetal retainer nut are removed and the drum is lifted off. If necessary, loosen the brake shoes following the same directions used for front drums in Section 2-5.

The broken stud can be removed from the axle flange with a small puller, press, or drift. A new lug bolt is lubricated and positioned in the hole. A thick washer or oversize nut is placed over the bolt to act as a pulling surface. A lug nut is threaded upside down over the new lug bolt. It is tightened until the lug bolt head is pulled against the axle flange. This procedure is illustrated in Figure 2-30. The nut and spacer are removed. The drum and wheel can then be reinstalled.

Front-Wheel Lug. If a lug bolt breaks on a front wheel, the brake drum and hub or rotor must be removed. The lug bolt must be pressed from the outside inward to remove it.

On drum brake vehicles, where the lug bolt is used to position and lock the drum to the hub, some preparatory steps must be taken before pressing the lug bolt from the hub. A special cutter, as shown in Figure 2-31, must be used to remove the swaged metal from the lug bolt at the drum face. Swaging is the process of upsetting a portion of the shoulder of the lug bolt to prevent movement between the hub and drum. The hub is then positioned on a press and supported around the bolt area so the lug can be pressed out without using the brake drum for support. Once removed, the hub and drum are turned over. The area around the lug hole is supported and the lug bolt is pressed into place. If necessary, the new lug bolt may be swaged to prevent movement of the drum on the hub. Reassembly of the hub and drum on the spindle and installation of the wheel complete the repair.

If the lug bolt needs to be replaced on a disc brake, the rotor must be removed from the vehicle. The rotor is supported around the broken lug bolt and the bolt is pressed from the hub as shown in Figure 2-32. The rotor is inverted and supported

Figure 2-30 Pulling a new lug into an axle flange with an oversize nut used as a spacer and an inverted lug nut.

Figure 2-31 A cutter used to remove the metal swage from around the lug.

around the new lug bolt as it was on for the drum brake. The new lug bolt is pressed into place. The rotor is replaced on the vehicle, the brake caliper installed, and the wheel tightened to specifications in the proper sequence. If a press is not available, replacement of the lug bolts in the hub should not be attempted because the hub or brake surface may be damaged.

2-7 TOWING THE VEHICLE

When a vehicle cannot be driven to a location where service can be performed, it must be towed. Professional towing companies and many service stations are equipped to tow vehicles.

The end of the vehicle requiring service will have to be raised from the road surface to prevent further damage. One concern when towing a vehicle is the drive line. If the drive wheels can be lifted, no problem exists. Precautions have to be taken if the vehicle must be towed with the drive wheels on the ground. When the drive wheels turn, they will turn the axles, differential, propeller shaft, and the transmission main shaft. The axle, differential, and propeller shaft work in a normal manner. The transmission does not. Both standard and automatic transmissions are not properly lubricated when only the main shaft turns. Severe transmission damage can be caused if the vehicle is towed for any distance without proper lubrication. The vehicle may be towed only for a short distance at a low speed when the drive wheels are on the ground.

The alternative is to have the tow rig operator place the drive wheels on a dolly. The dolly carries the vehicle wheels off the ground on a special set of wheels. The vehicle can be moved for any distance without damage to the transmission when the wheels are cradled on the dolly. If dolly wheels are not available, the drive shaft to the rear drive wheels can be disconnected. This will allow the vehicle to be towed with the drive wheels on the road with no transmission damage. Professional tow rigs have large rubber belts that are connected to the axle or suspension with chains. The large belts prevent sheetmetal damage as the vehicle is raised. Attempts to tow vehicles without the proper equipment should not be made if additional damage is to be avoided.

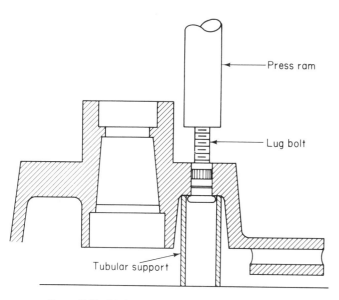

Figure 2-32 Method of using a press to remove a lug.

REVIEW QUESTIONS

1. What is required to give a vehicle maximum service life? [INTRODUCTION]
2. What is the goal of safety procedures? [2-1]
3. List the basic personal safety procedures. [2-1]
4. What safety precautions should be taken in the selection and care of tools? [2-1]
5. When should safety stands be used under a vehicle? [2-2]
6. List four types of jacks. When should each be used? [2-2]
7. What precautions should be observed when using a hoist? [2-2]
8. What precautions should be observed when selecting fluids for use in a vehicle? [2-3]
9. Where are oils used and where are greases used in vehicles? [2-3]
10. List the items serviced in chassis lubrication procedures. [2-4]
11. Describe the process used to pack wheel bearings. [2-5]
12. Describe the procedure used to adjust wheel bearings. [2-5]
13. How are broken lug bolts replaced? [2-6]
14. What precautions must be observed when towing vehicles? [2-7]

Tires and Wheels

Tires and wheels are important to vehicle ride, handling, and fuel economy. It is easy to overlook the many important functions the tires perform when viewing a vehicle that is not moving. The tires and wheels must first support the vehicle weight. This is referred to as the *radial load*. The tires are also required to change the vehicle direction while the vehicle is in motion. The side load placed on a tire in a turn is called the *lateral load*. If a tire could not sustain a lateral load, the vehicle would continue in a straight path, even though the wheels were turned at an angle. In addition, the tires and wheels have to transfer the engine and braking torque to the road surface for driving and braking the vehicle. This is called the *tractive load* of the tire. If a tire cannot develop a tractive load equal to or greater than the vehicle acceleration and braking requirements, wheel slip will occur.

3-1 TIRE REQUIREMENTS

All of the loads on a tire can be placed into three load groups or force components as illustrated in Figure 3-1. Note that the size of the force in each direction is not constant. The force values change as the vehicle is maneuvered, accelerated, and braked. While the vehicle is being driven along

Tires and Wheels

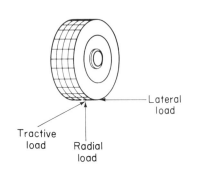

Figure 3-1 Directions of loads on a tire.

a straight road, the lateral load is near zero, the tractive load is only the force required to maintain a constant speed, and the radial load is nearly equal to the static load on the tire. As the vehicle is driven into a turn, the lateral load increases enough to cause the vehicle to round the corner, assuming the tire does not slip sidewise. The radial load is also changed due to the body roll of the vehicle. Body roll occurs when the vehicle body leans toward the outside of the turn. This will increase the effective weight on the tire that is on the outside of the turn and decrease the effective weight on the tire that is on the inside of the turn. The tractive load will be affected too, depending on whether the brakes are being applied or if the vehicle is being accelerated.

With a given load on a tire, the summation (vector sum) of the tractive and lateral loads cannot exceed the frictional force between the tire and the road (Figure 3-2). If it does, the tire will skid. Since the frictional force of a tire with the tread flat on the road surface is equal in all directions, any lateral load placed on the tire will reduce the amount of tractive load that can be placed on the tire at the same instant (frictional force = lateral load + tractive load). The reverse is also true; any tractive load on a tire will reduce the amount of lateral load the tire can handle at the same instant without slippage. This fact is most important in performance vehicles.

Tire Characteristics. The operator expects maximum mileage from each tire along with desirable vehicle handling characteristics. Maximum tire mileage with a given tire type is greatest with a tread rubber compound that is relatively tough and hard. Maximum frictional force or traction for maximum slip resistance, is best with a soft type tread rubber compound. It should be obvious that tread rubber compound is one area where a compromise must be made in tires to produce the desired operational characteristics.

Tires produce some noise as they are driven on the roadway. A tire with a smooth tread surface will be relatively noiseless as it rotates and it will work well on clean dry paved road surfaces. On the other hand, a smooth tire would be useless for vehicle control on wet highways because it would ride up on the surface water, a condition called *hydroplaning*. The tire is therefore designed with a tread that is a compromise between dry road handling, wet roadability, and noise.

Tires are also called upon to absorb many of the small road surface variations before they reach the suspension. This requires a relatively flexible

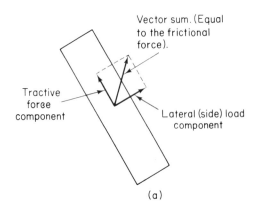 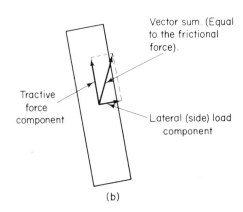

Figure 3-2 Maximum vector forces possible during a turn: (a) large lateral force while making a sharp turn; (b) less lateral force while making a mild turn.

tire sidewall and tread. The tire must also transmit operator input from the steering system to the road surface. This requires a sidewall and tread that is relatively firm. Again, a compromise must be made to make tires that are suitable for both ride quality and handling. These conditions must be met while the tire is rotating between 650 and 1000 rpm at 55 mph (88 km/h), depending upon the tire size.

3-2 TIRE CONSTRUCTION

Tires are made of many separate parts that are molded together to form the complete structure. These include the bead, cords, plies, rib, tread, sidewall, and liner as shown in Figure 3-3.

The *bead* is a bundle of wires in the section of the tire that fits around the wheel rim. It is wrapped with the inner edge of the plies. The *carcass* is the tire without either sidewall rubber or tread rubber. The *cords* are strands of material which are bonded together in a sheet to form the *plies* of the tire. These are made of textile when used for carcass plies. They may be either textiles or steel when used for *belts*. The *rib* is that section of the tire that is between the sidewall and the tread rubber. The *tread* is the part of the tire that is designed to run on the road surface. The tread rubber is grooved with a pattern that will provide maximum frictional force and minimum noise when used on the vehicle application for which it was designed. The *sidewall* is the flexible part of a tire that connects the bead to the tread. The *liner* is a thin air tight layer of rubber that covers the inner surface of the carcass.

Tire Design. There are three basic types of tires used on passenger vehicles. These three types are *bias ply, bias belted,* and *radial ply*. Each type has its own characteristics. They are considered advantages in some ways and disadvantages in other ways.

The bias ply tire, as shown in Figure 3-4, is constructed using layers of cord material running at an angle from bead to bead. The tire carcass has two or more body plies, each having its cord angle running opposite to the cords in the adjacent plies. They cross the tread centerline at 25 to 40 degrees, depending on the tire design. This results in body plies that crisscross throughout the tire carcass.

Low cord angles in the carcass plies are generally better for high-speed performance; they generate less heat and provide lateral stability, but they increase tread wear. High cord angles in the carcass plies will usually give a tire greater strength, increase its fatigue resistance, reduce its rolling resistance, and improve the contact patch when a belt is placed around the carcass under the tread.

The bias belted tire, as shown in Figure 3-5, is constructed much like the bias ply tire, with strengthening belt plies added in the tread area. These belts are wrapped around the circumference of the carcass with the cords in the belts running at an angle of 20 to 35 degrees to the centerline of the tire tread. The belts add to the strength and rigidity of the tire in the tread area.

The radial ply tire, as illustrated in Figure 3-6, differs greatly in construction from the bias ply tire. The cord material in the body plies runs from bead to bead. It intersects the centerline of the tread at essentially 90 degrees. There is no crisscrossing cord material in the carcass of the radial ply tire. Strength is added to the tread area by breaker plies or belts. The cords of these belts

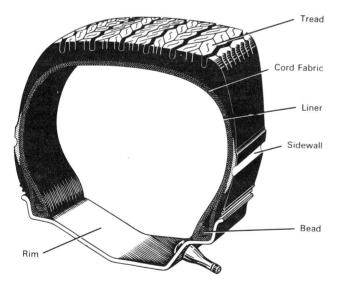

Figure 3-3 Section of a tire showing the various parts (Courtesy of The Firestone Tire & Rubber Company).

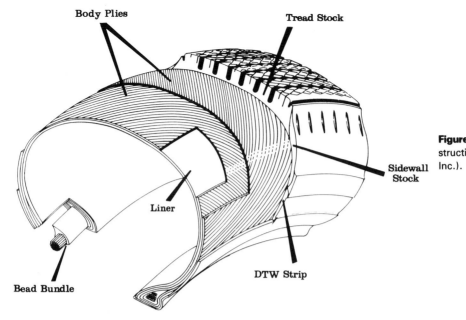

Figure 3-4 Bias angle tire construction (Courtesy of Uniroyal Inc.).

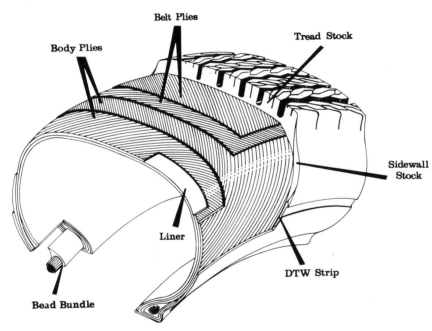

Figure 3-5 Bias belted tire construction (Courtesy of Uniroyal Inc.).

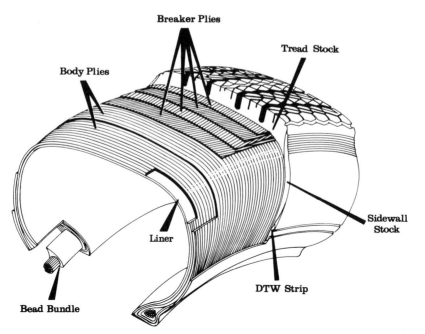

Figure 3-6 Radial tire construction (Courtesy of Uniroyal Inc.).

run at a low angle, usually 10 to 30 degrees to the centerline.

The bias type of construction forms a sidewall that is more rigid than the radial sidewall construction. The rigidity is apparent when viewing both a bias and a radial tire from the front at the same time when they have equal loads and equal air pressure. The bias angle of the cord material inhibits motion of the cord and this results in a more rigid appearing tire. With no bias angle present in the carcass plies, the cord material in a radial tire is more free to flex. This flexing may appear to be a disadvantage until it is completely understood. Friction between the cords of the plies on a bias tire causes heat to build up when the tire flexes as it rolls through the tire contact patch. The radial cords can flex through the tire contact patch without having friction between the cords of the plies and therefore they generate less heat. A tire that generates less heat while rolling will roll more freely than one that heats up. The tire with less friction requires less power to keep in motion, and this results in greater fuel economy.

We can see that no one type of tire is best for all applications. When the vehicle requirements are known, the tire type can be properly selected.

Tire Materials. Tires are constructed using a variety of materials. These include strands of textile, fiberglass, and steel along with natural and synthetic rubbers. The choice of the materials used in the tire is based upon economic considerations as well as the vehicle and operator requirements.

Tire Cord Materials. The cord used in the tire carcass provides strength and many of the tire operating characteristics. The cord materials in the tire carcass must be pliable to flex easily. This will allow the tire carcass to be flexible which, in turn, will provide the tire with smooth ride characteristics and long mileage life. The smooth ride results because greater tire deflections are possible with pliable material. The flexing tire is able to absorb many of the road surface irregularities. The more flexible the tire cord material is, the longer the tire life will be without failure and the flexibility will prevent distortion in the contact patch area.

Cord materials that are presently being used for the tire carcass are rayon, nylon, and polyester. Nylon and polyester have greater strength than rayon. After it has been run, nylon has a tendency to develop a "set" as it cools. This will produce a characteristic "thump" until the tire becomes warm again the next time it is used.

Tire cord material used for the belts is much stiffer and stronger than the cord that is used to make up the tire carcass. The characteristic stiffness in the tread area is helpful in preventing the tread from closing in the contact patch area. The greater stiffness and strength of the belt will also give the tire superb puncture resistance. A bias ply tire has a greater tendency to be pushed together or *squirm* as it rolls through the contact patch area than a radial ply tire. This occurs because the carcass is stiff and the tread flexes. The radial ply tire has less carcass stiffness and greater tread stiffness. This results in a tire with relatively no deformation or squirm taking place in the tread as it rolls through the contact patch area. This allows a softer rubber compound to be used in the tread. Any deformation that takes place in the tire contact patch area will produce wear, while also reducing water channeling when driving on water soaked highways.

Materials used for the belt cords are fiberglass, steel, or an organic fiber called aramid. These belt cord materials keep a tire from having squirm in the contact patch area, especially when used with a flexible radial sidewall.

Squirm in the contact patch area results from the sidewall stiffness and the load the tire carries. It must be noted that both the sidewall of the tire and the air pressure within the tire combine to support the load. As the tire sidewall is called upon to support a greater percentage of the vehicle weight, the tire has a tendency to push together in the footprint as shown in Figure 3-7. As a greater percentage of the load is carried by the air pressure, less deformation occurs in the tread contact patch. A tire that is underinflated or overloaded relies upon a greater percentage of the weight being carried by the sidewall than it would if it were properly inflated. This will result in excessive heat buildup in the carcass, squirm in the tread, and tire tread wear.

Tire Sidewall and Tread Rubber. Tire sidewalls and treads are made using different types of rub-

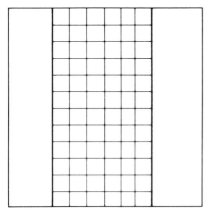
Normal tread ahead of contact patch

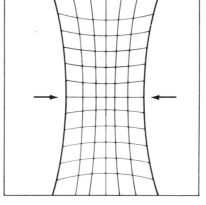
Contracted tread in contact patch

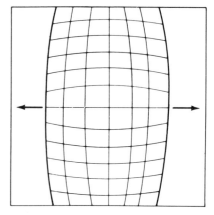
Expanded tread behind contact patch

Figure 3-7 Tire flexing through the tire contact patch.

ber. A large percentage of natural rubber is used in the sidewalls of the tire and in bonding the cords together to make up the tire plies. On the other hand, a large percentage of synthetic rubber is used in making the tire tread. The synthetic tread rubber is compounded so that it is superior to natural rubber in its wear and traction characteristics.

Synthetic rubber is compounded from a hydrocarbon called butadiene. Chemicals are combined to produce the desired vulcanizing effects to hold the tire together and carbon black is added to provide wear resistance. Chemicals are also added to slow the aging (oxidation) process that naturally occurs in rubber materials.

Although synthetic rubber is superior for wear and traction, natural rubber does flex easier in the sidewalls. This results in less heat buildup in the tire and allows greater fuel economy.

White sidewalls are produced by mixing titanium dioxide with the sidewall rubber. The white sidewall is extruded in the bulk sidewall rubber before the tire is built.

The various components of the tire are hand-assembled on a tire building machine. The assembled components are inspected and then subjected to high pressure called stitching to interlock the parts of the tire. This tire is then cured by vulcanizing with heat and pressure. The finished tire is cooled, inspected again, and made ready for shipment. Final inspection determines whether the tire will meet original equipment specifications, replacement specifications, blemish specifications, or if it is to be scrapped.

3-3 TIRE PERFORMANCE

Tire construction and materials will determine a tire's performance. Performance can be broken down into various areas, such as wet and dry road traction, wear characteristics, and ride and steering sensitivity.

The traction of a tire on a wet road is primarily a function of the tire's ability to channel water away from under the tire as illustrated in Figure 3-8. Channeling the water away from the tire contact patch prevents the development of a wedge of water that causes the tire to ride up off the road surface. This is called *hydroplaning* and is shown in Figure 3-9. If the tire contact patch flexing is minimal, the tire tread grooves will retain their

Figure 3-8 Water being forced from under a tire moving at low speed.

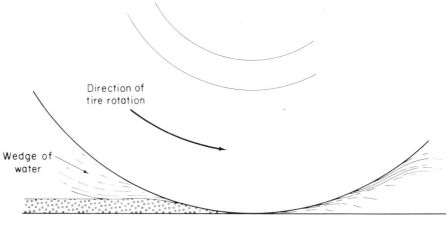

Figure 3-9 A wedge of water going under a tire.

shape. This provides a place for the water to flow from under the tread and so hydroplaning is minimized. The width and depth of the tread grooves are obviously of major importance for wet tire traction. The wet roadability decreases proportionally to the tire tread wear. This is due to the reduced ability of the water to channel from under the tire. Secondary factors in wet road performance of tires are the tread width and the softness of the tread rubber compound. A wide tire has a greater tendency to hydroplane than a narrower tire. The narrower tire exerts greater unit pressure, which improves its adhesion to the road surface. A soft rubber compound allows the tire to interact with the road surface to improve its adhesion quality. On the other hand, reducing the quantity of rubber in contact with the road surface (narrower tire) and using a softer compound will increase the rate of tire tread wear.

The dry road performance of a tire depends upon the type of tire construction, rubber compound, tread type and design, as well as the cord material used in its construction. Radial construction is superior for fuel economy, tread wear, and steering response. The softer the tread rubber compound, the greater will be the tractive force. However, the negative effect of a soft tread rubber compound is more rapid wear. The tread type and pattern is designed for the specific application and use of the tire. If a tire is designed for both tractive and lateral load, the tread must include grooves which run around the tire tread. The tread also has cross cuts, called *sipes,* that run across the tire. Sipes are pictured in Figure 3-10. They are mainly designed to help give the tire wet road traction. A tire that is perfectly smooth will provide traction as long as the road surface is perfectly clean and dry. The grooves that help to provide traction in wet and dirty road conditions do not improve dry road traction. The tire tread with grooves tends to fold under during cornering as shown in Figure 3-11. This increases the slip angle of the tire. Because of this folding, the slip angle of a new tire is greater than that of a used tire. A tire with little tread will, therefore, have better dry road traction than a tire that has deep tread grooves, providing the coefficient of friction is equal.

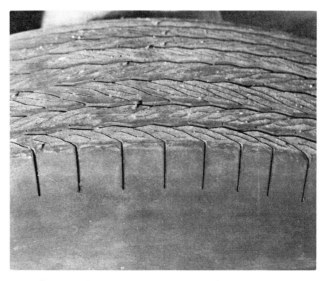

Figure 3-10 Unevenly spaced sipes on the edge of the tire tread.

Tires and Wheels

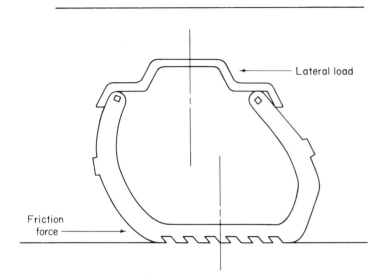

Figure 3-11 Typical flexing of a tire during a turn.

Flexing. The type of cord materials used in the tire construction affects the tire performance. The cord materials in the belt are a factor in determining the belt pattern that is used. The belt pattern affects the way the tire performs under various loading conditions. The stiffness of the cord materials in the belt also affects tire wear. Stiffer belt material allows less squirm and this reduces tread wear. The stiffer belt will also increase the amount of ride harshness, which may be objectionable to the operator. The belt stiffness may impair handling characteristics because it tends to reduce the length of the tire contact patch. The size of the contact patch is important for traction on irregular road surfaces.

Ride Quality. The quality of the ride is affected by the various types of tire construction and the materials used. The bias and bias belted tires have higher *radial spring rates* than do the radial tires. Radial spring rate of a tire is the amount of force that must be exerted on the tire to cause a one-inch deflection. The radial spring rate affects the resonant or rebounding frequency of the tire. The low radial spring rate of the radial tires contributes to what is known as *low-speed harshness*. A bias tire has a resonant frequency that peaks at about 160 Hz (cycles per second), or approximately 60 mph (96 kmph), while a radial tire has a resonant peak at about 60 Hz, or less than 20 mph (32 kmph). The resulting low-speed harshness and noise at the resonant frequency are some of the major disadvantages of radial ply tires. The radial ply tires are sensitive to wheel alignment settings, wheel balance, and construction tolerances. This sensitivity may lead to vibration and a wheel pull sensation when radial ply tires are used.

Steering Sensitivity. The area of steering sensitivity is one of the greatest apparent differences among the types of tire ply construction. The radial ply tire has the most steering sensitivity. This is one reason for its successful use on sports cars and for its delayed use on passenger vehicles. A sensitive tire requires more precise steering control because the vehicle tends to respond more rapidly to slight steering wheel movements.

Self-aligning Torque. A tire property that is an important part of handling and steering sensitivity is its self-aligning torque. The self-aligning torque is the tendency of the tire to twist between the footprint and the wheel rim. A radial tire has a low self-aligning torque value when a lateral force is applied. This produces a faster steering reaction and vehicle response from the operator input.

Slip Angle. The slip angle is another factor affected by various types of tire construction. The slip angle is the angle between the direction in which the tire is heading and the direction in which it is pointed, as illustrated in Figure 3-12. It is obvious that the slip angle can increase until the tire is actually sliding sidewise. The amount of slip that occurs as the lateral load is increased on the tire differs among the various types of tire construction. A radial ply tire typically has a low slip angle and this remains low until the tire suddenly breaks loose and slips sidewise. A bias ply tire has a slip angle that is nearly proportional to the lateral force being placed on

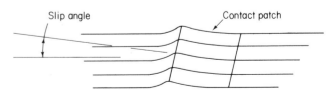

Figure 3-12 Twist in the tire tread causing slip angle.

the tire. The slip angle increases as the load increases. As a result, the bias ply tire gives the driver a gradually increasing warning that the limit of lateral force is being approached. This provides a large margin of warning between the time tire slip starts and the time when control is lost. The warning of slip is one of the reasons bias ply tires are still used in automobile racing. In racing, the vehicle must be run on the ragged edge of sliding sidewise. A great deal of skill is required to drive on the ragged edge between the start of slip and loss of control when using radial tires.

The low slip angle of the radial ply tire gives the operator more sensitive control during all speeds, until slip causes loss of control. This small margin between control and loss of control was one of the reasons the manufacturers were slow in changing from bias ply to radial ply tires on automobiles for the general public.

Because of the different slip angles, it is recommended that tires with different ply construction should not be mixed on a vehicle. Radial ply tires have the lowest slip angle. This is followed by bias belted and then bias ply tires which have the greatest slip angle. Mixing tire construction on one vehicle can produce dangerous handling characteristics. The different characteristics of the tires can produce severe unexpected steer reactions when entering a turn. These conditions are even more pronounced when driving on rain-slicked highways. If a vehicle had radial ply tires mounted on the front and bias ply tires on the rear, the different slip angles would cause the vehicle to oversteer (see page 217 for an explanation of oversteer and understeer). Having bias ply tires on the front and radial ply tires on the rear would produce a vehicle with understeer as a result of their different slip angles. If the steering wheel were held in a constant turn position and speed is increased, the vehicle is said to understeer when it increases the radius of the turn. It oversteers when it decreases the radius of the turn. It should, therefore, be obvious why tires of different ply construction should not be mixed on the same axle. Figure 3-13 shows what could be expected to happen when tires hav-

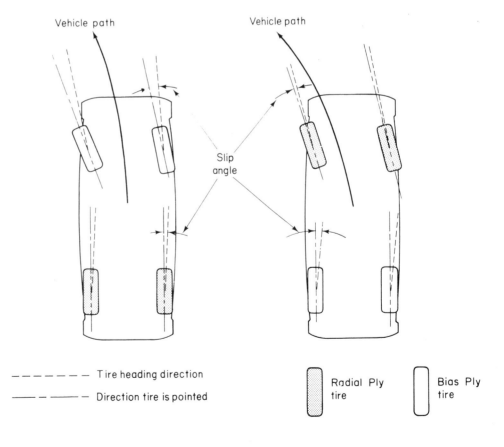

Figure 3-13 An illustration of how the mixing of bias and radial tire construction affects vehicle steering.

Tires and Wheels

ing different types of construction are used on the same vehicle.

Tires and Fuel Economy. With fuel economy being a major concern of everyone, research is being done on tires in an attempt to reduce their rolling resistance. This interest results from the fact that once a car is moving at constant highway speeds on a level road, approximately 20 percent of the fuel consumed is used for rolling the tires. This energy is basically lost in flexing the tire as it rolls through the footprint. The energy lost is due to the internal friction between the carcass cords of the tire. The lost energy is converted to heat. The friction of the rubber yields a hysteresis loss which is dependent upon the type of rubber used. Natural rubber has less hysteresis loss than synthetic rubber, so less energy is converted to heat. Hysteresis is the time required to return to the original shape after deflection.

The tire cord materials and belt cord angles affect internal friction. Radial ply tires have the lowest internal friction. Their use will give vehicles the maximum possible fuel economy. Unfortunately, when improving fuel economy through decreased rolling resistance, the tire ride quality may degrade. It can be seen that compromises must be made in each area of tire performance to achieve optimum tire operation.

Tire Deterioration. A great deal of discussion has been devoted to explaining how the different tire types and construction features affect the operation of new tires. Tires deteriorate with temperature and with age. The tire will respond as it was originally designed for only a short time before wear occurs. The tire compound continues to cure as it remains in the atmosphere after manufacture even if it is not placed on a vehicle. Higher temperatures increase the rate of oxidation, which is why tires will harden faster on the vehicle in use than they will when kept in storage.

As the tread rubber oxidizes, the rubber becomes harder, reducing the coefficient of friction between the tire and the road. A tire that runs cooler on the vehicle will not be subjected to as rapid a deterioration as a warmer running tire. Age can also cause deterioration of the cord material in the carcass. This can result in tire failure due to fatigue.

The water channeling that is so important on wet roadways is reduced considerably as the tread wears. This means that a vehicle with worn tire tread must be driven slower for normal tire to road adhesion.

All tires do not perform equally under similar conditions. It is difficult to judge whether a new tire is the best for a specific type of vehicle and type of driving. Performance during the life of the tread, as well as tire mileage, is one method used to evaluate a tire. Most consumers evaluate a tire solely on the mileage that can be obtained on a set of tires without consideration of other factors. It may be that the typical operator drives the vehicle well below the critical limits of the tire. The previous discussion shows that for satisfactory tire operation, a careful balance must be made between tire wear rate and tire performance as it adheres to the road surface.

3-4 REPLACEMENT OF TIRES AND WHEELS

The technician is often faced with the problem of recommending a good tire for replacing the original tires on a vehicle. Tire selection should be based on original equipment specifications, type of driving anticipated, weather conditions normally encountered, and the cost of replacement.

Tire Specifications. The manufacturers are presently using a series of numbers and letters to identify the tire size, load capacity, inflation limits, and type of service for which the tire is intended. The table in Figure 3-14 shows the popular tire sizes. A number of tire properties that must be considered when replacing tires are discussed below.

The first consideration when replacing tires is the wheel *rim size*. Two measurements are considered, the diameter across the wheel where the tire bead seats and the distance between the inner and outer bead flange. The diameter, which is measured in inches throughout the world, is the

Tire load range table

Bias and bias-belted series			Metric series	Radial series		Cold inflation pressure (psi)						
78	70	60		78	70	24	26	28	30	32	36	40
B78-13	B70-13		P175/75R13	BR78-13	BR70-13	980	1030	1070	1110	1150	1230	1390
D78-14	D70-14			DR78-14	DR70-14	1120	1170	1220	1270	1320	1410	1490
			P185/75R14			1160	1210	1260	1310	1360	1450	1540
E78-14	E70-14			ER78-14	ER70-14	1190	1240	1300	1350	1400	1490	1580
			P195/75R14			1270	1330	1390	1440	1500	1600	1690
F78-14	F70-14			FR78-14	FR70-14	1280	1340	1400	1450	1500	1610	1700
G78-14	G70-14		P205/70R14	GR78-14	GR70-14	1380	1440	1500	1560	1620	1730	1830
F78-15	F70-15	F60-15		FR78-15	FR70-15	1280	1340	1400	1450	1500	1610	1700
			P205/75R15			1370	1430	1490	1550	1610	1720	1820
G78-15	G70-15	G60-15		GR78-15	GR70-15	1380	1440	1500	1560	1620	1730	1830
			P215/75R15			1480	1550	1620	1680	1740	1860	1970
H78-15	H70-15			HR78-15	HR70-15	1510	1580	1650	1770	1770	1890	2010

Load range B extends through 32 psi; Load range C extends through 36 psi; Load range D extends through 40 psi.

Maximum pound load limit per tire at the cold inflation pressure.

Figure 3-14 A table of popular tire sizes.

most important of the two dimensions. Large wheel diameters require large tires which, in turn, will support more weight. Underfender clearance limits the maximum tire size that can be used on a vehicle. Oversize tires usually require wheels with wider rim widths, because the rim width must also match the tire size and tire design.

The *load range* is a term used to identify a given tire size, inflation limit, and maximum load the tire can safely carry. It is identified by letter codes. The codes presently being used are A, B, C, D, E, F, G, H, J, L, M, and N. As the size of the tire is increased, with a constant inflation pressure and aspect ratio, the load capacity of the tire will also increase. The load range is often compared directly to size, although it should be apparent that the important characteristic of a tire is its ability to carry a specified load.

The *aspect ratio* has to be considered when replacing tires. The aspect ratio, or *tire series,* is the ratio of the tire section height to the tire section width, as shown in Figure 3-15. Manufacturers are presently supplying 78 and 70 series tires on new vehicles. Replacement tires are available for passenger cars in 78, 75, 70, 60, and 50 series. A 50 series tire would have a section height one-half as large as the section width, while a 75 series tire would have a section height approximately three-fourths as large as the section width. The low aspect ratio or series tires tend to have a lower profile, giving the vehicle a sporty or performance look. Race cars have aspect ratios as low as 35, which provides better handling in turns. High-aspect-ratio tires typically will have better ride characteristics with poorer handling response than low-aspect-ratio tires. Vehicles using low-aspect-ratio tires must have wheels that have an adequate rim width to allow proper tire shape and tire-to-road contact.

Metric tire specifications and markings differ slightly from the customary system of tire identification as shown in Figure 3-15. The metric tire identifications will use the section width in millimeters, an "R" if it is a radial tire, the aspect ratio to the nearest 5 percent, and the wheel size in inches. The section width relates directly to load capacity of the tire. It is an accurate description of the maximum tire capacity.

Obsolete tire size and ratings are sometimes used by older people. They are briefly presented here to clarify the meaning of the terms. Prior to 1968, tire sizes were identified by their width in decimal inches and rim diameter. For example, two typical sizes were 6.70-15 and 7.35-14. The tire strength was specified in the number of plies used in the tire casing, such as two-ply, four-ply, or six-

Tires and Wheels

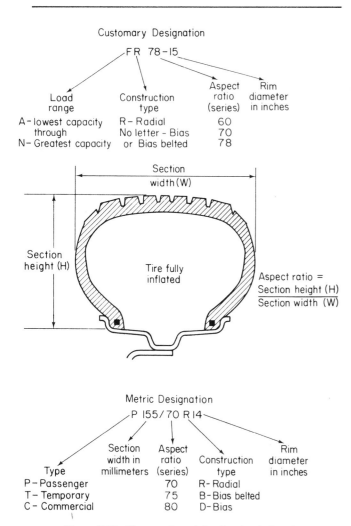

Figure 3-15 The meaning of the tire size designations.

ply. As a variety of cord materials became available with different strength properties, the ply rating became meaningless as a specification for tire strength. So this system was replaced with the customary and metric tire identification systems as described above.

Wheels. The wheels on a vehicle interact very closely with tire characteristics to provide ride quality, load capacity, and the handling desired. The wheel design will also affect front suspension design as related to vehicle control. Most passenger vehicle wheels are made from stamped steel. The stamped steel is made in two pieces and then welded or riveted to form the wheel assembly. This makes a true-running, airtight, and strong tire-holding device that can be attached to the automobile. The two wheel sections are the *rim* and the center *disc*. The rim is designed to provide the inner radius for the tire as well as a steel insert for attachement to the disc. The disc is the section of the wheel that bolts to the axle or wheel hub. Some automobile wheels that are designed for appearance and/or performance are cast in one piece. These are commonly referred to as "mag" wheels although they may be either magnesium or more commonly, an aluminum alloy.

Some heavy-duty trucks and trailers also use single-piece wheels. The rim, with the tire mounted, is bolted or clamped to the wheel hub. Other rims are designed with the bead seat separable from the rim to allow easy mounting and dismounting of the tires. This type of wheel does not have a drop center and it is referred to as a "split rim." Another wheel design is made by bolting two wheel halves together. This design would typically be used on small wheel diameters that would otherwise make tire mounting and dismounting difficult due to their physical size. Passenger car wheels are of the drop center type. This type allows the tire to be mounted and removed without disassembling the wheel.

Tire application determines the width of the rim that is required. This is the distance between the inner and outer bead seats. The rim diameter

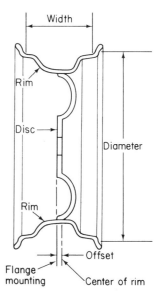

Figure 3-16 Location of wheel dimensions.

and offset are determined by the vehicle application. The rim diameter is the vertical distance between the bead seats on one side of the wheel. Wheel offset is another consideration that is selected to fit the requirements of the vehicle. Wheel offset is the horizontal distance between the centerline of the rim and the mounting face of the disc. The wheel dimension locations are shown in Figure 3-16. The offset is positive if the centerline of the rim is inboard of the mounting face and negative if outboard of the mounting face. The amount and type of offset is critical because changing the wheel offset will change the front suspension loading as well as the scrub radius as described in Section 15-1. This is important to front suspension alignment and handling characteristics.

3-5 TIRE AND WHEEL INSPECTION

Frequent inspection of the tires and wheels on vehicles will provide the operator with greater safety, reliability, and tire mileage. Inspection, from the automotive service technician's point of view, can result in a greater volume of suspension repairs and alignment jobs as well as the satisfaction of knowing that the customers are operating safe vehicles.

Inspection may be performed as routine preventive maintenance while the vehicle is raised for lubrication or other service. Abnormal tire wear is the most obvious condition. An inspection may also be the result of a customer complaint about the ride or handling of the vehicle. All that is normally done during preventive maintenance is a visual inspection and adjustment of the tire air pressure. When a customer complaint has led to tire and wheel inspection, diagnosis starts with a visual inspection. This most often leads to various measurements being taken to determine if the tire is the problem source. If the problem is found to be tire-related, the service technician has to find which one of the tires is at fault.

Tire Inspection. A tire inspection will disclose tread wear patterns that have developed due to poor driving habits, improper maintenance, mechanical problems, the need for suspension adjustments, wheel balancing, or a possible tire fault.

Frequent inspection will give the technician an opportunity to detect a problem and correct it before it becomes so serious that the tire must be replaced. The chart in Figure 3-17 shows typical tire wear problems and their usual causes.

Tire Wear. Poor driving habits, such as high-speed cornering, will result in the tires wearing on their outer edges. This wear is often similar to tire wear that occurs from underinflation. Figure 3-18 shows cornering wear. Improper maintenance will result in tire wear caused by underinflation and overinflation, neglecting to rotate tires, or tires that have been operated without being properly balanced. Inflation neglect results in predictable wear patterns, as pictured in Figure 3-19. Unbalanced tires will usually result in spotty or erratic wear patterns in the form of cupping or spot wear of the type shown in Figure 3-20. This could also be caused by loose, worn, or bent suspension components or a bent wheel. The need for adjustment of suspension alignment will usually result in tire wear that is easy to distinguish. Camber and toe (as discussed on page 224) are the most significant tire wear angles. Tire wear caused by incorrect camber is shown in Figure 3-21. Caster (as discussed on page 225) will only produce wear if it is extremely out of specifications. Toe-in, if excessive, will produce feathered edge wear on the outer edges of the tire tread grooves. Excessive toe-out will produce the same wear condition on the inner tire tread edges. Toe wear is shown in Figure 3-22.

Tire balance is necessary to produce an even wear around the circumference of the tire. Whenever a tire is reinstalled on the wheel, the assembly should be balanced to assure normal smooth tire operation.

Tire Failure. A faulty tire may show as ply separation, tread separation, bead separation, cord separation, or chunking. Ply separation is the parting of adjacent plies within the carcass or tread area due to failure of the rubber bond between the plies. Tread separation is the parting of the tire tread rubber from the tire carcass. Bead separation

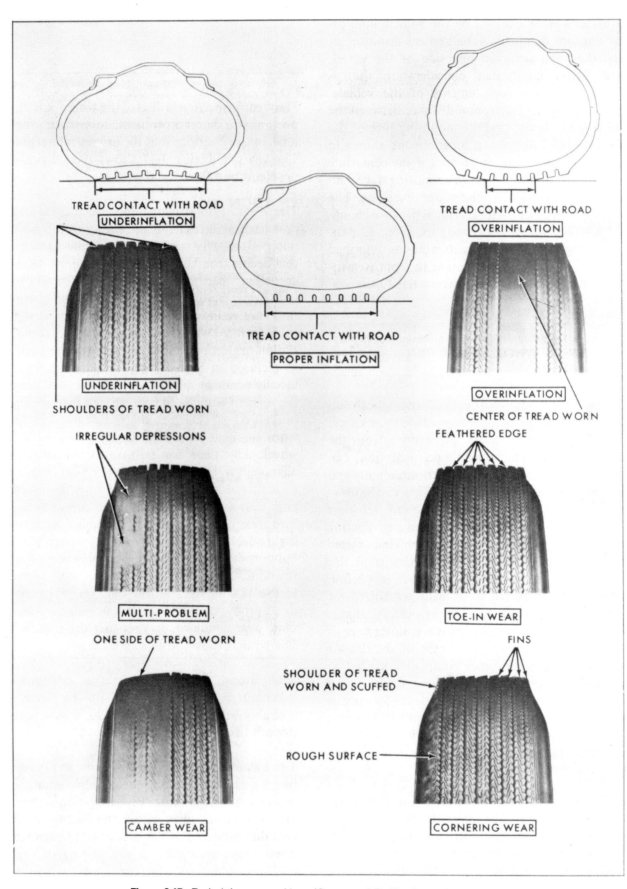

Figure 3-17 Typical tire wear problems (Courtesy of Cadillac Motor Car Division, General Motors Corporation).

Figure 3-18 Cornering wear on the edge of the tread.

(a)

(b)

Figure 3-19 Wear caused by improper inflation: (a) underinflation wear on the outer ribs; (b) overinflation wear on the inner ribs.

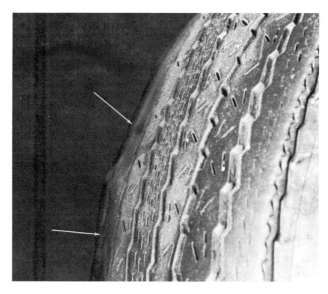

Figure 3-20 Advanced cupping wear.

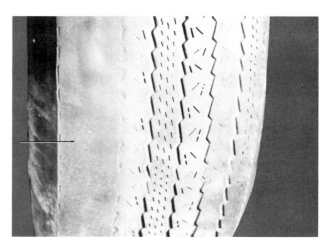

Figure 3-21 Wear on one side of the tread as a result of improper camber.

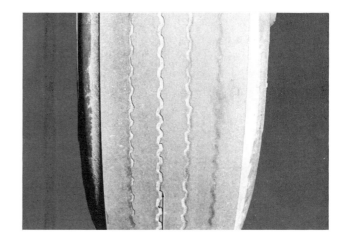

Figure 3-22 Feathered edge on the tread caused by incorrect toe setting.

Tires and Wheels

is a failure that occurs in the bead area of the tire at the wheel rim caused by a loosening of the carcass rubber and plies in that area. Cord separation is a failure of rubber bond between the cord material of each ply. Chunking is a condition where pieces of the tread rubber separate from the tire carcass and are thrown from the tire.

The failures of the type discussed in the previous paragraph are the result of *excessive heat buildup.* Heat buildup is a result of excessive tire flexing or scuffing. Most often the heat buildup occurs from operating with less than the air pressure specified for the loads being carried by the tire. Scuffing is the most usual cause of chunking. This results from high heat concentration on the tread surface which releases the bond between the tread surface and the tire carcass in localized areas.

Complaints that lead to a tire inspection usually involve vibration, noise, or wheel pull problems. Tire vibration is most often caused by an out of balance condition causing a shake that is traceable to one corner of the vehicle. A noisy tire could be the result of cupping that has occurred around the circumference of the tire. Extreme cupping could cause vibration that to the operator would feel similar to a failed wheel bearing. A slapping noise would result if separation of the tread, cord, or plies has occurred.

Wheel Pull. Often the wheels are aligned to correct wheel pull that originates in the tires. It is very difficult to determine if wheel pull is caused by a tire or by faulty wheel alignment settings. Over the years, it has been convenient to blame the wheel alignment settings for wheel pull. Most often this diagnosis was correct. Radial tires, which have become more popular, have a tendency to cause wheel pull or lead and they may develop unusual steer characteristics. It is usually difficult to determine if the alignment settings or tires are causing the wheel pull. If the alignment settings are known to be correct, then the diagnosis is rather straightforward. It simply involves changing the suspected tires from left to right. If this causes the vehicle to pull in the opposite direction, one of the tires is faulty. The tires are then restored to their original locations and the faulty tire replaced. If it is not known whether tires or alignment is at fault, the normal procedure is, first, to make sure the wheel alignment settings are correct and then check for tire problems. It may, in some instances, be more convenient to check the tires before checking the wheel alignment settings when diagnosing wheel pull.

The vehicle pull that originates in tires, especially radial tires, is usually the result of the tire *belt location* on the carcass. If during tire manufacture the layup of the plies varies, the tire may develop what is termed *conicity* or *ply steer*. This results in an effect like a rolling cone. Either belt location or ply steer will cause wheel pull if they are far enough out of specifications. If the ply steer is not equal around the tire, it will produce *waddle*. This is a periodic pull as the tire rolls producing a low-speed shaking of the vehicle. As a result of these problems, radial tires are extremely sensitive to production variations.

It is possible, after alignment, to have a tire produce a directional sensitivity. This could be caused by unequal wear across the tread surface that results from the previous misalignment. With realignment, the center of pressure is no longer located in the center of the tread, resulting in wheel pull. This problem is discussed in more detail in Chapter 15.

REVIEW QUESTIONS

1. Name three load groups that affect the tire. [INTRODUCTION]
2. What is meant by the summation of the loads on the tire? [3-1]
3. How does the tire tread rubber compound affect wear and traction? [3-1]
4. How does the design of the tire tread affect the tire operation? [3-1]
5. Make a sketch of a tire section and name the parts. [3-2]

6. Why does a properly inflated radial tire bulge more just above the road contact area than a bias ply tire? [3-2]
7. What are the advantages of a bias ply tire? [3-2]
8. What are the advantages of a belted tire? [3-2]
9. Describe hydroplaning. [3-3]
10. What are the advantages of wide tires and of narrow tires? [3-3]
11. What causes low-speed harshness in radial tires? [3-3]
12. Why is it important to use the same tire construction on all four wheels? [3-3]
13. What causes the change in operating characteristics as a tire is used? [3-3]
14. What precautions must be followed to match tire and wheel rims? [3-4]
15. Describe the tire sizing number system. [3-4]
16. List the types of visible tire faults and their cause. [3-4]
17. What tire problems apply only to radial tires? [3-4]

Tire and Wheel Service

Tires and wheels operate as an assembly. Malfunctions in either or both can cause an abnormal ride and poor handling. Most tire problems can be avoided with periodic inspection and preventive maintenance procedures. Faulty tires and wheels must be repaired properly or replaced when an inspection shows any part of the assembly to be faulty.

4-1 TIRE SERVICE

Tire service includes tire and wheel assembly rotation between vehicle axles, balancing, checking tire radial and lateral runout, repairing, replacing, and, in some instances, truing or trimming tires. When a problem exists, an analysis should always follow a logical sequence.

Tire Rotation. It is recommended by automobile manufacturers that tires be periodically rotated between front and rear to assure the operator maximum mileage from a set of tires. The recommended rotation period is usually 10,000 miles (16,000 km) or less. This practice results from the fact that rear tires are subjected to different operating conditions than the front tires. This is due to differences in suspension design and the differences in load on the tires. As a result, rotation will help to provide most even tread wear. Rotation also tends to correct slight abnormal wear conditions on the tire tread.

Tire rotation sequence is different, depending upon whether the vehicle is equipped with radial tires or bias ply tires. A radial tire is said to develop a "set" after operating on one side of the vehicle and rolling in one direction. Changing the radial tires from side to side may adversely affect either tire life or handling. Placing a tire on the opposite side will cause the tires to rotate in the opposite direction. The four- and five-tire rotation sequence patterns suggested by vehicle manufacturers are illustrated in Figure 4-1 for both radial and bias ply tires. Bias belted tires are rotated in the same sequence as bias ply tires.

A vibration that is suspected to be coming from tires that show no abnormal condition during the visual inspection should be checked for both runout and balance of the tire and wheel assemblies. Runout should be checked before balancing any tire and wheel assembly. Runout can occur in two ways: laterally and radially.

Lateral Runout. Lateral runout is the side to side movement of the tire as the tire and wheel assembly rotates on its bearing. It is measured with a dial indicator positioned close to the tread shoulder as pictured in Figure 4-2. Acceptable lateral runout is usually 0.090 inch (2.29 mm) or less. Although most tires fall within acceptable limits, excessive lateral runout will cause wheel shake. Excessive lateral runout can be caused by improper installation of the tire on the wheel, the wheel on the hub, a bent wheel, or a faulty tire.

If excessive lateral runout is found on the tire, the point of greatest *tire* runout should be marked

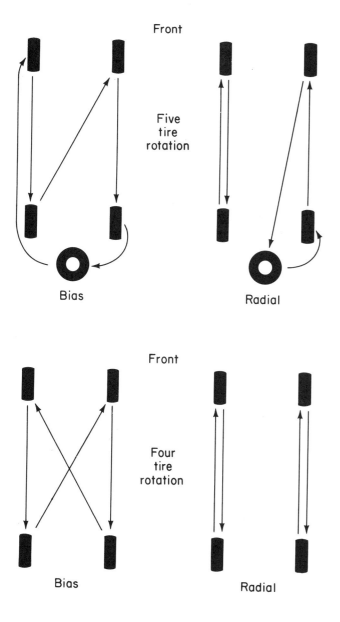

Figure 4-1 Tire rotation sequence patterns.

Figure 4-2 Measuring lateral runout of the tire.

Tire and Wheel Service

Figure 4-3 Measuring lateral runout of the wheel.

with chalk. The lateral runout of the wheel rim is then checked to make sure that it is within specifications. Wheel lateral runout, as illustrated in Figure 4-3, should not exceed half of the acceptable tire runout, or 0.045 inch (1.14 mm). The point of greatest *wheel* runout is also marked with chalk. If the chalk marks on the tire and wheel are close together, it is possible to deflate the tire, free it from the rim, and rotate it in relation to the wheel so the chalk marks are opposite each other, and then reinflate the tire. The assembly should then be reinstalled on the vehicle and the runout rechecked. If the tire runout is still out of specification, the runout readings must be examined. If the runout increases, the tire is at fault. If the runout decreases but it is still slightly out of specifications, it may not affect the operation. It may be worthwhile to balance the tire and test drive the vehicle. If rotating the tire on the wheel made no change in runout, the wheel itself is at fault.

Radial Runout. Tire runout can occur radially as well as laterally. Radial runout results from the tread having a varying distance from the wheel center. This causes the tire tread to move away from its axis at one point and closer to the axis at another point as the tire and wheel assembly is rotated. This can be thought of as a tire with a tread that does not form a true circle.

Radial runout is measured with the indicator positioned against the tread surface of the tire as shown in Figure 4-4. Excessive radial runout will contribute to noise and roughness that can be felt by the operator. Radial runout is more critical than lateral runout. The total runout, therefore, should not exceed 0.045 inch (1.14 mm). Tires will develop some "set" or flat spots when standing in one position as they cool. Whenever measuring runout, especially radial runout, the tires must be properly inflated and should be driven enough to warm them up prior to measurement. This will normalize the tire tread. If the runout is excessive at the tire tread, the point of greatest runout should be marked with chalk. The dial indicator is then positioned against the wheel rim, as shown in Figure 4-5, and the wheel runout checked. The high spot is marked. Radial runout of the wheel should not exceed 0.035 inches (0.89 mm). If the wheel radial runout is excessive, changing the position of the wheel on the hub two or three bolt holes from the original location may correct the runout. If the runout is then within specifications for the wheel and tire, mark one stud and the wheel so it can always be positioned

Figure 4-4 Measuring radial runout of the tire.

4-1 Tire Service

Figure 4-5 Measuring radial runout of the wheel.

in the same location. If the wheel runout remains excessive after repositioning the wheel on the hub, it is necessary to determine whether the wheel or the hub is at fault. If the hub is not at fault, the maximum runout chalk marks on the tire and wheel coincide, and the wheel is within specifications, it may be beneficial to deflate the tire and rotate the tire on the wheel so the chalk marks are opposite each other. After inflating the tire, reinstall the wheel on the vehicle, and torque the lugs in the proper sequence. The radial runout must again be measured. If the runout remains at the same location and is the same amount, the wheel is at fault. If the runout is greater than previously measured, the tire is at fault. If the runout decreases to an acceptable value, the tire and wheel assembly should be rebalanced and the vehicle test driven.

Loaded Runout. A tire falling within the preceding specifications for lateral and radial runout may still have runout when the tire is subjected to the vehicle load that it normally carries. This runout is called *loaded runout* or *tire force variation*. This type of runout is due to variations in the tire construction as it was built. If the tire has a stiff section, the tire will force the wheel spindle upwards as this section rolls through its footprint. The up-and-down movement of the spindle with each revolution of the tire will be felt by the operator. This type of runout cannot be located using the conventional runout measurement methods just described. Special equipment is needed to detect this problem. The equipment is basically a motored drum placed under the tire. It rotates the tire while an indicator measures the spindle deflection. This type of detector is illustrated in Figure 4-6. When this type of equipment is not available, the cause of the shake problem can sometimes be located by overinflation or substitution. The tire and wheel assembly must first be balanced and the runout checked. Using the overinflation method the tires are then inflated to 50 psi (345 kPa) and the vehicle test driven. If the problem has diminished, deflate one tire to the recommended air pressure and test drive. Repeat this procedure at each tire. When the faulty tire is returned to its original pressure, the problem should return. The faulty tire should be replaced to correct the problem. Substitution checking procedures require replacing the seemingly faulty tire with the spare and then test driving the vehicle. The spare tire and wheel assembly can be moved to each wheel location until the problem is corrected, thus identifying the faulty tire.

Every effort should be made to assure that everything affecting the tires is correct before

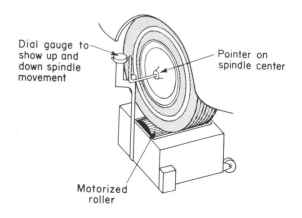

Figure 4-6 Measuring loaded runout with an indicator that measures spindle deflection.

Tire and Wheel Service

condemning tires and/or wheels. Tires should always be checked for proper inflation. The tires should be warm before measuring runout: lateral, radial, and loaded. The wheel bearings must be properly adjusted (as discussed in Section 2-5), and every effort must be made to assure that there are no defective wheel and axle bearings. The lug nuts must be properly tightened in the correct tightening sequence to the correct torque.

4-2 WHEEL SERVICE

The vehicle may be subject to wheel shake. If, during inspection, a wheel is found to have excessive lateral runout, the wheel may have to be replaced. Before condemning the wheel, it should be checked to make sure that the wheel lugs have been tightened in the proper sequence and tightened to the proper torque. A check should be made to make certain that the wheel runout is not a result of hub runout due to loose bearings, bearing races, or a bent hub flange. If the wheel is at fault, it must be replaced.

Wheel failure may be caused by many factors. These include improper maintenance, accident, overloading, and age. Figure 4-7 shows a rim

Figure 4-7 A rim flange that was bent when the wheel went through a pot hole in the road.

Figure 4-8 A wheel that failed because it was overloaded.

flange that was bent when the wheel went through a pot hole in the road.

Improper maintenance and accidents are the major causes of wheel failure. Improper maintenance will show up as a failure in the wheel disc bolt circle. This is generally due to the wheel lugs loosening during operation causing the bolt holes to wear. The bolt holes can also fail from fatigue in the bolt circle resulting from improper wheel lug tightening procedures that cause unequal bolt loading. Improper tightening can also damage wheel hubs and brake components.

Overloading wheels will usually show up as a failure in the rim bead flange and seat area. This area of the wheel will undergo fatigue due to continued high load and eventually fail. Overloading will also occur when tires designed for wide wheels are incorrectly placed on narrower wheels. Continued overloading with side loads on the tire will sometimes result in failure of the wheel disc in the bolt circle area as shown in Figure 4-8.

Vehicles that have been involved in accidents are probably the most common cause for wheel replacement. The wheel is usually impacted from the side. This will cause the rim and center section to bend. After an accident the technician must carefully inspect the vehicle suspension for other bent components.

Failure of the wheel due to age can occur in various ways. Corrosion can take place inside the wheel rim. In severe cases air leakage will occur. Air leakage can occur around the rivets or spot welds that hold the disc to the wheel rim. The raised bosses and bolt holes can become worn so the wheel will not accurately center or the lug nuts will not remain tight.

Careful wheel inspection should be undertaken each time a tire is to be mounted on a wheel. If the wheel is found to be damaged or defective, it should be replaced.

4-3 BALANCE

Tire balancing is one of the most common types of tire service performed. Whenever a tire is originally mounted or when it is remounted, the tire and wheel assembly should be balanced. Balancing can be performed using a number of different methods, although the type of imbalance or unbalanced condition falls into only two categories. These are *static imbalance* and *dynamic imbalance*.

Static Balance. The assembly is in static balance when there is equal weight distribution about the axis of rotation of the tire and wheel assembly. Therefore, static imbalance is corrected by adding weight to the light portion of the wheel rim until the distribution of weight is equal around the wheel center.

Static imbalance may be detected with the wheel stationary while supporting it in a horizontal position, with the wheel stationary in the vertical plane, or when the wheel is rotating. Accuracy of the final tire balance depends upon the operator of the balancing equipment and the type of equipment used.

Horizontal Static Balancing. Static tire balancers that balance the tire while it is held in a horizontal position with no rotation are usually referred to as *bubble balancers* (see Figure 4-9). This balancer has a hub over which the wheel fits. Imbalance is detected by viewing a sight glass in the center of the hub and locating the position of the bubble, similar to a carpenter's level. The bubble in the level shifts toward the light side of the wheel. Weights are added to the light side of the wheel rim until the bubble is centered. It is customary and acceptable to add half of the weight to the inner rim and half to the

(a)

(b)

Figure 4-9 A static bubble balancer: (a) in balancing position; (b) the bubble indicating imbalance.

outer rim, especially when the amount of imbalance exceeds two ounces (57 g). The purpose of adding weights to both inner and outer rims is to maintain whatever dynamic balance the assembly has. Once the required location of the weights is determined, the location is marked. The tire and wheel assembly is removed from the balancer, weights are tapped on the rim, and then the balance is rechecked.

Vertical Static Balancing. Some static balancing equipment places the tire and wheel assembly in a

Tire and Wheel Service

vertical position. They make use of the tendency for the heaviest spot in the tire to rotate toward the lowest position. With this type of balancing equipment, the tire and wheel assembly is mounted on a hub with the wheel in a vertical position as shown in Figure 4-10. The hub is supported with anti-friction bearings. The assembly is allowed to rotate freely until the heavy spot settles at the lowest point. Weight is added at the very top of the wheel rim flange, which is the lightest part of the assembly. The tire and wheel assembly is turned to various positions to determine if the assembly still has any tendency to rotate. If the section with the added weight rotates to the bottom, the weight is too great; it should then be decreased. If the added weight returns to the top, the weight is too small and it should be increased. If the assembly remains at any position in which it is placed, the assembly has equal weight distribution about its axis and it is, therefore, in static balance.

Detecting Static Imbalance While Rotating. Some tire balancing equipment will spin the wheel to detect static imbalance. This may be done with the wheel remaining installed on the vehicle or it may be done with the tire and wheel assembly removed and mounted on the balancing machine. There are advantages and disadvantages to each method.

Balancing the tire and wheel assembly while mounted on the vehicle allows all of the rotating mass—tire, wheel, brake drum, brake rotor, and wheel cover (which can cause shaking)— to be balanced at the same time. This may result in a better balance that is free of vibration. When balancing is completed on the vehicle, one lug and a corresponding location on the wheel should be marked to assure proper relocation if the tire and wheel assemblies have to be removed. As a result of the brake components being included in this type of balancing, any time the tire and wheel assemblies are rotated to another axle location, they must be rebalanced at their new location.

Balancing off the vehicle on a neutral spindle allows each tire and wheel assembly to be properly balanced as an assembly. As a result, the assemblies can be put at any wheel location on the vehicle. If imbalance occurs after mounting the assembly, the imbalance is caused by the rotating brake components. These should then be balanced separately.

Methods presently being employed to detect static imbalance on the vehicle rely either on the operator's "feel" or an electronic readout to determine when imbalance has been corrected. The equipment shown in Figure 4-11 has an attachment that allows the balancing head to be attached to the wheel. Inside the balancing head are movable weights. The weights can be moved while the wheel is rotating to position and change the amount of effective weight on the assembly. The weight and position are adjusted as the tire is spinning until no vehicle shake can be seen or felt. The tire is stopped and the amount of weight shown on the scale is added to the wheel rim in the position indicated. The wheel attachments are then removed and the wheel assembly is respun to check the balance. This type of balancing offers quick service while detecting vibrations that are annoying to the operator.

Figure 4-10 Static balance is done in a vertical position on the type of balancer shown.

Figure 4-11 Static balancing as the wheel is spun on the spindle.

Another type of equipment that detects static imbalance on the vehicle is of the type shown in Figure 4-12. This employs a vibration sensor that is positioned beneath a suspension component as close to the wheel as possible. The sensor detects downward suspension movement that is produced as the wheel spins. A strobe flash and weight indicator are triggered by the vibration sensor as the suspension is moved downward by the heavy portion of the tire and wheel assembly. The strobe flash is similar to an engine timing light. A chalk mark is placed on the tire, or the valve stem is used for a reference point. The tire is spun with the wheel spinner and the location of the reference mark is noted in the light of the strobe flash. Because the strobe light only flashes as the heavy spot rotates through the lowest point in its travel, the light flash will make the reference mark appear stationary. When the position of the location mark and reading of the weight indicator are noted, the wheel is stopped. The wheel is turned so the mark is positioned where it appeared in the strobe flash as the wheel was spinning. The amount of weight indicated is added to the wheel rim flange *at the top,* which is the lightest point of the assembly. The tire is spun again using the sensor and strobe light to check the balance. If the assembly is not in balance, the weight is increased, decreased, or its position changed, as indicated, to achieve balance. If on the respin the weight appears at the top, the added weight is too light. If it appears at the bottom, it is too heavy. If the weight appears anywhere between top or bottom, the *position* of the weight is incorrect. The weight position can be corrected by moving the weight *toward the lightest point* on the tire, a small amount at a time, and recheck the balance. If the weight appears at the top or bottom after repositioning, add or subtract weight as described above. Strobe balancers have the advantage of being able to balance any size tire,

Figure 4-12 Vibration sensor positioned on the lower A-arm to detect static balance as the wheel spins.

Tire and Wheel Service

although it usually takes more time than with other balancing methods.

Static wheel balancers that spin balance the wheel off the vehicle sense the force of imbalance about the axis of rotation of the balancing machine. The machine will indicate the amount and the location of the weight needed to balance the assembly. Weight is added and the tire respun to assure that proper balance has been achieved. This type of balancer is expensive but it eliminates much of the guesswork in balancing.

Dynamic Balancing. Dynamic imbalance, unlike static imbalance, can only be detected when the wheel is rotating. Dynamic balance is equal weight distribution about the center plane of the tire. Dynamic imbalance is thus caused by unequal weight distribution about the center plane through the tire as illustrated in Figure 4-13. The unequal side-to-side weight forces the tire to attempt to seek a new

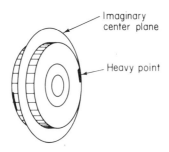

Figure 4-13 Dynamic imbalance shown as weight on each side of a center plane.

plane of rotation as the heavy section tries to move closer to the center plane as the rotational speed is increased. Figure 4-14 schematically represents a dynamic imbalance condition. As the tire is rotated, it can be seen that with every 180° of rotation, the tire would attempt to alternately turn in and out. This will result in *wheel shimmy*. It is possible to have static balance and still have dynamic imbalance if the heavy spots on the tire are 180° apart and on opposite sides of the tire. For this reason, the tire must be spun to detect dynamic imbalance. Dynamic imbalance on all balancers is

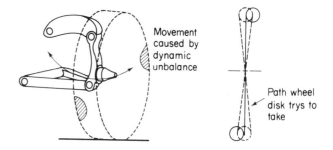

Figure 4-14 A schematic diagram of dynamic unbalance that causes shake.

corrected by placing two weights of equal amount on the wheel 180° apart, one on the inner rim and one on the outer rim. The use of two equal weights 180° apart does not affect static balance as it corrects dynamic imbalance.

Dynamic Balancing Equipment. Dynamic balancers that require removal of the tire and wheel assembly from the vehicle employ different methods to achieve static balance before dynamic balance can be checked. There are some machines that combine the static and dynamic balance test, allowing for correction of both imbalance conditions with one test spin and two weights.

Dynamic balancers of the first type require the tire to be in static balance before any dynamic balancing is attempted. One of these balancers is shown in Figure 4-10. Dynamic imbalance can then be checked as the wheel is spun. The weights of the size indicated are placed on the wheel rim in the proper position (180° from each other on opposite wheel rims). The position of the weights depends on the specific equipment design. The wheel is respun to make sure that the imbalance has been corrected.

Balancers of the second type, those that combine static and dynamic balance functions, use a sensing mechanism that is sensitive to the entire weight characteristic of the tire and wheel assembly. This allows imbalance to be corrected with one weight on each side of the wheel. This type of balancer is shown in Figure 4-15.

Dynamic balance may be checked and corrected on the vehicle. The strobe type balancer, as described on page 49 for static balance, is used. For dynamic balance the vibration sensor is positioned against the steering arm or brake splash shield as pictured in Figure 4-16. In this position it will

Figure 4-15 A balancer that combines static and dynamic balancing. The guard is up in this illustration to show the horizontal position of the tire.

Figure 4-16 Vibration sensor positioned on the steering arm to detect dynamic imbalance as the wheel is spun.

4-4 TIRE GRINDING

In some situations, tire balancing alone will not remove all of the shake or roughness problems associated with tires. The shake may be caused by a tire carcass that has unequal stiffness as it rolls through the footprint. This condition may be corrected by tire truing or tire trimming.

Tire Truing. Tire truing is done to make the tire perfectly round so that all points of the tire tread are at an equal distance from the wheel axis. Tire truing can be done where excessive runout cannot be corrected by repositioning the tire on the wheel or where some shake is still detectable after careful balancing. Tire truing is based on the theory that the closer to round the tire is, the smoother the ride will be. Truing is done with the wheel assembly removed from the vehicle. The tire and wheel assembly is mounted on the tire truing machine and the cutters positioned to conform to the shape of the tire tread. The tread is ground only enough to make it round. Grinding beyond the point where the tire runs true will reduce the tire life.

Trimming. Tire trimming is done to achieve equal *force* distribution about the circumference of the tire. This corrects for any loaded run-out that would be present as the tire rolls through its footprint under normal loads. During manufacture, tires may become stiffer in one part than in another. This concentrated stiffness is not detectable by conventional runout measuring methods nor can it be corrected by tire truing. Loaded runout can be measured by spindle movement or the resulting body movement as the loaded tire is rolled under load. This principle is illustrated in Figure 4-17 using different size springs between the wheel and tread in place of the tire carcass.

Loaded runout or force variation can be measured and corrected with the wheels and tires remaining on the vehicle. The vehicle is raised and the tire trimming machine (force variation grinder) is positioned under the tire to be trimmed. The

detect any tendency for the wheel to turn in or out. The reference mark is again used as it was for static balance. For dynamic balance, one weight is added inside the front of the wheel at the sensor height and one weight is added 180° from the sensor at the rear of the wheel on the outside. In dynamic balance, equal-sized weights are always used to maintain static balance. Static balancing should be done before dynamic balancing when using this type of machine.

Tire and Wheel Service

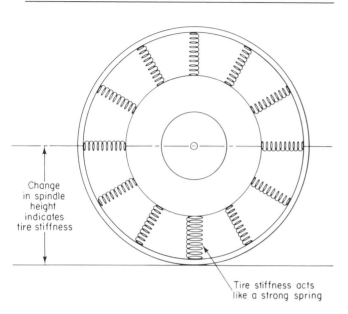

Figure 4-17 Springs used in place of the tire to indicate the effect of a stiff tire section.

motor-powered roller of the machine is placed directly beneath the tire. The vehicle is lowered until the normal weight is on the tire. The other three corners of the vehicle are supported to hold the

Figure 4-18 A typical force variation grinder with the cutter wheels backed away from the tire.

vehicle in its normal level position. A sensing mechanism is positioned to detect either spindle movement or the resulting body movement as the tire is rotated by the roller. A typical force variation grinder is shown in Figure 4-18. The motion of the spindle or vehicle body is used to trigger the machine's rasp cutters at the correct instant to automatically bring them into contact with the inner and outer ribs of the tire tread. Usually two ribs are trimmed on each side of the tire tread at the stiff section of the carcass. This automatic grinding of the tread only at the stiff section continues until the spindle or body deflection is within acceptable limits. The tire is then balanced to complete the operation. Force variation is the result of stiffness in the tire sidewall area which is corrected by grinding the outer rib beneath that sidewall area. Trimming has no negative effect on tire wear or noise, and it reduces ride roughness.

4-5 TIRE REPAIR

Tire repair is most commonly required because of punctures. If the repair is properly performed, the tire can be placed back in service, safely and without fear of the leak reoccurring. At present, very few vehicles use tube type tires and so their repair will not be discussed.

Tire repair begins with removing the wheel from the vehicle. Prior to removal, a wheel stud and corresponding lug hole should be marked so that the wheel can be reinstalled in the same location on the hub.

Locating a Leak. The tire is inflated to the maximum inflation pressure marked on the tire (usually 32 psi (221 kPa) as shown in Figure 4-19). To locate leaks the tire and wheel assembly generally will either have to be submerged in a water tank or covered with a soap and water solution applied with a brush or sponge. The water in the tank or soap will bubble at the exact location of the leak. A tire may leak in a number of locations. Each area of the tire should be individually checked. The tread area is checked first. If no leaks are found, the tire may have to be placed flat on the floor and the

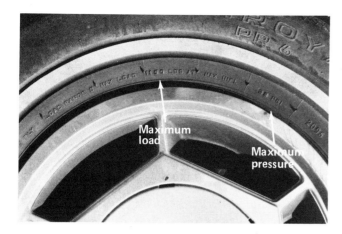

Figure 4-19 Tire load and pressure markings.

soapy solution applied to the sidewall and the bead areas. One side is inspected at a time. Slow leaks in these areas may not be detectable when the tire is in a vertical position. Leaks may also occur at the valve stem.

The leak in the tire is marked with crayon so the location can be identified after the tire is removed from the wheel for repair. An identifying crayon mark is also placed at the valve stem location so that the original tire balance and tire runout can be reestablished when the tire is remounted.

Leak Repair. Procedures used to make the repair depend upon the nature of the leak. Tire repair from the outside is *not* recommended. This type of repair is considered temporary, at best, and in some instances, it may even cause more damage to the tire. The outside repair is called *plugging*. An awl reamer is pushed through the punctured hole to clean it. A strip of rubber coated with tire cement is forced into the hole using a special inserting tool. When the tool is removed, the rubber remains to seal the leak. Problems exist with this type of repair. The puncture may not go straight into the tire and its path is not traceable with the awl. Forcing the awl through the tire in a straight line may actually cut part of the cord that was untouched by the original piercing object. Flexing of the repaired tire during operation may loosen the plug causing a new leak in the same location.

Repairs performed from the inside of the tire are considered permanent repairs when they are performed correctly. Puncture repairs should only be attempted in the tread area of the tire, not in the sidewall, bead, or shoulder area. A tire should be considered unrepairable if it is worn so much that the wear indicators show that it is considered unsafe for further use, as shown in Figure 4-20. More than two punctures may have damaged the tire cord enough that the tire should be considered unrepairable. Punctures larger than $\frac{1}{16}$ inch (2 mm) are not completely safe and so they should not be repaired.

Leaks sometimes occur in the sidewall or bead area. Sidewall leaks may be the result of a porous tire inner liner. This could develop if the tire were run with low pressure. Excess flexing will cause undue heat build-up, damaging the liner. The liner could leak if it were defective when new. The specific location of liner leaks cannot usually be detected and so they should be considered

Figure 4-20 Tire wear indicated by the wear bars.

Tire and Wheel Service

unrepairable. As a temporary expedient, a tube may be installed in the tire to prevent leakage. A tube will increase the operating temperature of a tire and should therefore only be used as an emergency repair. It should never be used in place of a good tire repair or replacement of the tire.

Bead area leaks usually result from corrosion build-up on the wheel rim in the bead area. Repair of the corroded area can usually be accomplished by thoroughly cleaning it with a wire brush and properly lubricating the bead of the tire with nonpetroleum-based tire lubricant.

Tire Dismounting. Removal of the tire from the wheel is best accomplished with the use of a tire changing machine. Such machines, using manual, electric, or pneumatic power will separate the bead from the wheel rim with little physical effort even if the tires have been mounted for long periods of time. First, the beads of the tire must be unseated from the wheel. This is done by forcing the tire beads inward from the wheel rim, as shown in Figure 4-21. After both beads are unseated, the tire can be removed by placing a bar or lever under the

Figure 4-21 Unseating the tire bead from the rim.

Figure 4-22 Bead of the tire lifted over the rim while the opposite side is in the rim drop center.

bead on one side of the tire while holding the opposite side of the same bead in the drop center portion of the wheel rim. When the bar or lever is rotated, the tire bead will be lifted to the outside of the wheel as shown in Figure 4-22. The process is repeated from the same side of the wheel for the opposite bead. The tire is then free from the wheel.

Some wheels are designed so that it is difficult to remove the tire from the front side of the wheel. The side of the wheel with the least distance between the bead seat and the lowest drop center is the side from which the tire must be removed. After the tire is removed from the wheel, a complete inspection of the carcass interior should be made. If no damage, other than the original leak, can be observed the tire repair can continue.

Repair of punctures is accomplished by first selecting a patch of the required size. The area around the leak is cleaned with a tire buffing solvent to remove all residue. The area is then buffed with a power-driven wire brush or a rasp to make sure proper bonding will take place. The size of the area to be buffed is determined by the size of the patch selected. The rubber removed during the buffing operation is cleaned from the tire. Always buff an area slightly larger than the patch, as shown in Figure 4-23. When buffing is complete, tire cement is applied to the buffed area. After the tire cement has dried, the patch is applied by hand. Then full contact is made using a specially designed roller to assure proper bonding. The roller is a

4-5 Tire Repair

Figure 4-23 Inner lining of the tire is buffed slightly larger than the patch to be applied, as shown.

corrugated wheel and the rolling process is called *stitching* the patch.

If the leak is in the bead area and wheel rim, careful inspection of the complete tire and wheel is necessary. If corrosion or dirt is at fault, the bead is properly cleaned and the tire prepared for remounting.

Reassembly. Reassembly of the tire on the wheel requires a few preparatory steps. The tubeless tire valve stem should be examined to assure no cracks or damage is present. It is advisable when mounting new tires to install new valve stems. The life of the tire rubber is close to the life of the valve stem rubber. The valve stem is replaced by drawing the new stem through the wheel from the inside to the outside until it is properly seated. The bead seat area of the wheel is cleaned with a wire brush. The bead area of the tire is then coated with either a soap solution or rubber lubricant. The tire is guided onto the wheel, lower bead first, with the use of the tire mounting equipment. One side of the bead must be held into the drop center portion of the wheel to allow the other side of the bead to be rolled over the rim flange onto the wheel. The outer bead is rolled over the wheel rim flange in the same manner.

The tire is then rotated on the wheel until the crayon mark is in line with the valve stem. The tire is then inflated. Inflation may require the use of a bead expander. This may either be a mechanical clamp or an expanding sleeve device that keeps the tread of the tire from expanding. As a result the bead of the tire is forced against the wheel as shown in Figure 4-24. This allows the tire to be inflated. Without the use of the expander, air will usually leak past the bead faster than the air is flowing in through the valve stem so the tire will not inflate.

Inflation of tires on passenger car wheels should always be done with the wheel fastened securely to the tire machine. This will hold the assembly safely as the bead seats against the rim. Always keep fingers clear of the bead area while inflating the tire.

Inflation will usually be easier if the valve core is removed from the valve stem prior to inflation. The tire should be inflated to approximately 40 psi (276 kPa) and then lowered to the proper operating pressure. This higher pressure will aid in seating the bead on the rim flange. The tire and wheel assembly should be rechecked to make sure that no leaks exist.

The final tire repair operations should involve balancing the tire and wheel assembly and installing it on the vehicle. With the tire properly marked

Figure 4-24 Expanding sleeve on the tire to help keep the bead against the tire rim as the tire is inflated.

Tire and Wheel Service

during disassembly, tire balancing may be omitted, provided the balance weights were not moved as the tire was removed and remounted. It is recommended, though, that the tire and wheel assembly be rebalanced. Replacement on the vehicle requires that previously marked lug and bolt holes be indexed together and the wheel nuts tightened in the proper sequence to the correct torque.

Retreading. Tire retreading is a means by which the good carcass with worn tread is fitted with new tread. This process is performed at specialized retreading shops.

One of the most important parts of the retreading process is the selection of tire carcasses that are in excellent condition. These will provide satisfactory service for the life of the new tread. Tire retreading is most commonly done on heavy-duty road trucks where high mileage is accumulated in a short period of time. Their tires usually have a good quality carcass.

The old tread rubber is ground down to the cord material. New tread rubber is placed around the carcass. The carcass is then placed in a tire mold and the new tread is molded and bonded to the old carcass under heat and pressure. This process, when properly performed, will provide an inexpensive tire with a good service life.

REVIEW QUESTIONS

1. What is the purpose of periodic tire rotation? [4-1]
2. How is tire runout checked? [4-1]
3. Why is it recommended to drive the tires before checking runout? [4-1]
4. What can be done to correct tire and wheel assembly runout? [4-1]
5. What problems can occur in wheels? [4-2]
6. How does static balance differ from dynamic balance? [4-3]
7. Where are weights attached in static balancing? [4-3]
8. Where are weights attached in dynamic balancing? [4-3]
9. What types of balance cause the wheels to bounce and what type causes wheel shimmy? [4-3]
10. How does correcting tire force variation differ from tire truing? [4-4]
11. What part of a tire cannot be repaired? [4-5]
12. Describe the procedure to be used to repair a puncture. [4-5]

Axles, Bearings, and Housings

The tire and wheel assembly is attached to a hub or flange. The hubs must be mounted on good wheel bearings if the wheel is to rotate smoothly. The wheel bearings, in turn, are mounted on a spindle in the front and in an axle housing in the rear of the typical drive train. Seals are provided to keep the lubricant in the bearing and keep the dirt out.

The operation of the brakes and steering is affected by the wheel bearings. Bearing life depends upon correct bearing fit and adjustment and the amount of lubrication and contamination present.

The technician will have to remove the front hubs when the brakes are serviced. The bearings are usually cleaned and relubricated, then new seals are installed before the hub is reinstalled. Wheel bearing adjustment is critical, for normal service life, and for normal operation of disc brakes. Rear axle bearings do not have to be removed when servicing the rear brakes, unless the grease seal has allowed lubricant to get into the brakes. In this case the axle must be serviced and a new seal installed. Of course, the bearing must be serviced any time a failure is indicated.

The wheels and hubs are often removed to gain access to other components when servicing the suspension. Routine servicing procedures for axles, bearings, and housings are, therefore, discussed here. They will not be repeated in the following chapters.

Axles, Bearings, and Housings

5-1 BEARINGS

Bearings are used to locate and support shafts and other components throughout the vehicle. Bearings minimize friction and wear between two surfaces where a difference in speed exists. Bearings used in automobiles may be divided into two major groupings, fluid film bearings and rolling contact bearings. The *fluid film bearings* are the type of plane bearings used in the engine, such as journal bearings. A typical fluid film bearing is shown in Figure 5-1. *Rolling contact bearings* are discussed in this

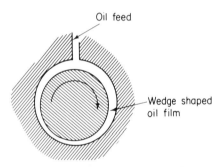

Figure 5-1 Typical fluid film plane bearing.

section. They are used where high loads and relatively high speeds exist. These bearings provide minimum friction and minimum clearance with a minimum amount of lubrication. Rolling contact bearings may be divided into three basic divisions, described by the shape of the rolling element within the bearing. These are ball, straight roller, and tapered roller designs as shown in Figure 5-2.

Ball Bearings. Ball bearings have many applications on vehicles. Their high-speed capability with minimum friction at a relatively low cost makes them a desirable type of bearing. The balls within the bearing assembly support the load. This makes a theoretical *point contact* between the balls and the races as shown in Figure 5-3. This point contact makes the ball bearing subject to early fatigue when it is required to carry heavy loads. Ball bearings can be called upon to support some axial (side) loads as well as radial loads. They can do this because the balls roll in grooves or raceways. Ball bearings are easy to assemble because they are self-contained assemblies. They have been used for front-wheel bearings. Presently they are being used on some vehicles as rear-axle bearings, although these are being replaced with roller bearings in new axle designs. Ball bearings also are used within transmissions, in the steering gear, and in various engine accessories.

Straight Roller Bearings. Straight roller bearings are used in a number of locations on vehicles. Roller bearings and needle bearings are similar in design. By convention, a roller element with a length over four times its diameter is called a needle bearing. Straight roller bearings have an advantage over ball bearings because they make a *line contact* between the bearing rolling element and the surface of the race. This is illustrated in Figure 5-4.

Figure 5-2 Rolling contact bearings. From left to right—ball, roller, needle, and tapered roller.

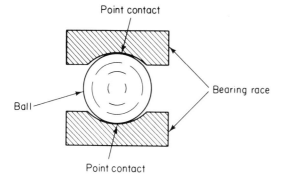

Figure 5-3 Ball bearing point contact.

Roller bearings are, therefore, capable of supporting greater loads than ball bearings. A line contact increases the rolling friction to over that of a ball bearing and, therefore, it limits the bearing to low- and moderate-speed applications. The roller bearing is relatively inexpensive. It can be designed to operate on a bearing journal surface that is ground on a shaft eliminating the need for separate bearing races. The roller bearing also has a low profile and it is, therefore, used where space is limited. One of the major drawbacks of a roller bearing is its inability to support axial side loads. Axial loads are retained by the differential case side bearings when straight roller bearings are used on the axle.

Straight roller bearings are used for radial load-carrying rear-axle bearings on many vehicles. They continue to find application, as either roller or needle bearings, in transmissions and other vehicle components that are subject to high loads and moderate speeds.

Tapered Roller Bearings. Tapered roller bearings are continually finding new applications on vehicles. Tapered roller bearings are more expensive to manufacture than either ball or straight roller bearings. They do incorporate some advantages of both the ball and straight roller bearings. The tapered roller bearing will accept radial loading and an axial load in one direction. This can be seen in Figure 5-5. These bearings have high fatigue resistance. The contact between the rolling element and the race is a line contact; therefore it has a high load-carrying capacity. Like straight roller bearings, the tapered roller bearing is limited to low and moderate speeds.

The tapered roller bearing is more complicated than the other two bearing types. They must

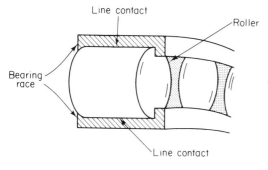

Figure 5-4 Roller bearing line contact.

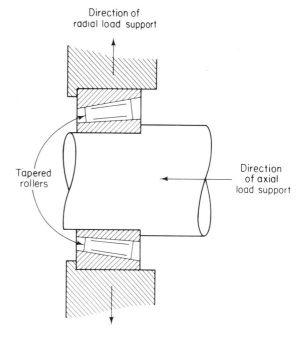

Figure 5-5 A tapered roller bearing will accept radial loads and axial loads in only one direction.

be used in pairs and they must have some means to adjust or retain axle end play. This complicates the assembly procedure.

The tapered roller bearing is used in differentials for case support bearings and for pinion bearings. It has replaced ball bearings for use as front-wheel bearings. Tapered roller bearings are also used in some rear axles. Their superior fatigue resistance makes them desirable bearings.

Axles, Bearings, and Housings

5-2 BEARING SUPPORT

Bearing support mechanisms for this discussion include axle housings, axle shafts, wheel spindles, and wheel hubs. A bearing is designed to provide low friction while maintaining minimum internal radial clearance. The bearing's ability to maintain the correct clearance is just as dependent upon its proper fit in the support as it is on the internal bearing clearances. For this reason, it is important to understand all the factors that influence the bearing fit. These include the support and the clearances, which ultimately affect the bearing life.

The driving axle, whether it be front- or rear-wheel drive, will have a bearing with the outer race stationary in the support housing. The inner race will rotate with the axle shaft. If the bearing is of the straight roller design, the axle shaft itself may have a ground surface that becomes the inner race for the bearing.

The nondriving axle of all passenger vehicles has the outer bearing race pressed into the wheel hub. The inner wheel bearing race is fitted over the wheel spindle. This type of assembly forces the outer bearing race to rotate with the wheel and the inner bearing race to remain nearly stationary.

Wheel Spindles. Wheel spindles, sometimes referred to as *steering knuckles* when used on the front of the vehicle, allow the wheels to be steered. When the vehicle has front-wheel drive, the rear wheels have a spindle or *stub axle* and the front wheels have a steering knuckle. The spindle, then is a nonrotating axle that provides support for the wheel bearings.

The axle shaft of the driving axle does not support any of the vehicle weight on medium- to heavy-duty trucks. The weight is supported by a tubular wheel spindle or housing. Only the torque for moving the vehicle is transferred through the axle shaft. This type of axle assembly, shown in Figure 5-6, is called a *full floating axle* because the axle shaft itself does not support weight.

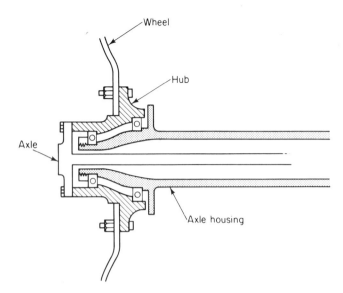

Figure 5-6 The basic design of a full floating axle that turns the wheel but does not support weight.

The wheel spindle is made of steel which provides the required wear properties as well as the toughness needed to resist impact. Because of the tremendous loads placed on the wheel spindles and the need for them to remain perfectly aligned, no attempt should ever be made to straighten them if they become bent.

The spindle is part of the front steering knuckle on vehicles using rear-wheel drive. The steering knuckle is also required to carry the disc caliper mounting and the steering arm for connection to the steering linkage. The steering knuckles have, in the past, been made of forged steel with the steering arm and the caliper mounting bolted to it. Manufacturers are now casting the steering knuckle, caliper mounting, and steering arm as one piece and then press-fitting the steel spindle into it. This unit, shown in Figure 5-7, is a rugged assembly.

The spindle provides support for the bearing races. The size and finish of the spindle area on which the bearing races are fitted is critical. The inner bearing races must rotate or *creep* slightly on the spindle to provide maximum bearing life. This is necessary to gradually rotate the small inner bearing race surface so the load is not always in the same location on the race. This will minimize bearing fatigue at any one area and increase bearing life. The creep is actually in the reverse direction to that of wheel rotation as shown in Figure 5-8. The

5-2 Bearing Support

Figure 5-7 One piece cast knuckle with pressed in spindle.

need for correct bearing clearance is essential. Wear does not occur on the spindle unless the bearing itself becomes seized. This will cause the inner race to rotate on the spindle at wheel speed. Excess inner race rotation can be recognized as grooving on the spindle at the race position. If wear has occurred, the spindle will have to be replaced.

The outer bearing race, usually called the bearing *cup,* is fitted into the wheel hub with a *press fit.* Care must be taken when servicing the bearing to make sure the cup is tight in the hub and that it is properly seated against the shoulder of the bore. Outer bearing races or cups are replaced by driving them out of the wheel hub with a hammer and drift as shown in Figure 5-9. The hub should be inspected to make sure the cup is tight in its bore. The cup is replaced in the hub using a bushing driver to properly position the cup without damaging it.

Axles and Axle Bearings. The engine torque is transmitted through the drive line and axles to the wheels to power the vehicle. The axle is also used to

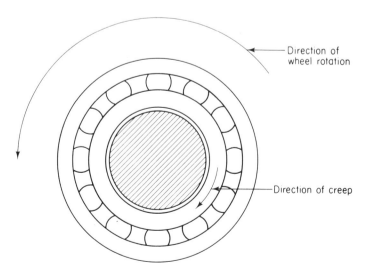

Figure 5-8 Wheel bearing movement to minimize fatigue.

Figure 5-9 A punch being used through a wheel hub to remove the outer tapered roller bearing race.

Axles, Bearings, and Housings

support the vehicle weight at the driving wheels of passenger vehicles and light trucks. This means that the axle shaft is constantly subjected to two loads. One is a *radial support load* that tries to *bend* the axle. The second is a *torsional load* which tries to *twist* the axle. Although both types of loading may result in axle failure, torsional loading will have no effect on axle bearing life. The axle bearings are only subjected to the radial load. Some axle bearings must also support a lateral load. This load tries to move the axle endways. The various loads placed on the bearing may cause premature bearing failure which will require replacement of the bearing. Bearing replacement must be completed before the brakes or other wheel systems can be properly serviced.

Axle shafts require a close fit of the bearing to the shaft. Ball and tapered roller bearings are press-fit on the axle shaft, along with a bearing collar as shown in Figure 5-10. A press fit of the bearing and collar on the axle is very important. The press fit is what securely retains the axle when side loads are applied.

The proper fit of the bearing on the axle shaft is essential on the straight roller bearing design as well as on other bearing designs. Side loads on the axle using this type of bearing are retained by a separate bearing system in the differential case. Fatigue of the bearing surface on the axle shaft may begin to appear. If this condition has occurred, the axle shaft as well as the bearing must be replaced to obtain normal service life from the replacement bearing.

Axle Housings. Vehicles equipped with typical dependent rear suspension and rear-wheel drive use the type of axle shafts and bearings discussed in the previous section. Those employing independent suspension at the driving wheels will use short axle shafts on each wheel hub with no bearing on the axle, as shown in Figure 5-11. Bearings are supported in the knuckle when used in a front-wheel drive or in the rear upright if a rear-wheel drive is used.

Any axle housing can become damaged in service. Problems such as a seized axle bearing may force the bearing to rotate in the housing. The resulting wear and scoring will cause a loose bearing fit in the housing.

Axle housings can become bent causing the tires to wear excessively. The wear can show up as

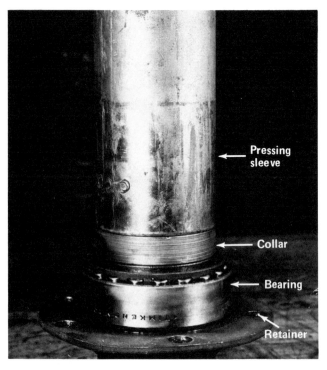

Figure 5-10 Set-up to press a bearing collar and bearing on an axle shaft.

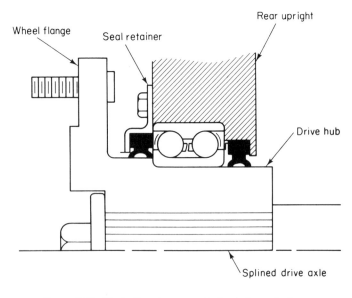

Figure 5-11 Typical bearing support of an independent drive axle.

Figure 5-12 A spalled outer race from a tapered roller axle bearing.

toe wear, as illustrated in Figure 3-22, if the housing is bent forward or backward. The bend will show up as camber wear, as pictured in Figure 3-21, if it has been bent in the vertical plane. The most common type of bend failure will result in a camber wear problem. This is most often the result of overloading the vehicle. Any bending of the rear axle will shorten bearing and axle life because the rotating axle is continually being forced to bend as it rotates. A typical example of this type of bearing failure from overloading, called *spalling,* is shown in Figure 5-12. Either type of axle failure will be cause for replacement of the axle housing. On independent rear suspensions, alignment can be adjusted. A worn or bent bearing support housing must be replaced.

5-3 SEALS

Seals used throughout a vehicle may take different shapes and be made of many different materials. The basic classifications of seals used in the suspension, steering, and brakes are radial lip seal, lathe cut seal, and O-ring seal. The materials used in the manufacture of the seals depend upon the temperature range in which the seals will operate, the qualities of the lubricant, the types of lubricant, the pressures to be sealed, types of contamination, and how effective the seals must be.

The *radial lip seal,* as shown in Figure 5-13, is a dynamic seal. It is used extensively to provide a seal between rotating axle shafts and hubs and the stationary spindles and axle housings. These seals

Figure 5-13 A typical radial lip seal.

may use either leather or synthetic rubber for the sealing element. Most synthetic rubber seals have a spring, called a *garter spring,* fitted to the seal element to aid in holding the sealing edge in contact with the moving member. The sealing element is usually encased in steel to provide a rigid supporting structure around the element. This aids in preventing distortion of the sealing lip. Synthetic seals may have double sealing lips. The secondary sealing lip of the double lip seal is designed to prevent contamination from reaching the primary sealing lip. Seals used at the wheels of the vehicle to retain lubricant in the bearing and seal out dirt are typical examples of the radial lip seal.

The *O-ring* and *lathe cut seals,* as shown in Figure 5-14, are used as either dynamic or static seals. They are desirable in applications where axial motion occurs and radial motion is minimal. Friction caused by this type of seal is greater than lip type seals. Both are low-cost seals. The lathe cut seal is the least costly of the two. These seals work ideally as static seals where thermal changes may occur between the parts or where vibration may make a gasket undesirable.

Seals that fail prematurely during service require replacement. Seal failure is most often caused by improper installation, inadequate lubrication, great temperature extremes, excessive contamina-

Axles, Bearings, and Housings

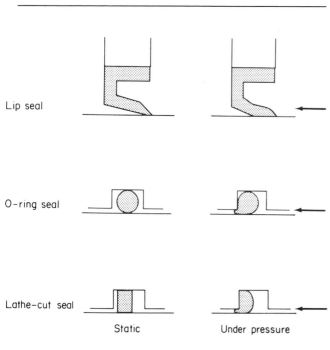

Figure 5-14 Shape and sealing principles of common seal types.

tion, excessive movement of the components, or misalignment of the shaft and seal. The technician must be aware of problems that cause seal failure to assure a proper repair that will provide normal service life of the seal.

Gaskets. Gaskets are a special type of seal used where movement does not occur between the surfaces to be sealed. A gasket, for this reason, is considered a *static seal*.

Gaskets are generally used where a low-cost seal is adequate to seal fluid-retaining components that are bolted together. Gaskets are generally used where there is a small pressure difference across the surface to be sealed. In special locations, with special types of gaskets, the gasket may be required to hold high pressure. High pressures are sealed by cylinder head gaskets in the engine while low pressure is sealed with gaskets on differential covers and brake backing plates.

Gaskets are often made of composition material that must be compatible with the fluid to be retained. They are composed of asbestos, cork, cellulose, organic, and inorganic materials with binders to give them the needed properties. In high-pressure sealing conditions, the gaskets may be stamped steel to provide the required sealing capability.

5-4 PROBLEM DIAGNOSIS

The first step in any repair procedure is to locate the malfunction. Bearing failures may be difficult to locate, depending on the degree of failure. Bearing failure is most easily detected when complete failure has occurred, although detecting a failing bearing in its early stages will usually prevent damage to other components.

Failure of a bearing may damage the bearing support and the seal if it is allowed to remain in operation. A damaged or failed seal is most easily detected prior to disassembly by a noticeable lubricant leak at the seal. A leaking seal at the wheel can contaminate the brake linings with lubricant. Contaminated linings must be replaced. In all instances, *early detection* and correction of the problem is the best policy for the automotive service technician to follow to minimize the cost of repairs for the customer.

Bearing Service. Bearing service may be done as periodic maintenance, during brake service, or when a customer has a noise, vibration, or even possibly an odor complaint. Bearing service is usually completed as periodic maintenance. The bearing and hub is often removed to repair another component. Bearing service is generally limited to the bearings on nondriving axles. This service includes disassembly of the bearing and races, cleaning and a careful inspection. Bearing service is completed by packing the bearing with grease and lubricating the required components.

Bearings at the driving axles may be sealed bearings. Sealed bearings are assembled with their own lubricant and do not require relubricating. Tapered roller or straight roller bearings may be lubricated with differential lubricant. Some tapered roller axle bearings require packing with

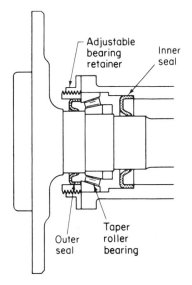

Figure 5-15 Typical seals on a grease-packed axle bearing.

grease. Grease-packed bearings are sealed from the differential lubricant and to prevent external leakage. This is illustrated in Figure 5-15.

Diagnostic Procedure. Bearing service that results from a customer complaint requires diagnosis. Although failure of one bearing may be a reason to inspect and service other bearings, the troublesome bearing should be located prior to disassembly. This will enable the technician to give an accurate estimate of the repair cost. The inspection may also show that a bearing noise complaint is actually tire noise. If possible, a road test with the operator should be used to determine the customer's specific complaint.

Road Test. Pinpointing the location of the faulty bearing can be difficult whether on a road test or in the service facility. The technician must understand conditions that will change wheel bearing loading while the vehicle is being driven. This will be an aid in locating the area of the problem.

During the test drive, the vehicle should be driven at constant speed, accelerated, decelerated, braked, and taken around left and right turns. The bearing noise should be noted during each mode of operation. A noise that changes in pitch and intensity during acceleration and deceleration may be either a differential or transmission noise. This can be traced further by selecting different transmission gear ranges and noting any changes in the sound. Acceleration and/or deceleration noise is most often traceable to the differential pinion bearings. Bearing noise generated at the pinion will increase at a faster rate than vehicle speed increases.

Bearing noise which changes intensity in a turn is most often a faulty axle or wheel bearing. The noise produced by the defective bearing will increase as the load on it is increased in a turn. During a left turn the noise level of a defective bearing on the right side will increase. This test should aid in locating a noisy bearing. A vehicle equipped with a C-washer lock axle controls the axial loads with the differential case support bearings. A faulty differential case support bearing can cause the noise level to increase during a turn and/or to increase during acceleration and deceleration.

If the location of the faulty bearing is difficult to trace, changes in the sound during braking may be an aid in identifying the faulty bearing. The noise is most likely coming from a front bearing if the bearing noise increases during light to moderate braking. The resulting change in noise level during braking is caused by the forward weight transfer to the front-wheel bearings.

Diagnosis in the Shop. Bearing diagnosis can also be done in the service area. Often it is not possible to road test the vehicle or possibly the noise level is so low that it cannot be heard during a road test. In the service area, the vehicle should be raised to allow an inspection of the tires. A severely cupped tire will make a noise similar to and barely distinguishable from a faulty bearing. If it is verified that the tires are not the noise source, the front and then the rear wheels can be spun to locate the defective axle or wheel bearing. A handy tool for rotating the nondriving wheels is the spin balancer used while balancing tires. This will turn each wheel at a speed high enough to locate any abnormal noise.

On front-wheel drive vehicles the engine can be used to spin the front wheels. One driving wheel is raised from the floor surface and the other wheels

Axles, Bearings, and Housings

are securely blocked to prevent vehicle movement. The raised wheel is driven at varying speeds not to exceed 30 mph (48 km/h) on the speedometer. It should be noted that when only one wheel is rotating, the speed of the raised wheel is twice the speedometer reading. Thirty miles per hour (48 km/h) on the speedometer is equivalent to sixty miles per hour (96 km/h) at the spinning wheel. If there is no distinguishable bearing noise, the opposite side should be raised and the test repeated.

The rear axle bearings of a rear-wheel drive vehicle can be checked by raising the vehicle from the floor surface. The parking brake can be applied on one wheel to isolate the noise to the other wheel. *Never* attempt to stop a rotating wheel by hand. Severe bodily injury can result. If the vehicle is equipped with a limited slip differential, both wheels drive at all times; therefore applying the parking brake will *not* stop one wheel.

If the preceding diagnosis procedure has not identified the faulty bearing, disassembly and inspection is in order. Once the defective bearing has been located, disassembly, inspection, and repair can be completed.

5-5 WHEEL BEARING SERVICE

Disassembly, whether to replace a defective bearing or service the used bearings, should follow an orderly procedure. This procedure will differ if the bearing serviced is on a driving wheel or on a nondriving wheel. Bearings used at a driving axle are usually referred to as *axle bearings,* while those at a nondriving wheel are called *wheel bearings.* The service procedure differs considerably for each type, so they will be discussed separately.

Disassembly. The wheel bearings are serviceable with the vehicle raised from the floor surface and properly supported. The wheel covers are removed to gain access to the wheel nuts. A chalk mark is placed on the wheel and hub or lug bolt to identify the original location so it can be replaced without disturbing the existing balance.

With the wheel removed, the brake components are accessible on a disc brake-equipped vehicle. On drum brake vehicles, removing the wheel allows the technician to work with the hub and drum while servicing the bearings. The drum and rotor removal procedures were discussed in Section 2-5.

If service is being performed because a bearing has failed and the bearing is seized to the spindle, the hub may have to be pulled off. It can be pulled using a rear hub puller attached to the front hub as shown in Figure 5-16. Use necessary precautions to prevent further damage to the spindle.

Inspection. After disassembly is complete, all parts must be thoroughly cleaned in a petroleum-based cleaning fluid, such as mineral spirits. The bearing cups, cones, rollers, cage, and spindle are carefully examined for any sign of wear or damage. When found, the reason for the failure should be determined.

Abnormal bearing roller and cup wear will usually show as spalling, brinelling, galling, or abrasive wear. *Spalling* is a fatigue-related phenomenon. At high mileage the bearing surface material loses its bond to the base metal. This leaves small cavities in the bearing surface that cause the bearing to run roughly. Spalling is pic-

Figure 5-16 Using a rear hub puller to remove a seized front hub.

tured in Figure 5-12. Both cup and cone are replaced when spalling occurs.

Brinelling is a wear condition that results from impact loading or vibrations. This type of bearing failure starts with bearing loads that occur while the bearing is not rotating. Rotation forces the bearing elements to move on a film of lubricant, spreading the load across the full bearing surface. If the bearing does not rotate while under load, the load will force the lubricant film from between the bearing element and the cup. This process, if continued, forms indentations in the cup or inner race. When the bearing is rotated, the brinelling will cause vibrations, roughness, and failure. For example, this problem could start on wheel bearings while the vehicle is being shipped by truck or rail. Brinelling requires replacement of both bearing cone and cup. The technician should determine the cause of the start of the brinelling process so it can be avoided on the new bearings. Brinelling is shown in figure 5-17.

Galling is another abnormal failure that occurs as a result of bearing overheating. This condition is usually traceable to inadequate lubrication, improper lubricant, or overheating. The actual transfer of material from the bearing race to the bearing element is evident as shown in Figure 5-18. Replacement of both the bearing cone and cup is necessary as well as a careful inspection of the related components for signs of wear and/or looseness.

(a)

(b)

Figure 5-18 Example of a galled bearing. (a) cup, (b) cone.

Figure 5-17 Brinelling shows up as lines across the bearing race.

Abrasive wear is not common in wheel bearings. This occurs when the lubricant is contaminated with foreign matter. An example is shown in Figure 5-19. This is more common in differential oil lubricated bearings where the oil may carry the contaminant from one part of the axle housing to another. Abrasive wear occurs on the complete bearing assembly and so the complete bearing must be replaced.

Smearing is caused by movement of the bearing cup or race in its bore. This shows up as shiny or polished areas around the outer circumference of the race or cup. When smear occurs, both wheel hub and bearing cup have to be replaced to satisfactorily correct the problem. While servicing wheel bearings, always inspect the cup for looseness in the hub. A properly fitting cup must be

Axles, Bearings, and Housings

Figure 5-19 Example of abrasive wear in a bearing: (a) cup; (b) cone.

pressed into the hub bore. Any cup looseness will allow looseness in the wheel assembly after bearing adjustment has been made. This looseness can result in premature failure of the bearing.

The spindle must also be carefully inspected for wear when servicing wheel bearings. Wear will most often show on the lower or load-carrying side of the spindle. The spindle will generally be damaged when a bearing seizes. This will show up as discoloration from heat or galling of the metal in the inner bearing race area on the spindle. The spindle must be replaced if it shows any damage, to assure maximum safety and service life. Spindle replacement is explained on page 253. If the spindle is in serviceable condition, the bearing cones should slide on the spindle by hand without looseness.

Assembly. Bearing cups that require replacement are carefully driven from the hub with a large drift and a hammer. New cups are installed in the hub with a tool that can push squarely on the cup to properly seat it in the bore. The best tool for this is a bushing driver. An improvised installing tool can be made by grinding the old cup to a reduced diameter. The old cup is fitted into the bore on top of the new cup. The new cup is seated by striking the old cup with a hammer. It is important to assure the bearing is properly seated against its shoulder in the hub bore or the cup will reposition itself. This will change the bearing adjustment while it is in service. When the cups are properly installed, the bearings are packed. Packing wheel bearings is discussed in Section 2-5.

5-6 DRIVE AXLE SERVICE

For the purpose of this discussion, axle service will consist of the axle, axle bearing, and seal replacement. Axle shaft removal and replacement is covered because it is necessary to remove the axle shaft to replace the bearing and/or oil seal. Axle service is divided into two major groupings: drive axles used with dependent suspensions and drive axles used on vehicles with independent suspensions.

Drive Axle with Dependent Suspension. Bearings and seals can be serviced with relative ease on standard vehicles using rear-wheel drive dependent suspensions. Bearing noise problems are diagnosed as described in Section 5-4. Seal failure will be indicated by lubricant leakage on the inside of the rear brake backing plate or as brake linings become saturated with differential lubricant.

The axle shaft must first be removed to enable the technician to gain access to the bearing and seal. The axle shaft is retained by one of two standard methods. A C-lock located on the axle inside the differential case holds most axles in place. A retainer plate outisde of the axle bearing holds the bearing on the other types. Each type of axle requires different disassembly and reassembly procedures. These are described in the following paragraphs.

Axle with C-lock. The tire and wheel assembly as well as the brake drum or rotor must be removed to allow the axle to be serviced. The brake shoes may require loosening to allow the brake drum to slide over the linings. If the vehicle has rear-wheel disc brakes, the caliper must be removed. This allows the brake rotor or drum to be removed. At this point, the axle will be identifiable as a C-lock type if there is no separate retainer plate. The C-lock type axle is shown in Figure 5-20.

Disassembly. The differential cover has to be removed to get to the C-locks. The lubricant is drained into a waste container as the cover cap screws are removed. The small differential pinion shaft lock pin, as pictured in Figure 5-21, is removed and the differential pinion shaft is slid from the differential case. Pushing the axle shaft inward slightly allows the C-lock to be removed from the

(a)

(b)

Figure 5-21 Removing the differential pinion shaft: (a) removing the shaft lock pin; (b) removing the differential pinion shaft.

axle inner end to free the axle shaft as shown in Figure 5-22. The axle is then carefully pulled out through the straight roller axle bearings and seal.

Service and Reassembly. With the axle shaft removed, the bearing surface on the shaft can be inspected. The bearing and the seal which remain in the housing may also be inspected. If there are any signs of spalling, galling, or brinelling, both the axle shaft and bearing must be replaced. Normal operating conditions show as a gray area on the bearing journal surface of the axle. Bearing and

Figure 5-20 An axle with a roller axle bearing held in place with a C-lock in the differential. There is no external bearing retainer.

Axles, Bearings, and Housings

Figure 5-22 With the axle shaft pushed slightly inwards, the C-lock can be removed.

seal removal is accomplished using a sliding hammer to pull the bearing and/or seal from the housing. One type of puller is pictured in Figure 5-23. Whenever the seal is removed, it must be replaced with a new seal. Never attempt to reuse an axle shaft seal.

The new bearing is coated with liberal amounts of differential lubricant, positioned in the axle housing, and driven in using a properly fitting bearing installation tool. The seal is then positioned

Figure 5-23 One type of puller used to remove the axle seal.

with the sealing lip facing inward. It is seated using a proper seal driver. The seal lip must be coated with differential lubricant. If a leather seal is used it has to be soaked 15 to 20 minutes in oil before it is installed.

The axle shaft is carefully inserted through the seal and bearing, turning it slightly until it slides into full mesh with the differential side gears.

The C-lock is installed in the recess groove at the inner end of the axle shaft inside the differential case. The axle is then pulled outward to seat the C-lock in its counterbore in the differential side gear. The differential pinion shaft is inserted with the lock pin hole aligned with the matching pin hole in the differential case. The lock pin is installed and torqued to the correct specification. The sealing areas of the differential cover are cleaned, a new gasket installed, and the cover properly tightened.

If seal failure allowed lubricant to get into the brake, the linings will have to be replaced. This procedure is covered in Section 9-8. If linings are to be replaced, they must be replaced in axle sets. This is due to the variation in lining coefficient of friction as discussed in Section 6-3. The brake drum or rotor is cleaned and installed along with any other components that were removed to make the repair. The brakes are readjusted and the tire and wheel assemblies are installed and the lugs properly torqued. The differential is filled to the required level with the proper lubricant. The brake fluid level is topped off. The job is completed when the wheel covers are installed and the vehicle is lowered to the floor.

Axle with Retainer Plate. Dependent rear-wheel drive suspensions that use either tapered or ball axle shaft bearings have a retainer plate located outside of the axle bearing. It is attached to the axle housing tube behind the axle flange. The retainer holds the bearing when axial loads occur. A typical retainer is pictured in Figure 5-24. The bearing and collar are held securely on the axle shaft by a press fit. Axle endwise movement is prevented because the bearing is held securely between a shoulder in the axle tube and the bearing retainer.

Disassembly. The axle shaft on this type of axle must be removed to service either bearing or seal. To facilitate removal, the vehicle is raised and pro-

Figure 5-24 An axle held in place with a bearing retainer.

Figure 5-25 A slide hammer axle puller used to slide the axle and bearing from the axle housing.

perly supported. The wheel and necessary brake components are removed as previously discussed. The axle retainer, located inside the axle flange, can now be loosened by removing the four or five attaching bolts. These can be reached through a hole in the axle flange. An axle puller (slide hammer with adapter) is attached to the axle shaft as shown in Figure 5-25 and impacted until the bearing moves from the housing. The axle must be pulled because the cup has a light press fit in the axle housing tube.

Service and Reassembly. With the axle shaft removed, the bearing and the seal can be inspected closely. Some ball axle bearings have their own lubricant sealed inside. An O-ring around the outside of the bearing provides the seal as pictured in Figure 5-26. Ball axle bearings that are lubricated with the differential oil have a separate axle shaft seal. Tapered roller bearings are often packed with wheel bearing grease and use two seals. One seal is placed in the bearing retainer (the outer seal) and one seal is placed in the axle tube (the inner seal). A gasket is used between the flange and axle housing tube. A shop manual should be consulted to identify which type of bearing and seal combination is being serviced.

The bearing is removed from the shaft by first removing the bearing collar. This is done by striking it with a hammer and chisel, as shown in Figure 5-27, to loosen it. This spreads the metal and loosens the collar. With the collar removed, a small bearing press, as shown in Figure 5-28, or a larger shop press, as shown in Figure 5-29, can be used to

Figure 5-26 A typical use of an O-ring seal on the outside of a self sealed ball axle bearing.

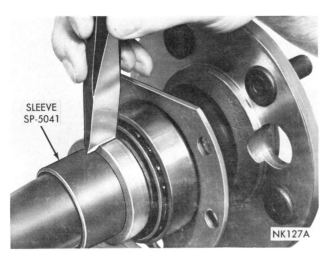

Figure 5-27 Using a chisel and hammer to loosen the bearing collar (Courtesy of Chrysler Corporation).

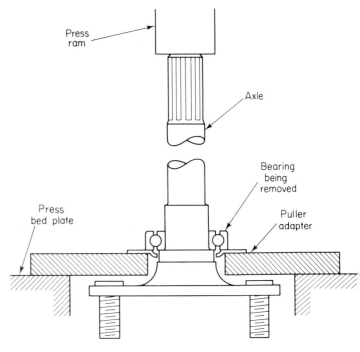

Figure 5-29 Set-up on a press to remove the bearing from the axle. The bearing will be damaged so it cannot be reused.

remove the bearing from the axle. In some cases it is necessary to grind a portion of the inner race flange to remove the rolling elements and outer race before the inner race can be removed from the axle. Never use an acetylene torch to heat or cut the

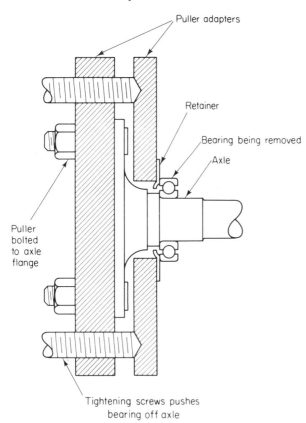

Figure 5-28 Bearing puller set up to remove the bearing from the axle. The bearing will be damaged so it cannot be reused.

bearing or collar to facilitate removal. The retainer can be removed after the bearing is off the axle.

With the bearing removed, the axle shaft can be inspected. Any damage to the axle shaft will require replacement of the complete axle shaft. A new seal is placed in the bearing retainer with the sealing lip facing the center of the vehicle. The retainer is placed over the axle shaft with the proper side inward. This is followed by the new bearing. If an adjustable tapered bearing is used, it is packed following the same procedures used for wheel bearings. The cup and cone are both placed on the axle with the small end of the cone facing outward toward the seal and flange. The bearing may now either be pressed on as shown in Figure 5-30 or pulled on using a smaller bearing tool, as shown in Figure 5-31. Sometimes the collar is pressed on with the bearing and sometimes it is pressed on separately. Care must be taken to fully seat the bearing against the shoulder. Depending on the design, the axle roller bearing outer race may remain in the axle housing bore after removing the bearing. If the bearing has been replaced, the outer race should also be replaced.

The axle housing is carefully inspected for wear or burrs and then thoroughly cleaned. Using a seal driver, the inner seal is installed to the proper depth with the sealing lip facing inward. Never reuse any axle seal. If the bearing outer race is fitted into the housing separately, it is coated with wheel bearing grease. It is driven into the housing bore with a properly fitting tool until it is seated. If the roller bearing is lubricated in service with differential lubricant, the bearing and seal area should be lubricated with liberal amounts of differential lubricant. If the bearing is of the sealed ball design, the surface of the outer bearing race should be coated with wheel bearing grease to prevent seizing in the axle housing bore.

The axle shaft is now carefully inserted through the seal and engaged in the side gear of the differential. The bearing retainer is positioned and properly tightened. Vehicles using tapered roller bearings may require an adjustment to set the proper bearing freeplay. This is accomplished through the use of shims at the retainer plate or by an adjuster placed in one of the retainer plates. The adjuster is only required on one side of the vehicle. It will control the total axle bearing freeplay. The

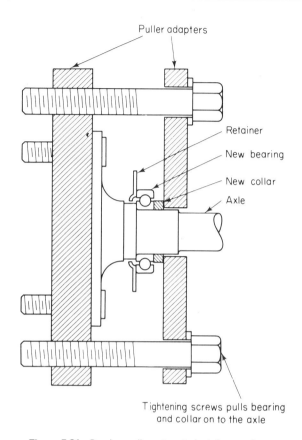

Figure 5-31 Bearing puller set-up to install a new bearing and new collar.

shop manual should be consulted to determine the type of adjustment, the proper method of adjustment, and the proper setting.

The brakes can now be serviced as discussed in Chapter 9 if they have become contaminated with differential lubricant. The brakes are assembled and properly adjusted. The tire and wheel assemblies are installed and properly tightened, and the wheel covers put in place. The differential lubricant and brake fluid are inspected and filled, as required. The vehicle is then lowered to the floor to complete the repair.

Drive Axle with Independent Suspension. The drive axle of an independent suspension uses two bearings at each wheel. Disassembly procedures differ, depending upon the particular suspension design. For disassembly, it is necessary that the

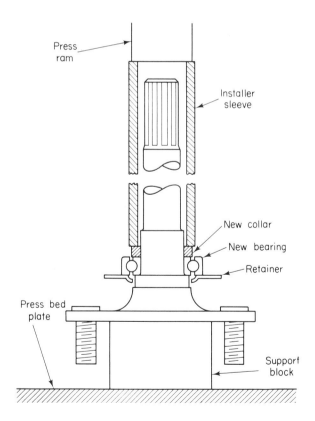

Figure 5-30 Set-up on a press to install a new bearing and a new collar.

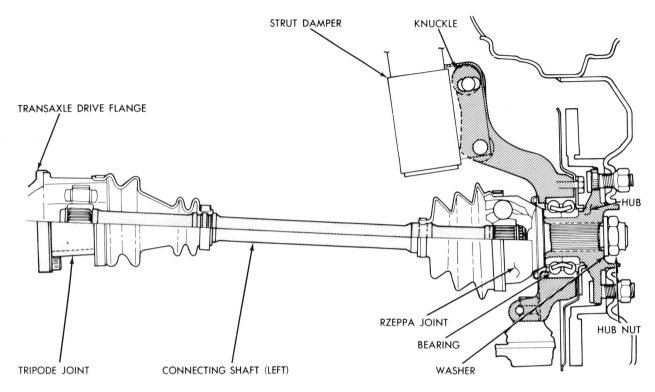

Figure 5-32 Typical front-wheel drive axle assembly (Courtesy of Chrysler Corporation).

vehicle be raised and properly supported. The tire and wheel, along with interfering brake components must be removed; either brake caliper or brake drum. The drive shaft, sometimes referred to as the *half shaft,* is removed from the vehicle by unbolting or loosening the clamps holding the shaft in place. A typical front wheel drive half shaft, called a connecting shaft, is shown in Figure 5-32. An axle shaft nut or stub shaft nut is commonly used to retain either the hub flange or the companion flange to the short axle. The axle shaft can now be either pushed or pulled from the bearings. This will allow access to the bearing in the hub carrier. The bearings used with this type of mounting are usually sealed and do not require replacement of seals or packing with grease. Reassembly with new parts is accomplished by reversing the disassembly procedure. It is most important that proper torque be applied to the flange retaining nut to prevent any relative motion. The axle shaft is installed, removed brake components remounted, and the wheel and tire properly tightened.

REVIEW QUESTIONS

1. List the advantages of each of the following: ball bearing, straight roller bearing, and tapered roller bearing. [5-1]
2. How does each type of bearing support shaft end thrust? [5-1]
3. What mountings on automotive wheel and axle bearings are light press fit and what mountings are a slip fit? [5-2]
4. What kind of loads will cause premature bearing failure? [5-2]
5. What axle housing faults can cause bearing failure? [5-2]
6. Name three types of seals used in suspension, steering, and brakes. In what application is each used? [5-3]

7. Describe the process used to locate a faulty bearing during a road test. [5-4]
8. Describe the process used to locate a faulty bearing in the service area. [5-4]
9. Why is it important to carefully inspect a used bearing before reinstalling it? [5-5]
10. How are bearing cups removed and installed? [5-5]
11. Describe two general procedures to be followed to remove axle shafts for bearing service. [5-6]
12. Where are seals located on axle bearings? [5-6]
13. Describe the procedure used to remove and replace a pressed on axle bearing. [5-6]
14. What axle bearings need to be adjusted? How is this done? [5-6]

6

Fundamentals of Brakes

Automotive service brakes must be able to stop the car, prevent excess speed when coasting, and hold the vehicle in position when stopped on grades. They are designed so that the braking effort can be varied by the driver to keep the vehicle under control.

To reduce the rotational speed of the wheel and slow the automobile, the service brakes force a stationary friction material, called a *lining,* against a rotating member, either a *drum* or *rotor.* The action of the friction material on the moving cast iron surface changes the kinetic energy of the moving vehicle into thermal energy or heat in the brakes. This heat is removed by moving air past the brakes.

The driver controls the force of the brake lining against the moving cast iron braking surface through mechanical, vacuum, and hydraulic mechanisms. Braking effect increases as the force transferred from the pedal pushes the stationary linings against the moving braking surface. Maximum braking occurs just before the wheel locks and the tire begins to slide on the road surface. Maximum braking, therefore, depends upon the adhesion between the tire and the road surface. When the tire slides on the road, braking effort is reduced and directional control of the vehicle may be lost.

6-1 BRAKING REQUIREMENTS

The brake system must provide smooth, predictable stopping power that is controlled by the operator. The required driver effort must not be excessive during a panic stop. To meet these braking requirements, minimum braking standards have been set for automotive brakes.

Braking Standards. These standards require the automobile to stop from a speed of 60 mph (96 km/h) with a minimum deceleration rate of 18 ft/sec^2 (5.5 m/sec^2) without skidding as the brakes are being applied using a maximum service brake pedal effort of 120 lb (534 N). The standards also include a high-speed stopping test made from a maximum speed of 100 mph (161 km/h) without skidding as a maximum pedal effort of 200 lb (890 N) is held. The brakes must provide straight stops with no wheel pull, regardless of the road surface on which the automobile is being driven. Anti-skid braking (covered in Section 7-8) of the type used on the heavy trucks and some automobiles will stop the vehicle in a straight line even though the road surface adhesion may change from tire to tire. The brakes are required to function normally when they operate in environmental extremes that include hot and cold climates as well as dry or humid climates. The environment in which the brake operates is constantly changing—not only the different climatic conditions, but also different local weather and road conditions. The brakes are subjected to dampness while being driven over rain-slicked highways and to dirt and dust while being driven over unpaved roadways. They must have the ability to recover their normal braking action after being wet. They must also be able to function in a dusty and dirty environment without adverse effects on the braking action.

Heat Load. The brakes are subjected to high heat loads. Heat results from the large amount of kinetic energy that is changed to thermal energy during braking. Brake temperatures may exceed 500°F (260°C) during a hard stop from cruising speed. The brakes must have the ability to work in these high temperatures and still exhibit little fade.

Fade is a condition that occurs when there is little braking effect with full brake pedal force. This condition results from the temperature rise in the brakes that becomes excessive within the wheel brake as a result of a number of frequent severe stops.

Brake Design Conditions. Wheel brakes are designed to occupy a minimal amount of space at the wheel so that small wheels and tires can be used on the vehicle. At the same time, enough space must be present to allow the heat to transfer away from the brake system to keep the brake temperature low enough to brake without fade. If the braking surface were not cooled as quickly as possible, the temperature of the brake would rise to the point that brake fade would occur.

Brake components are designed to function well at high temperatures. When the heat transferred to the atmosphere surrounding the brake is less than the heat developed while braking, the temperature of the brake components will rise. Brake drums, which absorb a great amount of heat during braking, may rise in temperature to more than 600°F (315°C). This high temperature will cause an increase in diameter of the drum braking surface, due to thermal expansion, as illustrated in Figure 6-1. This results in a greater required movement of the brake shoes to maintain contact with the braking surface. The effect of the added shoe movement is a low brake pedal height. The pedal height with the brakes applied is called the *pedal reserve*. The heat absorbed by the brake

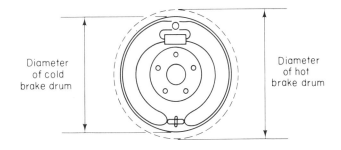

Figure 6-1 Drum diameter increases as it gets hot, which increases the lining-to-drum clearance.

Fundamentals of Brakes

linings increases their temperature to a point where the binders and saturates (page 86) liquefy within the lining itself and are forced to the surface by the pressure of the lining against the drum. The liquefied lining ingredients serve as a lubricant between the lining and brake surface resulting in reduced braking effect. Excessive overheating may also cause the brake fluid to boil within the hydraulic system which results in the vaporization of the brake fluid and this, in turn, causes a soft or "spongy" pedal feel.

Brake Fade. Fade is more common on drum brakes than on disc brakes. Disc brake rotors expose the braking surface directly to the atmosphere which results in rapid cooling. The rotor, when heated during braking, does not expand in the direction of lining travel. This can be seen in Figure 6-2. The expansion of the rotor in the radial direction does not affect the brake pedal height. Lining temperatures of the disc brake may rise to the point where the binders and saturates liquefy but the rapid cooling cycle of the rotor prevents this from causing a loss of braking action.

The close proximity of the lining to the hydraulic caliper cylinder in a disc brake as compared to a drum brake causes some of the heat to be absorbed by the cylinder and hydraulic fluid. This heat will increase the temperature of the fluid. Excess temperature will cause the brake fluid to boil and this will produce a soft pedal feel, just as it does in systems using drum brakes. The disc brake requires a high-temperature type brake fluid to prevent the fluid from boiling while under the high heat loads that are normal for disc brakes. During the early stages of disc brake development, boiling brake fluid was a problem. Changes in brake caliper design and brake fluid boiling points have nearly eliminated this problem in modern automobiles.

6-2 AUTOMOTIVE BRAKE TYPES

Various types and combinations of types of brakes are employed to provide safe braking on current vehicle designs. Each vehicle must, by law, have two independent brake systems for safety. The main braking system is hydraulically operated and is called the *service brake system*. The secondary or *parking brake system* is mechanically operated. It is designed to hold the vehicle on a 30% (16.7°) slope, and it must be capable of stopping the vehicle if the service brake system should fail.

The automotive brake systems are divided into three types of service brake combinations: drum brakes, disc brakes, and disc-drum combinations. All three combinations of brake systems are in production. Automotive drum brakes may be either the dual-servo type or the leading-trailing shoe type while some trucks use the double leading shoe type. Disc brakes are divided into two main types: the fixed caliper and the floating caliper design. Disc-drum combinations employ nearly every possible combination of disc and drum brake types.

Drum Brake Design. Drum brakes use an internal expanding *brake shoe* with the *lining* attached, working within the confines of a rotating brake surface called a *brake drum*. Typical drum brake parts are shown in Figure 6-3. The brake shoe diameter is expanded to contact the brake surface by a hydraulic cylinder that is referred to as a *wheel cylinder*. The brake drum is fastened between the wheel-tire assembly and the hub assembly on the nondriving wheels. It is fastened to the axle flange on the driving wheels of the vehicle. Fluid pressure from the *master cylinder* supplies fluid to the wheel cylinders causing them to expand. The expansion of the wheel cylinder through mechanical linkage forces the brake linings into contact with the rotating brake drum to provide braking action.

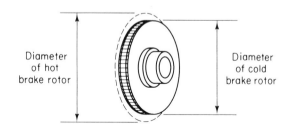

Figure 6-2 Rotor diameter increases as it gets hot but this does not change the lining-to-rotor clearance.

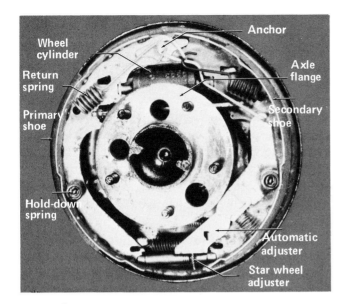

Figure 6-3 Parts of a typical drum brake.

The many components within the wheel brake are used to support, apply, adjust, release, and transfer the braking torque to other suspension members. A *backing plate* is attached to the wheel spindle or axle housing to provide support for the brake shoes and to transfer torque from the brake shoes to the suspension member. As the brake linings come into contact with the rotating brake drum, they attempt to rotate in the same direction as the rotating wheel. The brake shoes, with the linings attached, are prevented from rotating by the *anchor* which is securely attached to the backing plate. The wheel cylinder is also fastened to the backing plate and this provides a secure mounting location. On dual-servo brake systems, a *screw adjustment* placed between the two brake shoes, opposite the wheel cylinder, provides a means for adjustment to compensate for brake wear. This adjuster may be automatic (self-adjuster) or manual. Nearly all dual-servo brake systems used in conventional passenger cars employ an automatic adjuster. The automatic adjuster may not be used on vehicles employing heavy-duty brakes, such as police cars or trucks. The automatic adjuster makes use of the slight rotation of the brakes that occurs as the brakes are applied while backing the vehicle to make the automatic adjustment.

Leading-trailing (Figure 6-4) and double-leading brake shoe arrangements use a cam or screw to manually adjust for lining wear. One exception is a leading-trailing configuration used on the rear of some passenger vehicles that adjusts when the parking brake is applied and released as shown in Figure 6-5. These types cannot use an automatic adjustment as used in the dual-servo brake because two anchor points are used to prevent even the slight rotation of the brake shoes that is needed to actuate automatic adjusters.

The dual-servo brake shoes are held to the backing plate with *pins, hold-down springs,* and *retainers*. These parts are pictured with other brake parts in Figure 6-6. These parts allow the brake shoes to move when the brakes are applied. The additional fluid that has been forced into the wheel cylinder to expand the brake shoes and linings must

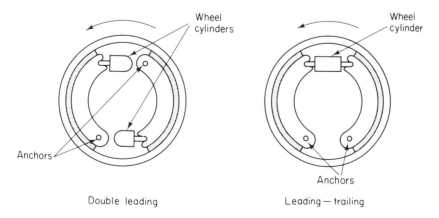

Figure 6-4 Alternative drum brake shoe arrangements.

Figure 6-5 A drum brake design that adjusts when the parking brake is set.

voir. The return springs may be used in a number of different arrangements, although their purpose, that of contracting the expanded shoes, is the same.

Drum Brake Operation. Self-energizing drum brakes are the most commonly used brake design. A self-energizing brake is one in which the drum rotation increases the brake shoe application force supplied by the wheel cylinder. Conversely a non-self-energizing brake design would be of the type in which drum rotation decreases the application force of the wheel cylinder. Figure 6-7 compares self-energizing with nonself-energizing brake action with a trailer that comes off from a hitch. Some wheel brakes incorporate one brake shoe that is self-energizing and a second shoe that is non-self-energizing. Other designs have both brake shoes self-energized or neither self-energized when the wheel is rotated forward.

be returned to the master cylinder to allow the brakes to release. The *return springs* provide the force necessary to pull the shoes away from the drum and to contract the wheel cylinder. This action will force the surplus fluid from the wheel cylinder and return it to the master cylinder reser-

Dual-Servo Brake Design. The most popular drum brake design currently in use employs self-energization. In this brake the front brake shoe (primary shoe) is applied by the wheel cylinder.

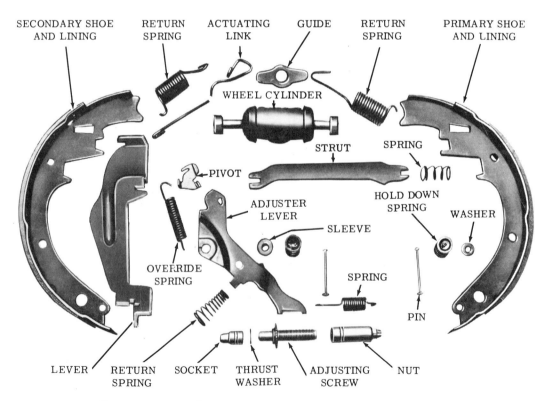

Figure 6-6 Parts of a self-adjusting dual-servo brake (Courtesy of Oldsmobile Division, General Motors Corporation).

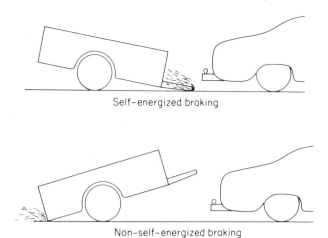

Figure 6-7 Comparison of the effect of self-energizing and nonself-energizing braking.

The force of the primary shoe, being pulled by the drum, is used to apply the rear (secondary) shoe. This type of brake is referred to as a *dual-servo* design and it is self-energized during both forward and rearward wheel rotation.

Self-energization occurs any time the drum rotates from the point of force application (wheel cylinder end) toward the fixed or anchored end of the shoe. In drum brakes as shown in Figure 6-8 the portion of the shoe that is forced into the drum is referred to as the *toe* while the end of the shoe near the anchor is termed the *heel*. If a point on the drum rotates from the toe to the heel, the shoe is a *leading* shoe. If rotation is from the heel to the toe, the shoe is a *trailing* shoe.

The self-energizing force on any brake is dependent upon a number of design limitations.

These limitations include the location of the anchor pin, the arc formed by the brake lining, the location of the lining on the shoe, and the coefficient of friction of the lining. The anchor pin location and coefficient of friction can be selected and designed so the brake will lock and not release until the drum is turned opposite to the direction of rotation. The locking effect of the self-energizing force will increase as the anchor position is selected closer to the center of the backing plate and as the coefficient of friction of the lining is increased. The location of the lining on the shoe and the arc of the lining will modify the self-energizing force. If the lining arc exceeds approximately 120° of the brake surface, as shown in Figure 6-9, the energizing force tends to increase near the ends of the linings to the extent that brake squeal may develop. As the lining location on the shoe is moved from the toe toward the heel of the shoe, the self-energizing

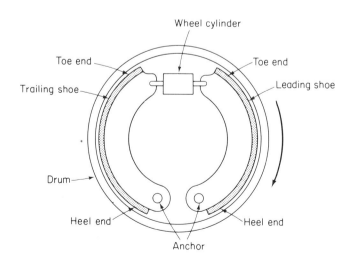

Figure 6-8 Brake shoe type identification.

Figure 6-9 Maximum lining length on a brake shoe.

force of the shoe is decreased. If the lining is positioned too near the toe, the energizing force is greatly increased and the brake will tend to grab.

Advantages and Disadvantages of Dual-Servo Brakes. Because of its efficient self-energizing system, the dual-servo brake requires less operator effort to stop the vehicle than other types of drum brakes. This makes it possible for the wheel cylinder to be made smaller than those on other types of brakes. This results in less pedal travel for the operator. The dual-servo system, because it employs a double-ended wheel cylinder and a moveable link between the two shoes, provides self-energized braking in both the forward and rearward directions.

The dual-servo brake, as shown in Figure 6-3, is not without its drawbacks. The secondary shoe application results from the applied forces of the primary shoe. Anything that affects the coefficient of friction of either lining will alter brake performance drastically. This brake design suffers more from brake fade than the other brake designs as a result of the loss in coefficient of friction due to increased temperatures. Water on brake linings also changes the coefficient of friction. When one brake is wet, on dual-servo brakes, violent pull will occur as the brakes are applied. This system, due to self-energizing, relies on the secondary shoe to provide approximately 75% of the total braking effort.

Alternative Brake Designs. Other designs of drum brakes (including the leading-trailing, double-leading, and double-trailing types) do not rely on the force of one shoe to apply the other shoe as in the dual-servo brake. This feature allows each brake shoe and lining to function independently of the other within each wheel brake. Because of their independent action, the effects of fade and dampness are not as great as in the case of dual-servo brakes. This results in more predictable braking action under adverse driving conditions.

The independent action of each brake shoe using the shoe arrangements above produces less stopping power for the same pedal force when compared to the dual-servo brake. Consequently, the wheel cylinder and brake diameters must be made large enough to provide the greater required braking power with a reasonable amount of driver effort.

The double-leading and double-trailing brakes require two one-ended wheel cylinders to provide the required operating characteristics. The double-leading brake has two self-energized shoes operating independently of each other in a forward direction. In a rearward direction the shoes are not self-energized and so braking will require considerably more pedal force. The double-trailing brake design operates without self-energizing action in the forward direction in the same way as the double-leading brake does in the reverse direction. Because these brakes do not have a self-energization force the braking action is directly *proportional* to the input pedal force. Due to a forward weight transfer of the vehicle during a stop, the front brakes require greater braking power and the rear brakes require a reduced or at least a constant braking power to prevent premature rear wheel lockup in a hard stop. Brake proportioning in some lightweight vehicles with disc-drum brakes uses a leading-trailing rear brake shoe configuration. The double-leading brake shoe configuration used on the front in conjunction with the double-trailing brakes used in the rear has, in obsolete brake systems, provided the required brake *proportioning*.

Disc Brakes. Disc brakes employ a brake disc that rotates with the wheel. The brake disc is usually referred to as a brake *rotor*. A hydraulically operated *caliper* is used to force the lining friction material against the braking surface of the rotor for stopping wheel rotation. The brake *shoes* fit inside the caliper housing to prevent end movement of the shoes when they are being forced into contact with the rotating brake disc. A typical disc brake is pictured in Figure 6-10.

Disc brake shoes move perpendicular to the face of the brake rotor to provide a clamping action on the rotor to slow the vehicle motion. The clamping action of the shoes against the rotor produces a force that is proportional to the driver pedal effort.

Disc Brake Design. Disc brakes are made in two basic configurations. The two types differ mainly in the number of pistons and the type of mounting

Figure 6-10 Disc brake assembly showing how the caliper can squeeze the brake shoes against the rotor.

used for the caliper. This is required because different application methods are used for the outboard shoe. The *floating caliper* design uses an adapter bracket mounted to or cast as a part of the steering knuckle. The caliper is mounted on this bracket. This allows the caliper to slide in and out with little resistance, while still preventing the caliper from rotating with the rotor as the brakes are applied. The piston forces the inboard shoe or pad into contact with the inboard side of the rotor. At the same time the reaction force from the piston moves the caliper housing inward, bringing the outboard shoe or pad into contact with the outboard side of the rotor. This action–reaction force of the caliper provides an equal force to both the inboard and outboard shoes to provide the clamping action of the linings on the rotor that is necessary to produce the required stopping power.

The floating caliper type of disc brake, as shown in Figure 6-11, usually employs a single piston when used on passenger vehicles. Heavier vehicle applications may use two or more pistons in the inboard side of the caliper to increase the applied force on the lining and to create more application points along the length of the shoe for more even lining wear.

The *fixed caliper* type of disc brake uses a securely mounted caliper, centered over the brake rotor with application pistons for each brake shoe. The caliper has an adapter plate mounted to the steering knuckle. The caliper is mounted on the plate in position over the brake disc. The secondary purpose of the adapter is to provide a link that will allow for some movement of the caliper housing. A small amount of deflection is required to prevent damage to the caliper and rotor during brake application.

The fixed caliper brake has two or more apply pistons. A minimum of one piston for each brake shoe is required. Many fixed caliper brakes use more than two pistons. Because the caliper movement is not used to equalize the application force on each side of the rotor, as in the floating caliper design, the separate pistons applying each shoe are required to equalize the application force on each side of the rotor. Equal application force prevents the rotor from being pushed to one side as the brakes are being applied. The use of more than one piston per brake shoe makes it possible to use a longer, more slender caliper. With more application points per shoe there can be greater total application force for each shoe. Drawbacks of the multiple piston caliper include an increased machining cost and the increased number of contact points provide more paths for heat to travel into the brake fluid, thus raising the temperatue of the brake fluid more rapidly, which in turn can lead to earlier brake fade.

Disc Brake Operation. Disc brakes offer better heat dissipation than drum brakes because they have direct contact with the moving air that surrounds the brake. The direct air cooling shortens the cooling cycle to provide quick recovery if the linings are heated to a point where fade could occur. The brake rotor will quickly free itself of any water and foreign matter from its braking surface during rotation. This occurs as the centrifugal force throws the contaminants outward, away from the vertical brake surface.

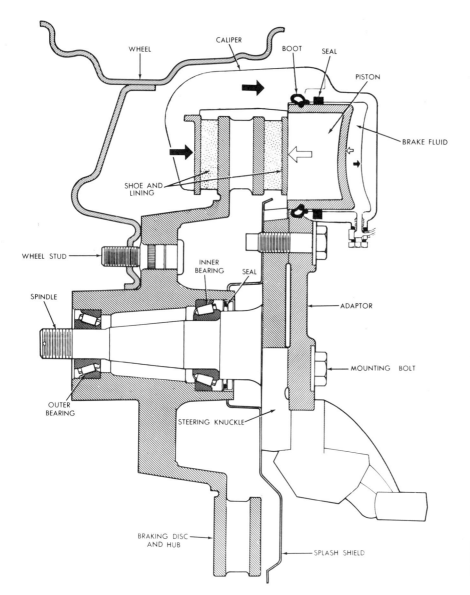

Figure 6-11 A section view of a typical floating caliper disc brake (Courtesy of Chrysler Corporation).

Disc brake linings are not self-energized because their movement is perpendicular to the disc rotation when they are being applied. This lack of self-energization reduces the tendency for adverse braking that occurs in self-energized brakes. This is especially noticeable when the brakes are overheated. Because the linings are not self-energizing, a greater application force is necessary to provide the required braking action with moderate driver pedal effort. The high application force on the rotor is achieved by using large diameter hydraulic cylinders and, most often, includes the use of a power assist to increase the force produced by the brake pedal. Because the disc brake calipers have larger diameter pistons than those used in drum brakes, the disc brake requires a larger quantity of brake fluid to move the pistons an equal distance. The service brake pedal must travel farther to provide this increased volume of brake fluid. The requirement for increased pedal travel may be overcome by using a larger diameter master cylinder. This, in turn, will decrease the mechanical advantage of the brake system. A power brake booster is then required to increase the mechanical advantage sufficiently to allow stops with a reasonable brake pedal force.

The brake caliper operates much like a vise as the brakes are being applied. The clamping action of the caliper on the rotor places a force on the caliper that tries to make the caliper spread apart. Any distortion of the caliper will result in a greater pedal travel and a "spongy" pedal feel. For this reason most domestic vehicles use calipers made of cast iron. Calipers on some lightweight vehicles are made of aluminum to lessen the weight of the brake components at the wheel.

6-3 BRAKE LINING

The *brake lining* is a composition friction material that is fastened to the *brake shoe*. It comes into contact with the brake surface as the brakes are applied. The brake lining is carefully compounded to give the required coefficient of friction when it is in contact with the brake surface, so that the brakes will provide smooth, predictable braking without being overly sensitive.

The brake lining is fastened to the brake shoe by *riveting* or by *bonding*. A picture of shoes using these fastening methods are shown in Figure 6-12. Riveted linings use brass rivets to attach the lining to the shoes. This makes a solid mounting that is soft enough to dampen vibrations allowing it to be quieter than a bonded lining. The brass rivet prevents damage if the lining is so worn that the rivets come into contact with the brake surface. Bonding is a method of cementing the lining to the brake shoe. The bonding adhesive cement is applied and then the shoe and lining are clamped together under high pressure. The assembly is heated at a temperature of approximately 1000°F (535°C) to cure the adhesive. Bonding provides a good solid method of attaching the lining to the shoes that lends itself to assembly line manufacturing. Bonded lining tends to be nosier in use than riveted lining.

Brake Lining Indentification. Brake linings are coded to provide the purchaser with a vendor compound and coefficient of friction identification. An examle of one lining code is FF-20-AB:

FF   Friction Identification
20   Compound Identification
AB   Vendor Identification

This is called an "edge code" and it is found on the edge of the lining. The friction identification must be uniform from manufacturer to manufacturer. The vendor and compound identifications may be any combination of numbers or letters. An example of edge coding is shown in Figure 6-13. The friction identification will indicate to the technician how a lining will perform when it is used at normal temperatures and when it is used at high temperatures. The first letter in the friction code identifies the normal coefficient of friction while the second letter indicates the hot coefficient of friction.

Code Letter	Coefficient of Friction
C	Not over 0.15
D	Over 0.15 but not over 0.25
E	Over 0.25 but not over 0.35
F	Over 0.35 but not over 0.45
G	Over 0.45 but not over 0.55
H	Over 0.55
Z	Unclassified

Most automotive linings have a normal temperature coefficient of friction between code E and F, or 0.25 to 0.45. The high-temperature coefficient of friction would normally be between D and F, or 0.15 to 0.45.

Organic brake linings will usually have a normal coefficient of friction code higher than their

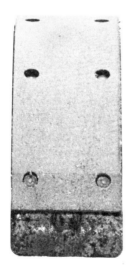

Figure 6-12 Bonded brake lining (left) and riveted brake lining (right).

Fundamentals of Brakes

Figure 6-13 An example of brake lining edge coding.

high-temperature coefficient of friction code. This means that the coefficient of friction of the organic lining decreases as the temperature increases. Metallic linings may show a high-temperature coefficient of friction higher than the normal temperature coefficient of friction.

The coefficient of friction is *not* a measure of a brake lining quality. When replacing brake linings, the coefficient of friction should be matched with that of the original brakes. The vehicle manufacturer selects the proper coefficient of friction for the lining so that it will provide the best braking under all normal operating conditions.

A high coefficient of friction may result in an overly sensitive brake that could cause abnormal wheel locking while operating on slippery road surfaces. A low coefficient of friction would require greater pedal-apply pressures to produce normal braking. When replacing only one axle brake lining set at a time, the technician should *always* match the coefficient of friction of the replacement linings to the original equipment linings. This will prevent altering the *brake bias* from the front to the rear wheels. Brake bias is the increase in braking effectiveness of the front wheels over the rear wheels that occurs as a result of the normal weight transfer to the front wheels during a stop. If the brake bias is altered, premature brake lockup of either the front or the rear wheels may occur. Brake bias is designed so that the front wheels will lock just prior to the rear wheels locking. This situation will provide maximum control of the vehicle under all braking conditions. If the rear wheels were to lock prior to the front wheels, the loss of rear wheel traction would allow the vehicle to spin or skid uncontrolled. With a bias allowing the front wheels to lock just prior to the rear wheels locking, the driver can adjust the force on the brake pedal to vary the brake application pressure and thus prevent tire skidding to maintain vehicle control. With the front brakes locked and the rear wheels turning the vehicle will brake going straight ahead and not spinout as occurs when the rear wheels lock up.

Lining Surface Area. The ratio of lining surface area to vehicle weight is an important consideration in brake designs. If the lining area for a particular vehicle were too small, the service life of the lining would be abnormally short. The use of small lining areas will increase the possibility of fade due to overheated linings. Excessive lining and drum areas require larger brake drums and thus increase the space required at the wheel. If the lining area is increased by making the lining longer, within a specified drum size, cooling will be reduced due to the area occupied by the linings on the drum surface. This can also cause brake squeal from "pinching" the linings near the toe and heel ends.

Passenger car linings are usually limited to about 25 pounds of vehicle gross weight per square inch (1.75 kg/cm^2) of lining for drum brakes. Disc brakes are limited to about 50 to 75 pounds of vehicle gross weight per square inch (3.5 to 5.25 kg/cm^2) of lining. Lining wear rate is accelerated as the temperatures are increased. The increase in vehicle weight per unit area of lining for disc brakes is possible because the disc brakes have a much greater cooling rate than drum brakes.

Lining Materials. Organic linings are most popular due to their wide operational temperature ranges. Metallic linings on the other hand have excellent high-temperature characteristics with poor low-temperature performance. Organic brake linings are compounded from mixtures of asbestos, resin, and filler materials, formed into the proper shape for attachment to the shoes. Organic lining material is composed of approximately 50 percent asbestos, with the remaining 50 percent taken up by seven or eight ingredients, including one which is metallic, to give the lining the desirable temperature characteristics. Other ingredients include compounds to hold the lining ingredients together called *binder,* and compounds to prevent the lining from absorbing water called *saturates.* The remaining ingredients build desirable and necessary operating characteristics into the lining material.

The mixture of materials which make up the lining compound are blended together in lots. Each lot has a manufacturer's tolerance for each ingredient as well as for the lot as a whole. These tolerances often lead to linings made from different lots having slightly different braking characteristics. For this reason manufacturers package linings in sets that originate from the *same lot.* Every measure is taken to prevent mixing linings from different lots to make sure that the same lot is placed on both brakes on the same axle. These are placed together in the same packaging units. The result of a lot mix may result in abnormal braking such as wheel pull, making the vehicle dangerous to drive and producing a brake problem that is difficult to diagnose.

6-4 PARKING BRAKES

The parking brake system is used to hold one or more of the vehicle brakes in an applied position for an extended period of time. This brake system must be capable of holding the vehicle on a grade and bringing the vehicle to a stop if the service brakes fail. The parking brake system used on most current model passenger vehicles operates by applying two rear-wheel brakes through a mechanical system of cables and levers. A typical parking brake system is shown in Figure 6-14.

Parking Brake Operation. The parking brake is applied, by the operator, by depressing a foot pedal or pulling on a hand-operated lever. The parking brake must remain effective even though brake fluid leakage and/or engine failure may have occurred. For this reason the parking brake is either completely independent of the service brakes or uses a mechanical system to directly apply the service brake linings.

The parking brake must meet rigid specifications to be considered safe for public use. For passenger vehicles the parking brakes must be capable of holding the vehicle on a 30% grade with an application force of no greater than 150 pounds (667 N) on a foot pedal or 100 pounds (445 N) on a hand-operated lever.

When the parking brake is used in conjunction with either dual-servo or leading-trailing service brakes, the service brake shoes and linings are also used for the parking brake. A cable is attached to the parking brake mechanism at the driver's controls. The cable travels beneath the vehicle to

Figure 6-14 A schematic of a typical parking brake system.

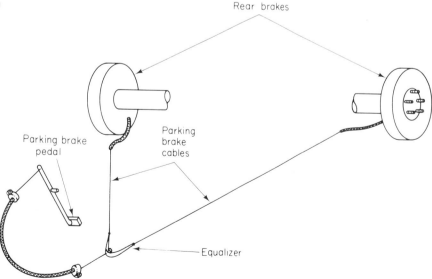

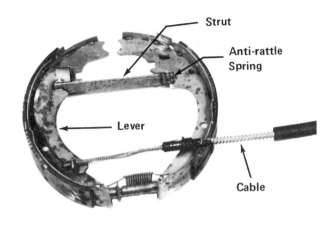

Figure 6-15 Parking brake parts within a rear drum brake.

the brake cable equalizer. From the equalizer a separate cable runs to each rear wheel. The equalizer allows equal tension in the brake cable to each wheel for equal brake effectiveness. The brake cables leading inside the wheel brakes are attached to a wheel brake lever which in turn works the brake strut to mechanically force the shoes to expand (Figure 6-15). It is common for the parking brake system within the wheel to have a mechanical advantage of approximately 3:1. If a hand-operated dash-mounted parking brake lever is used, an intermediate lever is often used between the front cable and the equalizer to increase the mechanical advantage (Figure 6-16). The intermediate lever usually produces a mechanical advantage of at least 2:1. If a foot-operated or floor-mounted hand lever is used, the intermediate lever is not necessary because the mechanism itself produces the required mechanical advantage. The overall mechanical advantage encountered is approximately 6:1 (3 × 2). With a mechanical advantage of 6:1 an input force of 100 pounds (445 N) on the parking brake lever will move the brake shoes out with a 600 pound (2670 N) force.

Parking Brake Systems. A parking brake used in conjunction with rear-wheel disc brakes may take a number of different forms. On domestic passenger vehicles two types are used. The first type of parking brake system used with rear-wheel disc brakes employs a completely separate brake system for the parking brake. The inside surface of the rear disc brake rotor is machined to form a small (usually 6 to 7 inch (15–18 cm diameter) drum. Inside the drum is a dual-servo cable-operated brake system, as shown in Figure 6-17. The system operates in the

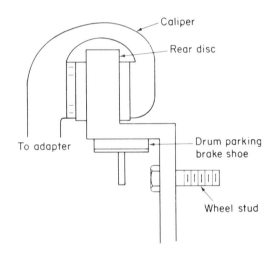

Figure 6-17 Drum parking brake in a rear disc brake.

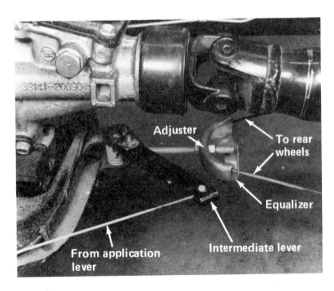

Figure 6-16 Location of the intermediate lever in a typical parking brake system.

same manner as that previously discussed for dual-servo brakes. This system does not feature automatic adjustment so periodic manual adjustment is required when the linings wear.

The second type of parking brake system used with rear disc brakes provides a mechanism for the brake to be mechanically set through the disc brake caliper piston. A caliper of this type is shown in Figure 6-18. A ball riding on an inclined ramp within the caliper causes the rotational movement of the caliper lever to produce an axial movement

6-4 Parking Brakes

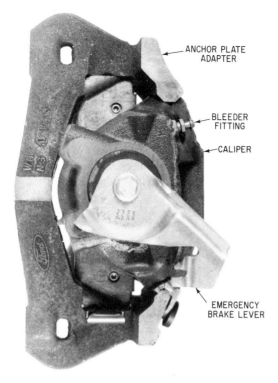

Figure 6-18 Parking brake built into a rear caliper.

Figure 6-19 Parts of the parking brake mechanism in a rear caliper rotor.

of the caliper piston. As the parking brake is applied, the balls are forced up the inclined ramps forcing the ramps apart. This can be seen in Figure 6-19. This separation is transferred along the parking brake adjuster mechanism to force the piston outward clamping the lining against the rotor. A screw and high-helix-angle external threads on a nut provide for automatic adjustment to compensate for lining wear. The adjustment mechanism is pressed into the caliper piston. Clearance between the external threads on the nut allow for normal hydraulic brake application and release. If the piston movement exceeds that clearance, the threads force the nut to turn and this provides automatic adjustment.

Other types of parking brakes have been used in the past. One type that was popular was fitted into a drum connected to the transmission output shaft. Setting the parking brake would then prevent rotation of the drive shaft to prevent vehicle movement. Similar drive-shaft parking brakes are sometimes used on medium-size trucks.

REVIEW QUESTIONS

1. What energy change occurs during braking? [INTRODUCTION]
2. What is the minimum braking deceleration rate without skidding? [6-1]
3. When might brake fade occur? [6-1]
4. What limits the minimum size of the brakes? [6-1]
5. Describe how heat expansion affects drums and rotors. [6-1]
6. What conditions cause the brake pedal reserve to change during braking? [6-1]
7. Name three types of service brake combinations. [6-2]
8. Identify the parts of a drum brake assembly using terms in this chapter. [6-2]
9. Describe self-energizing as it applies to dual-servo brakes. [6-2]
10. What is the advantage of using self-energizing brake shoes? [6-2]

11. What type of brake shoe is affected most by brake fade? [6-2]
12. How does floating caliper design differ from fixed caliper design? [6-2]
13. Why are disc brakes considered superior to drum brakes? [6-2]
14. How is brake lining fastened to the shoe? [6-3]
15. What do the lining codes indicate about the lining? [6-3]
16. Why should brake linings be replaced in axle sets? [6-3]
17. What is meant by brake bias? [6-3]
18. How is the required brake lining area determined by the brake engineer? [6-3]
19. Why is the lining lot number important? [6-3]
20. Why is an equalizer required for the parking brake? [6-4]
21. What applies the rear brake shoes in the drum parking brakes? [6-4]
22. What applies the rear brake shoes in disc parking brakes? [6-4]

Brake Operating Systems

Automotive brakes are applied through both mechanical linkages and hydraulic systems. The driver controls the vehicle braking with a brake pedal. The service brake pedal is usually suspended from a channel attached between the steering column-to-dash attaching point and the fire wall. A mechanical linkage from the pedal moves the master cylinder piston which then converts the mechanical pedal force to hydraulic pressure. Each wheel cylinder receives the same hydraulic pressure. The pressure pushes each wheel cylinder piston outward. The wheel cylinder pistons force the brake linings against the braking surface of the drums or rotors. The harder the driver pushes the pedal the greater the stopping force of the brakes. In modern automotive brake systems the hydraulic pressure in the brake system is modified with a number of specialized hydraulic valves to produce the required brake bias for the vehicle.

7-1 HYDRAULIC PRINCIPLES

A review of hydraulic fundamentals is necessary in a discussion of the hydraulic brake system. Hydraulic system operation is based on Pascal's law, which states that pressure on a confined fluid is transmitted equally in all directions. As this principle applies to the automotive hydraulic brakes, the pressure in any one part of the brake fluid in

Brake Operating Systems

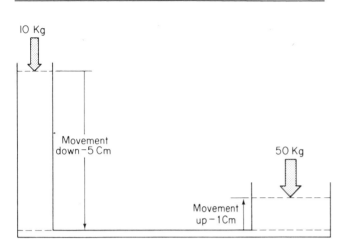

Figure 7-1 Pressure remains equal but the moving distance and force change between two interconnecting piston-in-cylinders of different sizes.

the enclosed hydraulic brake passage is equal to the pressure in any other part of the same passage.

A Simple Hydraulic System. The classical example used to illustrate Pascal's law is to show two different-sized interconnected cylinders. The two-cylinder system is filled with fluid and pistons are placed on top of the fluid in each cylinder, as shown in Figure 7-1. A small weight on the small piston can balance a large weight on the large piston, when the areas of the pistons are scaled properly. This is the same as balancing two weights on a mechanical beam lever system. Increasing the weight on the small piston will allow an increase in the weight that can be supported on the large piston. The force on the two pistons in a balanced system is directly proportional to the ratio of the piston areas. Assume that the small piston has an area of one square centimeter and the large piston has an area of five square centimeters. The weight that can be supported on the large piston is five times the weight that is on the small piston when the system is in balance (large piston area/small piston area equals the ratio of the force difference).

Nothing can be obtained free. In the hydraulic system just described, the movement of the small piston is five times the movement of the large piston. This is true of the mechanical lever system, too. The amount of movement of the pistons is the reciprocal of the force the piston produces (small piston area/large piston area = ratio of the distance moved). When the one square centimeter small piston moves down ten centimeters, the five square centimeter piston moves up two centimeters. *The volume of the fluid displaced by each piston is equal when movement takes place.* When one piston moves down, the fluid displaced by that piston is forced into the other cylinder to push the piston up. To summarize: The force on the small piston times the distance it moves is equal to the force on the large piston times the distance it moves.

In both cases just described, when the pressure in one part of the system changes, the pressure in all parts of the system change. The best example of this principle in automobiles is in the brake system. In the brake system the fluid is trapped between the small area of the master cylinder piston and the large area, made up of the combined areas of all of the wheel cylinder pistons.

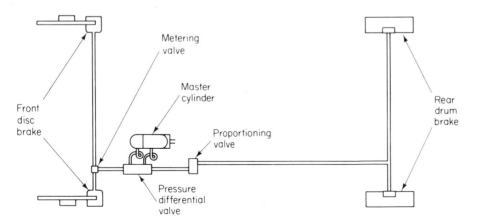

Figure 7-2 Schematic diagram of a typical front-rear split hydraulic brake system.

The hydraulic brake system illustrated in Figure 7-2 consists of the *master cylinder, wheel cylinders, hydraulic valves* (that either alter, delay, maintain, or compare pressure), steel *brake tubes,* and flexible *brake hoses* (that deliver the brake fluid under pressure to the wheel components). The pressure is evenly distributed through the brake lines, independent of the size or length of the lines. Through the use of different size areas on hydraulic components in the master cylinder and wheel cylinders, a force multiplication takes place. The use of a hydraulic system provides ease of connection between the service brake pedal inside the vehicle and the brake assemblies in the wheels that move up and down as well as turn as the wheels rotate.

7-2 THE BRAKE SYSTEM

The primary purpose of the master cylinder is to move hydraulic fluid under pressure in the brake system. This is accomplished by converting the mechanical input force from the service brake pedal to hydraulic pressure within the master cylinder. The mechanical advantage produced by the linkage arrangement at the pedal, as shown in Figure 7-3, multiplies the operators effort to increase the input force delivered to the master cylinder. This mechanical advantage is a ratio between 3:1 and 7:1. An operator-applied effort of 150 pounds (668 N) on the service brake pedal is considered maximum effort for the average operator. With a mechanical advantage of 4:1 and using 150 pound (668 N) force the mechanical input force to the master cylinder would be 4 × 150 pounds (668 N) or 600 pounds (2670 N).

Master Cylinder Size. Master cylinder bore sizes of 1 inch and 1.125 inches (25.4 mm and 28.6 mm) in diameter are commonly used on domestic passenger vehicles. The larger-size cylinder bore would most commonly be used with disc or disc-drum brakes in conjunction with a power booster. This large size is required to provide the fluid displacement necessary for the large brake cylinders used in disc brake calipers without having excessive pedal travel. Drum brakes will usually have a 1 inch (25.4 mm) diameter master cylinder because the volume of fluid required for brake application is less than that required for disc brakes. Disc brakes require a higher pressure than drum brakes. This is due to the nonself-energizing feature of the disc brakes. At the same time a greater fluid displacement is required because the disc brakes have larger cylinder bores. *Greater hydraulic pressure,* with the same operator input force, can only be accomplished by increasing the mechanical advantage of the pedal or decreasing the master cylinder bore size, both of which will result in *increased pedal travel. Greater fluid displacement* can only be achieved with a decreased mechanical advantage or an increased master cylinder bore size, and this will result in a *reduced hydraulic pressure.* For these reasons, a power booster is generally used to multiply the pedal force while a larger diameter master cylinder is used to increase the fluid displacement. In this way both demands of the disc brake will be fulfilled while maintaining normal operator pedal force and pedal travel. Some of the lighter vehicles of today do not require the use of a power booster. The operator of a lightweight vehicle can push the pedal just as hard as he can on a heavy vehicle. The applied force on the pedal of a light vehicle is a larger percentage of the required stopping force than that put on a heavy vehicle. This is enough to adequately stop the light vehicle.

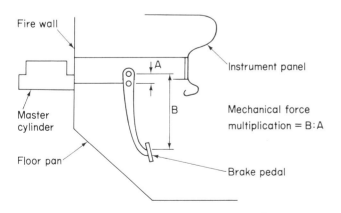

Figure 7-3 Typical service brake pedal linkage arrangement.

Brake Operating Systems

Figure 7-4 A dual master cylinder with the cover removed for fluid level inspection.

Dual Master Cylinder. A dual master cylinder (Figure 7-4) is used on conventional passenger vehicles. It is designed with two separate hydraulic pressure areas, each connected to one-half of the brake system. This will provide emergency braking in the event that one part of the brake system fails. With the hydraulic system divided at the master cylinder, a hydraulic failure at one wheel will result in only one-half of the system becoming inoperative. The other half of the system will still provide braking.

When using a dual master cylinder, the hydraulic brake system may be split in a variety of different ways. The most common method of splitting the system on domestic passenger vehicles is to split the front from the rear. The front-wheel brakes connect to one-half of the master cylinder, and the rear-wheel brakes connect to the other half. A diagonal split as shown in Figure 7-5 is employed on some vehicles, especially front-wheel drive types. In these systems the left front and right rear brakes are split from the right front and the left rear brakes to make up the two halves of the split system. A three-wheel system split is another alternative. It is connected such that the front wheels and one rear wheel make up half the system. Each half is connected to the separate halves of the master cylinder. A four/two-wheel system can be used. In this system the front brakes will always work in the event of partial system failure. The front and rear split is quite effective on front-engine rear-wheel drive vehicles, but is is considerably less effective on front-engine front-wheel drive vehicles. In the latter case, if the front brakes fail, the rear brakes would have little stopping effect and this would allow the vehicle to spin. A diagonal split is effective on front-wheel drive vehicles but it suffers in complexity. The three-wheel and four/two-wheel systems are so complex that early failure could be due to the number of hoses, cylinders, and valves required to make the systems operative, and cost of the systems is a major drawback.

7-3 MASTER CYLINDER

The dual master cylinder is designed with a primary and a secondary piston. The *primary* piston is in direct mechanical contact with the input push rod from the brake pedal linkage. The primary piston

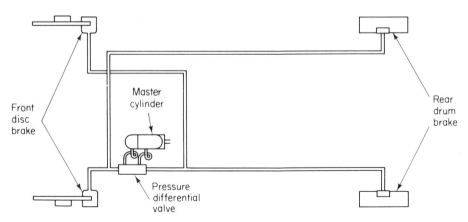

7-5 Schematic diagram of a typical diagonal split hydraulic brake system.

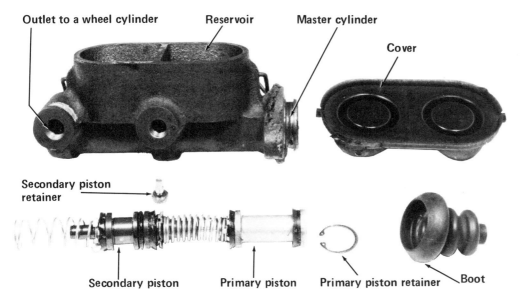

Figure 7-6 Parts of a disassembled master cylinder.

is fitted with a primary cup seal and a secondary seal. In normal operation the *secondary* piston is moved by hydraulic force from the primary piston. The secondary piston is fitted with a primary cup seal, a secondary cup seal, and an inverted secondary cup seal (these may be O-ring seals). The parts of a typical master cylinder are shown in Figure 7-6.

Master Cylinder Design. Each half of the master cylinder includes a fluid reservoir, compensating port, breather port, outlet port, and a residual check valve at the outlet port (except when used with disc brakes).

The residual check valve at the outlet port (Figure 7-7) is used to hold a static pressure ranging from 6–18 psi (41–124 kPa) with the pedal released. This *static pressure* helps maintain contact between the wheel cylinder cup lips and cylinder bore while the brake is released. This will prevent air from entering the system at the wheel cylinders. The residual check valve is not required on disc brake systems because the caliper seal is not a lip seal and no piston return springs are used. Static pressure held by a residual check valve would be enough to hold the caliper pistons out. This would keep the pads in contact with the rotor and cause brake drag. Drag is critical on disc brakes because it uses power, the brakes overheat, and they wear rapidly.

Master Cylinder Operation. In master cylinder operation, as the push rod begins to move, the primary piston seal first passes over the compensating port which opens to the reservoir. When the compensating port is closed, pressure begins to build between the primary piston seal and the inverted secondary piston seal. As this pressure between the pistons increases, it will begin to move the secondary piston until the primary seal on the secondary piston passes over its compensating port. This position is shown in Figure 7-8. There is no pressure increase in either section of the master cylinder until the compensating port for that section of the cylinder is closed off. As the primary piston is pushed in the bore by the brake linkage

Figure 7-7 A residual check valve being removed from below the brake line seat at the master cylinder drum brake line outlet.

95

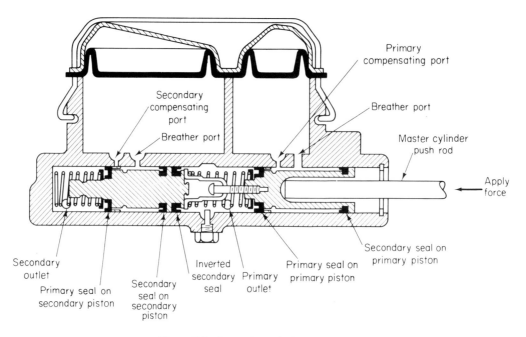

Figure 7-8 Parts of a typical master cylinder.

and push rod, the pressure increases. This pressure will continue to move the secondary piston in the master cylinder. The fluid will flow past the residual check valve (on drum brakes) and into the brake lines to apply the brakes. The secondary seal of the primary piston is used to carry a supply of brake fluid as it moves. This fluid can pass around the back side of the primary seal during pump-up as described in the next paragraph. The main function of the secondary seal on the primary piston is to prevent external leakage from the cylinder. Any leakage that occurs around the primary seal will return to the reservoir through the breather port located between the primary and secondary piston seals. The secondary seal on the secondary piston performs the same functions except that it prevents the fluid from passing into the primary chamber rather than preventing external leakage. It is important to note that at no time during the movement of the master cylinder piston is either breather port closed. This is important because the fluid carried by the secondary seals must not exceed atmospheric pressure.

Pump-up. The phenomenon of pump-up occurs when the brake pedal is rapidly cycled. Pumping forces fluid into the brake lines and wheel cylinders faster than it can return to the master cylinder. The additional fluid that is forced into the brake lines causes an increase in the brake pedal height. This added fluid will take up any excess clearance that exists between the wheel cylinder piston, lining, and braking surface. Brake pedal cycling, therefore, will create a temporarily high pedal to provide some braking. Pump-up is possible because the pistons in the master cylinder can return faster than the brake shoe springs can collapse the wheel cylinders to force the fluid back to the master cylinder. This creates a negative pressure in the master cylinder where normally a positive pressure is maintained. This negative pressure causes the primary seals to move away from the walls of the master cylinder to allow the extra fluid that has been carried by the secondary seals to fill the space ahead of the primary seal. As the pistons again move into the cylinder, the additional fluid will raise the pedal height. This is a safety factor in service brakes. When the brakes are released the master cylinder pistons also move to the release position. The returning brake shoe springs will force the wheel cylinders in. This pushes the additional fluid back to the reservoir through the compensating port. The compensating port also functions to allow fluid to enter or to return from the hydraulic system as the fluid expands and contracts from changes in the fluid temperature.

Safety Considerations. In the event of a partial system failure, one portion of the dual master cylinder will continue to function to prevent total loss of braking. If pressure is suddenly lost in the secondary chamber of the master cylinder (due to a line or seal failure), the pressure in the primary chamber will move the secondary piston to a stop at the end of its travel. With the secondary piston resting against the end of the cylinder bore, pressure will again rise in the primary section of the master cylinder to apply that portion of the brake system. This is shown in Figure 7-9. If pressure is lost in the primary chamber, the primary piston will continue to move until an extension on the end of the primary piston comes into contact with the secondary piston. Once in contact the secondary piston is mechanically moved to develop hydraulic pressure to provide braking from that portion of the system. This is shown in Figure 7-10. A failure of either portion of the system will result in a reduction both in pedal height and in the pressure difference between the two halves of the systems. The pressure difference will cause the illumination of the brake warning light to indicate to the operator that there is a brake system malfunction.

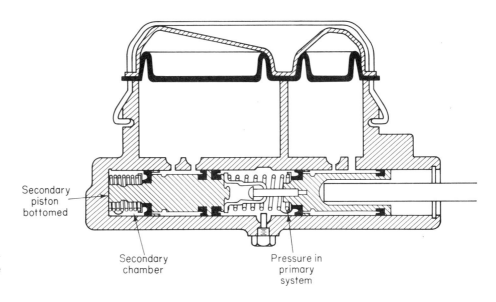

Figure 7-9 Master cylinder operation with loss of pressure in the secondary chamber.

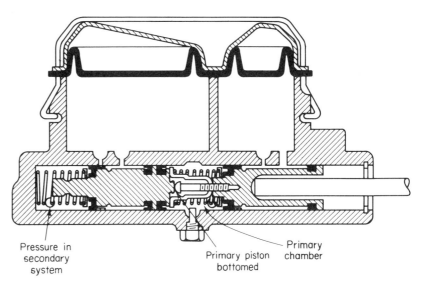

Figure 7-10 Master cylinder operation with loss of pressure in the primary chamber.

Brake Operating Systems

An alternative to the tandem master cylinder used on production vehicles is the use of two separate cylinders as used on racing vehicles. This system uses two cylinders and a bias bar connected to the brake pedal. This provides a fail-safe system with adjustment for the desired amount of braking front to rear, or bias. Since high speeds, and maximum controllable deceleration rates are necessary, this system provides the fine adjustment that is necessary due to changes in track surfaces, tires, and/or local weather conditions.

7-4 WHEEL CYLINDERS

The wheel cylinders change the hydraulic line pressure to mechanical motion to bring the brake linings into contact with the braking surface. Once the linings contact the brake surface, they provide the force required to accomplish the braking that is desired by the operator. The brake system converts the *mechanical input force to hydraulic pressure* at the master cylinder. The hydraulic pressure is transferred to the wheel cylinders where the *hydraulic pressure* is converted back to a *mechanical force*. The diameter of the wheel cylinder and the brake line pressure determine the amount of mechanical force the cylinder will generate. When either the diameter of the wheel cylinder or the brake line pressure is increased, the mechanical force on the linings will also increase. Because the area of a circle is equal to πR^2 or $0.785 D^2$ (R = radius of the cylinder and D = diameter), it can be seen that the area of the wheel cylinder piston is dependent upon the diameter of the piston squared. Because of this relationship, mechanical force will increase as the square of the diameter. If the diameter of the wheel cylinder were increased by a factor of *two,* the mechanical force it could apply with the same pressure would increase *four times.* Pressure has a linear relationship with force, thus as the pressure is doubled the force is also doubled.

In domestic passenger vehicles using drum brakes, the wheel cylinders' diameters range from approximately 0.875 inch to 1.125 inch (22 mm to 29 mm). If the brakes are of the dual-servo type, the front cylinders are typically about $\frac{1}{8}$ to $\frac{1}{4}$ inch (3 to 6 mm) larger than the rear cylinders on automobiles having a front engine with rear-drive wheels. If an automobile has the engine in the rear, larger wheel cylinders are required on the rear brakes. The wheel cylinder size difference helps to compensate for the weight bias that results from the engine location and from the weight transfer that occurs during braking.

Drum Brake Wheel Cylinders. The typical drum brake wheel cylinder consists of the cylinder housing, spring, expanders, two cups, two pistons, two boots, and a bleeder valve as illustrated in Figure 7-11. Brake systems of the double-leading or double-trailing type require two single-ended wheel cylinders, rather than one double-ended cylinder because each wheel cylinder only activates one brake shoe. Depending upon the cup design, the expanders may not be used by some brake manufacturers.

The spring holds the expanders and cups firmly against the pistons. The expanders hold the sealing edge of the cup against the wheel cylinder bore to prevent air from entering the cylinder as the brake pedal is released. The wheel cylinder pistons provide a backing for the rubber cups as they convert the brake fluid pressure to the mechanical force on the brake shoes. A bleeder screw provides a place where the air can be removed from the fluid system when the hydraulic system is being serviced. It is

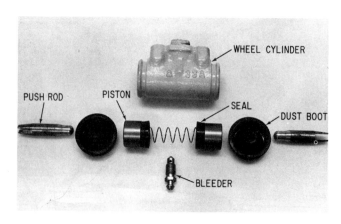

Figure 7-11 Parts of a typical wheel cylinder.

7-4 Wheel Cylinders

Figure 7-12 A typical wheel cylinder on the backing plate.

always necessary to mount the cylinder with the bleeder screw at the top of the fluid chamber or air will be trapped in the wheel cylinder with no means to expel it. The two boots, one at each end of the wheel cylinder, prevent dust and water from entering the cylinder bore.

The drum brake wheel cylinder is attached to the backing plate. A typical installation is shown in Figure 7-12. Fluid enters the wheel cylinder from the brake line that is attached to the cylinder housing. As pressure builds in the master cylinder, fluid flows into the wheel cylinder forcing the cylinder cups and pistons to move further apart (Figure 7-13). The pistons act through a connecting link or directly on the brake shoe forcing the lining into contact with the drum.

Disc Brake Caliper. The typical disc brake caliper assembly consists of the *caliper* housing, one or more *pistons*, a *seal* on each piston, a *dust boot* for each piston, and two *brake shoes* or *pads*. A typical disc brake caliper is shown in Figure 7-14.

The disc brake caliper may be either fixed or floating. The *fixed caliper* uses pistons on both the inboard and outboard side of the caliper. The piston force is applied directly to each shoe as shown in Figure 7-15. The *floating caliper* design uses pistons on only the inboard side. The reaction force of the caliper then applies the outboard shoe. This can be seen in Figure 7-16.

Caliper Piston Force. The mechanical application force of a caliper must always be equal on both inboard and outboard shoes to prevent excessive side loads on the rotor. The floating caliper accomplishes this by applying Newton's first law of motion: *For every action, there is an equal and opposite reaction.* This is done by using a caliper that is free to move in or out on bushings or guide surfaces. The outside shoe is pulled toward the rotor as the inside shoe is pushed into the rotor by the caliper piston. The fixed caliper design has the caliper rigidly mounted. The pistons themselves move toward each other to apply the equal opposing forces on the shoes. It should be noted that the pistons in the fixed caliper apply the forces in op-

Figure 7-13 Operation of a typical wheel cylinder.

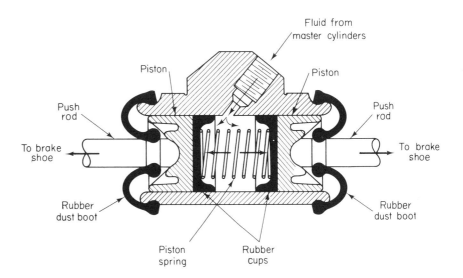

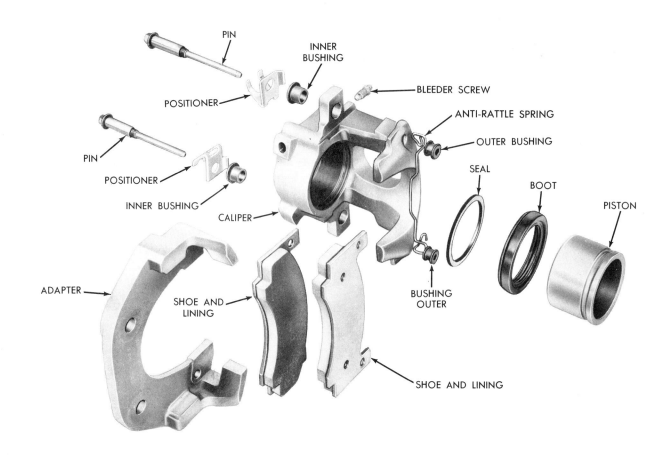

Figure 7-14 Parts of a typical single piston floating caliper (Courtesy of Chrysler Corporation).

posite directions, both toward the rotor. The total application force for each shoe can be calculated using the cross-sectional area of the pistons that directly apply that shoe. The application force for each shoe of a floating caliper can be calculated for both shoes by using the cross-sectional area of the pistons in the caliper. This can be done because one side of the caliper supplies both opposing forces, one on each side of the rotor. A floating caliper that has one piston would apply the same force to each shoe that a fixed caliper with a piston on each side of the rotor would apply, providing the cross-sectional areas of all the pistons are equal.

Caliper Design. The fixed caliper requires fluid to be directed to the pistons on each side of the caliper. This means that the caliper must have either an internal passage or an external tube to transfer fluid between the two caliper halves. The fixed caliper provides a greater cooling surface for the caliper and fluid than the floating caliper. This results from the increased number of cylinder bores that extend into the air stream moving past the brake. The fixed caliper design is more expensive to manufacture because it requires more machining operations and has more components. The fixed caliper is more sensitive to rotor runout (wobble)

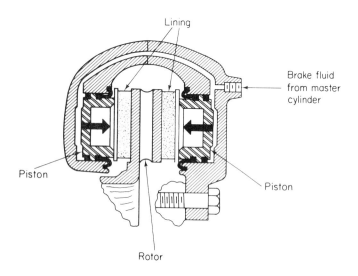

Figure 7-15 Operation of a fixed caliper.

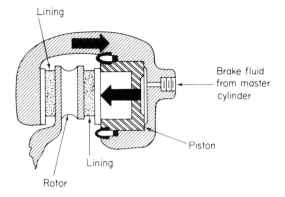

Figure 7-16 Operation of a floating caliper.

than the floating caliper. Runout will produce excessive clearance between the pad and rotor called *knockback*. This results in a low brake pedal height.

The high forces generated on the caliper require the caliper to be heavily constructed to prevent it from flexing during severe brake applications. Any flex that occurs at the caliper will feel like a "spongy" pedal to the operator. Most domestic disc brake calipers are, therefore, made of cast iron to prevent flexing. The cast iron provides a strong rigid housing with minimal distortion or flexing from the high hydraulic pressures, torque forces, and reaction forces to which it is subjected. Some vehicles use cast aluminum calipers with strengthening ribs and thick wall castings to prevent distortion. Experimental caliper designs have been produced of fiberglass-reinforced plastics. The objective of those designs is to reduce the weight and cost of the caliper.

Caliper Pistons. The caliper pistons are most often made of steel with chromium plating to prevent corrosion on the sealing surface. An alternative material being used is a compression-molded phenolic. The phenolic material resists corrosion, is strong, and provides an excellent heat dam that keeps the high lining temperatures from reaching the brake fluid. Aluminum alloy pistons have been used on various caliper designs with the piston seal attached to the piston with the seal sliding in the caliper cylinder bore.

Piston Seal. Most floating-type calipers have the piston seal mounted stationary in the housing and the piston moves in the seal. This design requires the piston itself to have a surface finish that will provide maximum seal life. Mounting the seal in this way produces the force required to pull the piston away from the rotor as shown in Figure 7-17. This design provides "self-adjustment" to automatically maintain the correct lining clearance in the disc brake system. Some fixed-type calipers have the seal mounted on the piston and the seal slides in the cylinder bore. This design requires the cylinder bore to have a smooth surface finish that will provide maximum seal life. In this caliper design the linings maintain slight contact with the rotor at all times. This slight amount of contact has very little effect on pad wear or on brake temperatures but it does prevent knockback so the pedal height remains constant.

The brake linings used on disc brakes must be "tougher" than the linings used with drum brakes. The lining must be made harder to prevent compression of the lining material from the high force

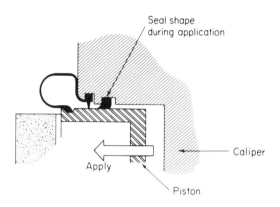

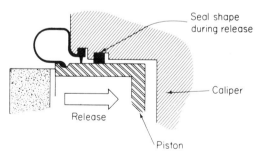

Figure 7-17 Seal mounting that pulls the piston from the rotor.

Brake Operating Systems

placed on them by the piston. The disc brake lining must be made thicker to give a lining life that closely matches that of drum brakes. The lining-to-rotor interface, which is the point where the surfaces contact, operates at a much higher temperature than that of the drum brakes. The linings must therefore be compounded to function at these temperature extremes.

Brake Application Forces. The disc brake wheel cylinder is referred to as the *disc brake caliper bore.* Because the disc brake is nonself-energizing, the application force for the lining, during a moderate to rapid stop, must be much greater than that required for self-energized brakes. For this reason the disc brake wheel cylinder piston cross-sectional area is much larger than the area of the wheel cylinder piston used in a drum brake. In the following example, to minimize confusion, only customary measurements are used to describe forces in the brake system. In a typical full-sized domestic passenger vehicle the dual-servo wheel cylinder diameter would be 1.125 inches. That same vehicle equipped with disc brakes would have a disc brake piston diameter of approximately 3.0 inches. Comparing the areas of the two different cylinders, the cross-sectional area of the drum brake wheel cylinder would be 0.994 in^2, while the cross-sectional area of the disc brake piston would be 7.069 in^2. These values can be calculated by squaring the diameter and multiplying by 0.785 [1.125^2 inches $\times$ 0.785 = 0.994 in^2 and 3.0^2 inches $\times$ 0.785 = 7.065 in^2]. This results in the disc brake piston cross-sectional area being 7.065/0.994 = 7.1 times, or 710%, larger than the drum brake cross-sectional area. If the same brake line pressure was applied to both the caliper and the drum brake wheel cylinder, the caliper piston would apply 7.1 times the mechanical force to the brake lining. Using the example mentioned earlier with a service brake pedal having a mechanical ratio of 4:1 and an application force of 150 pounds, the mechanical force applied to the master cylinder would be 600 pounds. If the master cylinder had a 1.125 inch diameter piston (0.994 in^2 area as shown above), the line pressure in this example would be 604 psi (600 lb/0.994 in^2). The line pressure equals the application force divided by the cross-sectional area of the cylinder. The drum brake wheel cylinder mechanical force would then be 600 pounds. The mechanical force applied to the shoe is the line pressure times the wheel cylinder piston area (604 lb/in^2 $\times$ 0.994 in^2 = 600 lb). The comparative mechanical application force for the disc caliper would be 4270 pounds. Using the same formula: 604 lb/in^2 $\times$ 7.069 in^2 = 4270 lbs.

The self-energizing brake actually tends to increase the effective application force from the wheel cylinder, while the nonself-energizing brake tends to decrease the apply force. A self-energizing factor must be considered where this factor increases the cylinder application force to produce the total effective lining apply force. The self-energizing factors are dependent upon lining coefficient of friction, temperature, brake design, and application pressure. For the purposes of the following discussion the forward wheel rotation self-energizing factors will be considered to be as follows:

Dual-servo brakes $\cong$ 5.0
Double-leading $\cong$ 3.5
Leading-trailing $\cong$ 2.5
Double-trailing $\cong$ 0.7
Disc $\cong$ 0.7

This implies that the total mechanical force applied to the lining in dual-servo brakes is approximately 5 times as great as the mechanical force supplied by the wheel cylinder piston. In contrast, the total effective mechanical lining force for the disc brake is only 70% of the caliper piston mechanical application force during a moderate stop. Using the self-energizing factors above, it can be determined that the disc brake piston cross-sectional area must be approximately 7.1 times the area of a dual-servo wheel cylinder piston for a similar braking response (self-energizing factor of 5.0 divided by the effective mechanical force of 0.7).

7-5 HYDRAULIC VALVES AND THEIR FUNCTION

Brake systems need hydraulic valves to regulate, control, and compare brake fluid pressure in different portions of the system. Depending upon design, the hydraulic system may incorporate a residual check valve, a pressure differential valve, a metering valve, a proportioning valve, or a combination valve.

Residual Check Valve. The residual check valve is located in the master cylinder at the outlet fitting. Fluid enters the brake lines as the brake pedal is depressed. Upon release of the brake pedal, fluid will return through the check valve until 6–18 psi (41–124 kPa) of line pressure remains in the lines. Once the pressure has dropped to this value the valve will close, holding a static or residual pressure in the lines. The action of the residual check valve is illustrated in Figure 7-18.

The residual pressure is high enough to keep the wheel cylinder cup lips expanded tightly against the cylinder walls. This will prevent air from entering the system as the brake is released. The residual pressure is low enough so that the return springs on the brake shoes are able to fully return the brake shoes against the pressure.

The residual check valve is not used with disc brakes. The caliper seal is the only force used to return the disc brake piston. The piston would not be returned against residual pressure because it has no return spring and the caliper has a large piston area. The residual pressure would be great enough to keep the disc brake linings in tight contact with the rotor, causing brake drag and greatly shortened lining life. A disc-drum combination brake system uses a residual check valve in the drum brake side of the master cylinder outlet. The check valve is omitted from the outlet side going to the disc brakes.

Pressure Differential Valve. A pressure differential valve or brake warning light switch is used on vehicles with a tandem master cylinder. The valve senses the outlet pressure on each side of the

Figure 7-18 Operation of the residual check valve below the outlet seat.

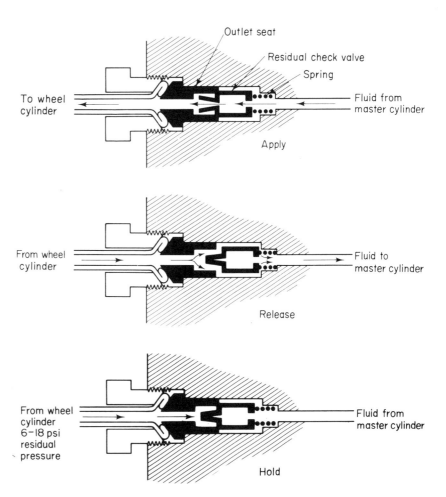

Brake Operating Systems

tandem master cylinder. It compares these pressures and turns an indicator light on to warn the operator of a pressure failure in one half of the brake system.

The valve assembly consists of an electrical switch, or contact, centered between a small two-headed spool valve. This is shown in Figure 7-19. One end of the valve is exposed to pressure from one half of the master cylinder while the other end of the valve is exposed to pressure from the other half of the master cylinder. A centering spring on each end of the valve will prevent the valve from shifting and turning the warning light on when normal pressure fluctuations occur in the brake line.

The pressure differential switch remains open or off when the valve is in its centered position. When the pressure rises evenly in both halves of the brake system during brake application the switch will remain centered. The spool valve moves toward the low-pressure side if the brake system has a malfunction that causes the pressure to drop in that part of the system. The shift in valve position closes the switch, connecting the switch terminal to ground to complete the indicator light circuit. This activates the brake warning light on the instrument panel to warn the operator of the existing malfunction in the brake system. The malfunction can be

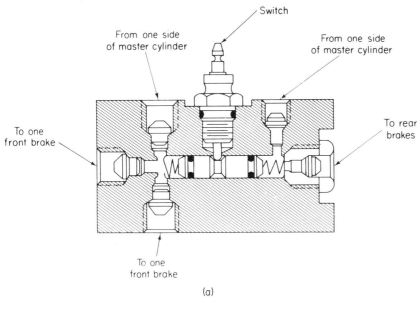

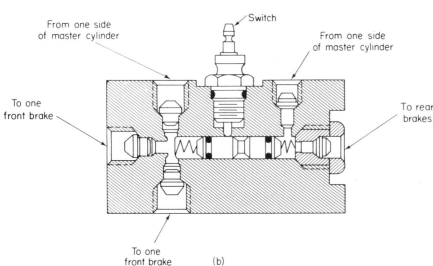

Figure 7-19 Operation of the pressure differential valve and warning switch: (a) equal pressure and valve centered; (b) unequal pressure and valve off center to close the warning light switch (valve shown with rear system failure).

either hydraulic or mechanical. If the malfunction is mechanical, it is serious enough to affect the pressure in one half of the system. It could be the result of an improper brake lining adjustment. The lining-to-drum clearance could be so great that the volume of fluid displaced in one half of the master cylinder is not enough to expand the linings into full contact with the brake surface. The pressure in the normally operating half of the master cylinder would continue to rise. The difference in pressure of the two halves of the brake system will cause the valve to shift from its centered position to illuminate the brake warning light.

Metering Valve. The metering valve, or hold-off valve, is used on disc-drum combination brake systems to coordinate the application times of the drum and the disc brakes. This is necessary because the disc brake shoes have no return springs while the drum brakes have rather stiff return springs. Without a metering valve, the disc brakes would normally start to apply long before the drum brakes with their stiff return springs would have started to function. This type of braking could be hazardous on slippery road surfaces. The front wheels could lock when the driver applies the brakes lightly, causing a loss of steering control.

The metering valve is located in the front brake line. It prevents hydraulic pressure from going to the front disc brake calipers until a minimum brake pressure has been reached. This minimum pressure is approximately the same pressure that is required to overcome the force of the drum brake return springs. The metering valve begins to open when the minimum pressure is reached, allowing the disc brakes to start to apply at the same time as the drum brakes. The metering valve prevents pressure from reaching the disc brake calipers until there is approximately 125 psi (861 kPa) of line pressure. The pressure to the front calipers trails behind the pressure to the rear drum brakes until

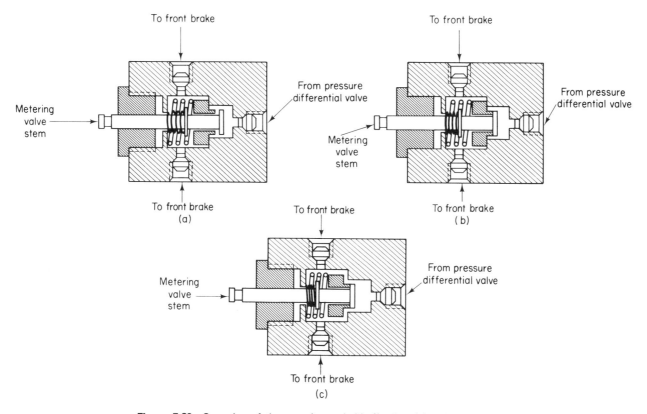

Figure 7-20 Operation of the metering or hold-off valve: (a) position with the brakes released; (b) position with light pedal apply force; (c) position with heavy pedal apply force.

the metering valve reaches its fully open position. When the brakes are released, the metering valve will close on its seat and a second valve located inside the metering valve assembly, will open, allowing free fluid return to the master cylinder. The metering valve operation is illustrated in Figure 7-20.

Proportioning Valve. The brake proportioning valve or balancing valve was developed to distribute the braking effect of front-wheel disc and rear-wheel drum brakes and thus provide the required brake bias. Two factors make this necessary. First, with a disc-drum brake combination, the disc brakes require much higher pressure to achieve the same braking force as the drum brakes during moderate to hard brake application, especially with systems using dual-servo drum brakes. Second, during a hard brake application the effective weight is transferred from the rear to the front of the vehicle. This causes the effective weight on the front to become heavier and the rear to become lighter. These two factors make rear-wheel brake proportioning a requirement on many vehicles.

The proportioning valve is located in the rear brake line. It proportions rear brake pressures as a percentage of front brake pressure, after a minimum brake pressure is reached. This allows equal line pressure at both front and rear wheels during normal braking. During hard brake application the pressure going to the rear wheels has a smaller increase than the front brake line pressure. This will prevent the rear wheels from sliding prematurely.

The proportioning valve functions by sensing pressure on different size surface areas of a spool valve. The larger surface area senses the drum brake application pressure. The smaller surface area is acted upon by the master cylinder pressure and a spring as shown in Figure 7-21. During normal braking, the spring force is great enough to oppose the force developed by the application pressure. The valve allows fluid to flow freely to the rear brake cylinders. During hard application the pressure from the master cylinder forces the valve to move against the check valve spring, closing the center passage through the valve. From this point on, the master cylinder pressure and spring force are applied to the smaller end of the spool valve while rear-wheel brake pressure is applied on the larger end of the spool valve to balance the forces. A low pressure on the large surface will balance a

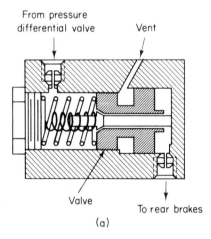
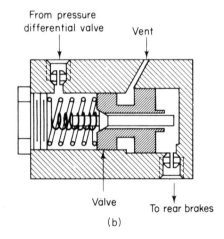

Figure 7-21 Operation of the proportioning valve: (a) release and lift braking; (b) heavy braking.

high pressure on the small surface. This allows rear-wheel line pressure to increase as a proportion of master cylinder pressure.

Proportioning valves are not limited to use solely on disc-drum combinations. They are used in brake systems whenever rear-wheel sliding, brake effectiveness, and/or weight transfer is a problem. It is important for the driver's control of the vehicle to prevent the rear wheels from sliding prior to the front wheels. If the rear-wheels slide, a vehicle

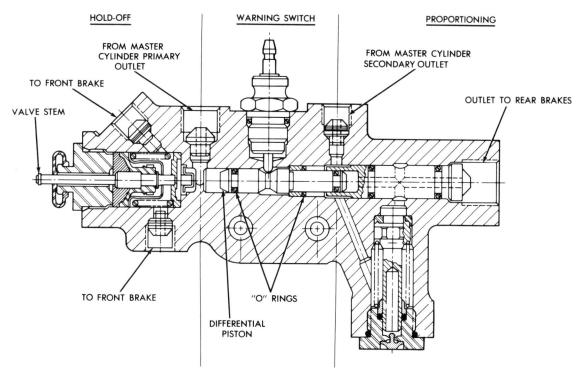

Figure 7-22 A cutaway view of a combination valve including a metering or hold-off valve, pressure differential valve or warning light switch, and a proportioning valve (Courtesy of Chrysler Corporation).

is unstable and may spin out. Maximum braking and good control can be accomplished if the front wheels lock just prior to the rear wheels locking.

Combination Valve. A combination or pressure control valve is used on many disc brake-equipped vehicles. The combination valve combines the pressure differential valve, metering or hold-off valve, and the proportioning valve in one valve housing. This is shown as a section view in Figure 7-22. Each portion of the valve assembly functions in the same manner as the independent valves would function. By combining the various pressure controls into one housing, a number of fittings and much additional line routing is eliminated. The combination valve is most often located on the vehicle frame directly beneath the master cylinder.

7-6 BRAKE LINES

Brake lines in the brake hydraulic system provide a means of directing and transporting the brake fluid from the master cylinder to the wheel cylinders. Brake tubing and brake hose are made in various dimensions. The common tubing sizes used for passenger vehicles have either $3/16$-inch or $1/4$-inch diameters. The required inside diameter of the tubing and hose is determined by the time required to apply the brakes. The smaller hose or tube will deliver the same pressure as the larger. It will expand less under pressure and thus requires less pedal travel than the larger sizes. The smaller diameters, however, require a greater application and release time than the larger size tube or hose.

Brake Hose. Brake hose is used to connect the brake system components that are mounted on the steering knuckle or rear axle housing to the steel brake tube on the vehicle frame. These units move in relation to the frame. A brake hose is used at each front wheel to allow the brakes to be applied as the wheel assembly turns for steering and as it moves up and down during normal suspension travel. One brake hose is normally used at the rear in the case of a Hotchkiss drive line. Two brake hoses are required in the rear when independent rear suspension is used. The number and length of brake hoses are kept to a minimum. The hoses expand during braking and this takes more fluid,

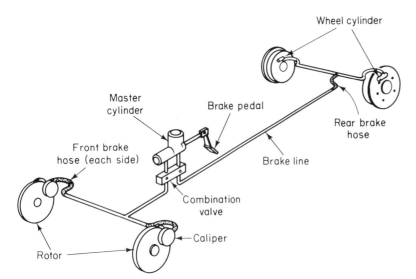

Figure 7-23 Location of typical brake lines and hoses in a hydraulic brake system.

which results in greater pedal travel for the operator. Steel brake tubing is used in all parts of the hydraulic brake system where movement does not normally take place. The location of brake hoses and lines is illustrated in Figure 7-23.

The brake hose is tested for bursting strength. The minimum acceptable burst strength is 5000 psi (35,000 kPa). Along with burst strength, the hoses are also tested for expansion at both 1000 and 1500 psi (6900 and 10,000 kPa) and the amount of expansion must stay within the specified limits. The maximum expansion limits are dependent upon hose size and type, but hoses are limited to an expansion of less than 1.3 cc increased volume per foot (0.30 m) of hose. Other tests to which the hose is subjected are water absorption and tensile strength.

Brake Tubing. Brake tubes are used to carry the fluid beneath the vehicle. Tubing is used to interconnect the brake valving and route the fluid as close to the moving wheels or axle as possible. The brake hoses make the final connections to the front-wheel cylinders and between the frame and rear axle. Brake tubes are connected to the rear-wheel cylinders. The material used in brake tubing must be metallic and have high strength properties. The primary mechanical properties for which it is tested are bursting strength, fatigue resistance, impact resistance, and heat resistance. The bursting strength test requires the tubing to withstand an internal pressure of 8000 psi (55,000 kPa) without failure. The steel tubing used for brake lines may be either seamless tubing or multiple-ply tubing in which the plies are bonded by a metallurgical bond. If the plies have seams, the seams must be separated 120° from each other. The seamless steel tubing is the most widely used type in automotive applications. The most common sizes are ³⁄₁₆ to ¼ inch. The ³⁄₁₆ inch is the most widely used size. The tubing connections on brake line must be double-flared as shown in Figure 7-24 to provide the strength and durability required of a brake line.

Brake Tubing and Hose Installation Guidelines. When installing brake tubing on a vehicle, there are a number of guidelines to follow. The tubing

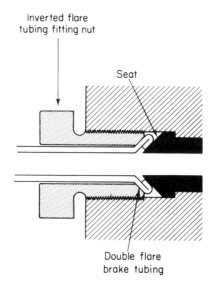

Figure 7-24 Section view of an assembled double-flare brake line fitting.

should be routed so it is protected from impact by gravel and stones thrown up by the tires as well as lifting hoists, towing, vibrating components, areas of extreme heat, and areas that are subject to salt and water buildup. Some systems are equipped with guards to prevent damage from gravel and stones. Tubing usually has loops to minimize the effects of vibration. The loops can be seen in Figure 12-1.

Brake hose installation is highly critical. The most important consideration should be to avoid *wear* due to the hose rubbing against a moving component. Brake hoses must be properly routed and be the correct length so they will neither stretch nor kink. The brake hose must not contact anything as the wheels are moved up and down and as they are turned to the extreme limit of travel.

Brake hoses are often manufactured with a white stripe running the length of the hose. The stripe is used to show hose twist. It should have less than a 15° twist from end-to-end after installation to provide normal service life. The stripe makes the technician aware of any hose twist that has taken place during installation. The hose should always be routed away from areas of extreme heat.

7-7 BRAKE FLUID

Brake fluid is one of the brake system's most important elements. The fluid has a major effect on the operation of the complete brake system and must meet government and industry quality-control standards. Brake fluid should be handled with care and stored properly, as described below, to assure that it remains a high-quality product.

Brake Fluid Requirements. The fluid must satisfy a number of requirements before it can be used in the brake hydraulic system. A high *boiling point* is necessary to provide braking at high brake component temperatures. A loss of braking will occur if the fluid boils and becomes vapor in a hydraulic cylinder. The *viscosity* or thickness of the fluid is important to provide good application and release times when the air temperatures surrounding the brakes are low. If the viscosity of the fluid is too great at low temperatures brake response is difficult and slow. The fluid must have good *lubrication* properties to prevent wear of the rubber and metallic components in the system. The fluid must contain *inhibitors* to prevent swelling of the rubber components and to prevent the formation of rust.

Brake fluid is susceptible to contamination by water. It will absorb moisture if the container is left open to the atmosphere. Moisture in brake fluid will lower the fluid boiling point and cause corrosion in the system. Because of this *hygroscopic* or water absorbing property, fluid from a container that has been left open should never be used in a brake system.

When handling brake fluid, there are a number of precautions one should follow. Always store brake fluid in its original container to be sure that it is properly identified. The container should always remain tightly capped and free from dirt to prevent contamination. Never reuse brake fluid that has been used for bleeding brakes. Always throw it in the proper waste container when the brake repair is completed. Brake fluid will damage painted surfaces. If brake fluid should accidentally get spilled on a painted surface, immediately flush it off with water. Never use any petroleum products, such as naphtha to clean brake system parts.

Silicone Brake Fluid. Silicone brake fluid is currently on the market and, therefore, it should be discussed. This fluid has a number of distinct advantages over the glycol-based fluid, which is the common brake fluid. The boiling point of silicone brake fluid is considerably higher than the conventional brake fluid and for this reason silicone brake fluid has gained popularity. The silicone fluid retains its low viscosity in extremely low temperatures, and thus it provides consistent operation in cold weather. This fluid does not attack painted surfaces if accidentally spilled on them.

Some of the disadvantages of the silicone fluid also deserve discussion. The cost of this fluid is usually high. The high cost is not worth the benefits to the average automobile owner. Silicone

Brake Operating Systems

fluids do not mix with other brake fluids and thus they cannot be added to glycol-based fluids but require a complete change of the brake fluid in the system. Some silicone fluids absorb water at a rate that may equal or exceed that of glycol-based fluids. Other silicone fluids absorb much less water.

Always observe all of the handling and storage precautions when working with brake fluids. Cleanliness practiced while working on all of the brake system components will always extend the service life of the repair being made.

7-8 WHEEL SLIP BRAKE CONTROL SYSTEMS

Brake control systems that provide the operator with maximum controlled braking are continuously being tested. Some of these systems have been applied to passenger vehicles. The wheel slip control systems or antilock systems, commonly but incorrectly called anti-skid systems, provide the operator greater control during panic braking. This control is used to prevent spinouts while allowing steering control in some systems. In others it will only prevent spinouts.

The wheels must continue to rotate to retain directional control of the vehicle. Steering control is lost if the brakes are applied hard enough to slide the tires on the road surface. The tires will follow the path of least resistance when sliding. If all four wheels are locked, the vehicle will usually spin out of control. The braking distance is increased when the wheels slide because the maximum friction force between the tire and the road surface is achieved just before total wheel slip. During wheel slip, braking is reduced by 20 to 30 percent from the maximum possible nonslip braking.

Wheel slip control systems sense wheel lockup. They then automatically and rapidly cycle the brakes through release and application. This occurs as often as four or five times a second on the brakes being controlled. The resulting effect is very similar to a driver pumping the brakes to stop. However, it is done faster and at a controlled rate with the wheel slip control. The automatic pumping action allows the wheels to cycle from locked to rotating in a rapid sequence so that control is not lost. A vehicle can generally be brought to a controlled stop in the shortest distance using this type of system. This might not be true when making a panic stop on gravel or dirt. The braking distance may actually be slightly increased when using one of the braking control systems on these surface.

Brake control systems have been offered on luxury-type vehicles by all of the major vehicle manufacturers at one time or another. Customer acceptance of the systems have met with limited success. This is mainly due to the units' cost and the rare occasions when maximum braking is required. All heavy-duty trucks of current manufacture are equipped with wheel slip controls, commonly called antilock controls. FMVSS121 (Federal Motor Vehicle Safety Standards No. 121) is a law requiring all air-brake-equipped vehicles to stop at an average deceleration rate of 18 ft/sec^2 from 60 mph (5.5 m/sec^2 from 97 km/h) without loss of control. This requirement is so severe that the only way in which the requirement can be met is through the use of the wheel slip brake control systems. The systems used on trucks are similar to those offered on passenger vehicles.

Control System Variations. The wheel slip brake control systems are made in different combinations. They are individual wheel control, front and rear axle control, rear axle control or any combination of these. The systems that control both front and rear axles can be steered while preventing spinout as the brakes are fully applied. A typical system is illustrated in Figure 7-25. Steering control will be lost in a panic stop when the control system is only used on the rear. The vehicle will not spin out of control with rear axle control but will travel in a straight path with the front wheels sliding.

Each type of system offers some distinct advantages and often some disadvantages. The factors considered are the vehicle suspension type, desired levels of braking and handling, and the cost of the system.

The individual four-wheel braking system is the most complex but it gives the operator maxi-

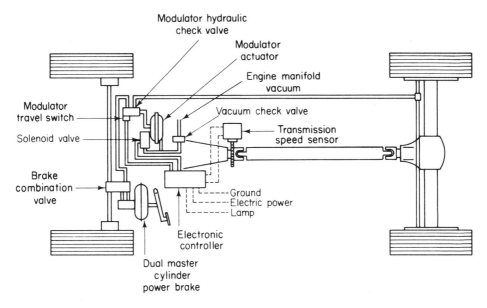

Figure 7-25 A schematic diagram of a wheel slip brake control system.

mum controlled braking. This results from the system's ability to respond to individual friction force variations that are produced at each wheel by a changing road surface friction (wet, dry, or ice). This configuration sometimes produces undesirable steering feedback. It is caused by instantaneous wheel pull when the frictional force changes at each front wheel. The amount of steering feedback depends upon the front suspension design.

Systems offering brake controls on both front and rear axles sense wheel speed at each wheel location. Control during hard braking occurs simultaneously at the two wheels of an axle. Each axle set operates independently of the other. These systems offer good stability and steering control during a panic stop. The cost of the system is less than the four-wheel independent control system. Braking effectiveness is not as great as the four-wheel independent system on surfaces where different friction coefficients exist from side to side.

Wheel slip brake control systems that offer control of only the rear wheels are more popular because they cost less. There are a number of different ways these systems are made. Some are made with individual wheel speed sensors having individual wheel control. Some are designed with individual wheel sensors that control both wheels on the axle. Other systems have one axle sensor to control both wheels of the axle. The only advantage of this system is stability during maximum braking at the lowest cost. The two-wheel control system does not allow steering control when the brakes are applied during maximum braking. The front wheels will lock and slide.

Systems may employ combinations of the systems described. One combination that has been used is the individual front-wheel control used with rear axle control. Although this system is a compromise, it works well. Steering control can be maintained while braking effectiveness is only slightly compromised in this system. It is less costly than the independent wheel slip brake control system.

Each system has its own particular advantages. Cost is the factor that has the greatest effect on the type of system used. In the past, only the two-rear-wheel control and the individual front-wheel control coupled with rear axle control types have been produced by major domestic vehicle manufacturers.

System Components. The system components fall into three major groupings. These are the speed sensors, the controllers, and the modulators. *Speed sensors* are electromagnetic devices mounted to sense wheel or drive line speed. Systems using independent front-wheel control or front axle control have the sensors mounted at each wheel position. Rear axle control systems may locate the sensors at

Brake Operating Systems

each wheel or one end of the drive shaft. Use of the drive line sensor reduces cost.

Speed Sensors. The sensing device consists of a toothed ring and a magnetic pickup coil. The toothed ring is mounted on the rotating component. The magnetic pickup coil is mounted on a stationary member close to the rotating toothed ring. The rotating toothed ring induces a voltage in the pickup coil. As vehicle speed increases, the rotating speed of the toothed ring also increases. The voltage produced in the coil is proportional to the speed of the wheel. The voltages from each pickup coil are sensed by the electronic control unit.

Controllers. The *electronic control unit* is a solid state "black box" that is usually mounted inside the drivers compartment. Its purpose is to interpret the voltages produced by the wheel sensors and to relay the appropriate signal to the brake modulator that controls the brake application.

The electronic control unit continually senses output voltage produced by the speed sensors. If the output voltage from the speed sensors reduces too rapidly to near zero, the control unit senses this as wheel lockup. It then signals the modulator to release the brake at that wheel. As the wheel begins to rotate the voltage from the wheel sensor coil increases. With the brake pedal still applied, the control unit again directs the brake at the released wheel to be applied. This cycling continues at a rate of four or five times each second. When the vehicle speed drops below apprpoximately 5 mph (8 km/h) cycling stops so the brakes lock. Systems that have a sensor at each wheel provide a voltage input from each sensor to the control unit. The control unit then senses the faster or slower rotating wheel depending on the system type, by comparing the voltage produced by each sensor. Vehicles using one drive line magnetic pickup coil actually sense the average wheel speed. If the average wheel speed slows too rapidly, both rear brakes begin to cycle.

Brake Modulators. The *brake modulators* are the working part of the system. On an independent wheel control system the modulators are mounted in the brake line going to each wheel. On an axle control system they are placed in the line going to the two wheels. Depending on the type of system, one to four modulators are used in wheel slip brake control systems.

The brake modulators operate using hydraulic pressure, air pressure, or vacuum to control brake fluid pressure. Basically, the modulator contains a solenoid valve, a chamber for the working pressure or vacuum, and a chamber of varying size that is used to control the wheel braking.

When the signal is received from the control unit, the solenoid moves to close the passage between the brake master cylinder and the brake. With this passage closed, the cycling pressure will not be felt in the brake pedal. At the same time, the working pressure—air, hydraulic, or vacuum—is admitted to increase the volume of the chamber of varying size that is holding brake application pressure. By increasing the volume of the chamber the pressure is reduced at the wheel(s) to release the brake(s) and the wheel(s) begins to rotate. This action is seen in Figure 7-26. When the wheel starts to

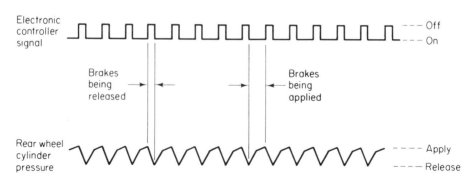

Figure 7-26 Drawing of oscilloscope traces of the electrical signal and brake pressure as the brakes are being applied on a wheel slip brake control system.

rotate the working pressure or vacuum is released, decreasing the volume in the brake chamber and reapplying the brakes. This cycle is continuously repeated until the wheel stops sliding or the speed has dropped to about 5 mph (8 km/h). This speedup and slowdown cycling of the brakes holds the wheels at the point of impending skid where braking force is greatest. It allows some wheel rotation for stability and directional control.

Problem Diagnosis. Most brake control systems work in conjunction with the brake warning light switch. If a malfunction occurs in the system a fuse will blow. This will cause the brake warning light to turn on. In this way the driver will be alerted that there is a malfunction in the brake system. The fuse has a time delay action so that it will blow if the modulator remains activated too long. The service brake system operates as if the vehicle had no wheel slip brake control when the fuse is blown.

To check the operation of the wheel slip control system, the wheels must be rotated with the vehicle raised from the floor. On a vehicle employing independent wheel control, each nondriving tire can be rotated with a wheel spinner of the type used in wheel balancing. This must be done with the engine running and the wheel rotating at a minimum speed of 35 mph (57 km/h). When the brakes are applied hard, the wheel should cycle three or four times before it comes to a complete stop. A vehicle using rear axle control may require both wheels to be rotating to make this check. This is due to the location of the sensor. Engine-driven rear wheels can be rotated with the drive line. They are operated above 35 mph (57 km/h), the transmission placed in neutral, and the brakes applied. Again, cycling should take place.

Problems encountered should be diagnosed using the instructions in the correct shop manual. Components of the various slip control systems cannot be repaired. Faulty units must be replaced. Wheel slip control systems have no effect on repair of the regular service and parking brake systems.

REVIEW QUESTIONS

1. How does Pascal's law apply to brake systems? [7-1]
2. Name the parts in a typical hydraulic brake system. [7-1]
3. Why are power brakes usually installed on medium and large vehicles that have front disc brakes? [7-2]
4. What combinations are used to provide safe braking with split brake systems? [7-2]
5. Why is the master cylinder outlet going to a drum brake equipped with a residual check valve? [7-3]
6. Describe the operation of the master cylinder compensating port. [7-3]
7. Name all of the master cylinder parts that could cause an internal leak if they were faulty. [7-3]
8. What happens in the brake system during pump-up? [7-3]
9. Why are the wheel cylinders used on front drum brakes often larger than the rear wheel cylinders? [7-4]
10. Why is the bleeder screw opening at the top of the wheel cylinder bore? [7-4]
11. How does the floating caliper differ from the fixed caliper? [7-4]
12. What could cause a spongy pedal feeling on disc brakes? [7-4]
13. How is the caliper kept from turning with the rotor when the brake is applied? [7-4]

14. Why is no adjustment required on disc brakes? [7-4]
15. Why are caliper pistons larger than drum brake wheel cylinders? [7-4]
16. Why is a pressure differential valve needed? How does it operate? [7-5]
17. Why is a metering valve needed? How does it operate? [7-5]
18. What does the proportioning valve do? [7-5]
19. What special design characteristics must brake tubing and hoses have? [7-6]
20. What brake hose installation precautions should be followed? [7-6]
21. What requirements must brake fluid meet? [7-7]
22. From what materials are brake fluids made? [7-7]
23. How does the wheel slip brake control system allow both maximum braking and vehicle directional control? [7-8]
24. Why have wheel slip braking control systems only met with limited customer acceptance? [7-8]

Brake Inspection

Brake service begins with the owner bringing the vehicle into the repair facility for either routine inspection or because, in his opinion, the brakes are not functioning properly. Possibly the reason the vehicle owner feels a need for a brake inspection is that something abnormal happens as the brakes are applied. If the complaint is abnormal braking, the vehicle should be test driven. Before moving the vehicle, make sure that the brakes will operate sufficiently for a safe test drive. The test drive will aid the automotive service technician in determining if the complaint is actually brake related, or is a fault of the steering, suspension, or tires. When test driving, use extreme care and drive only fast enough to produce the customer's complaint, always staying within the legal speed limit. If the road test proves the brakes to be at fault, the complete brake system should be inspected.

A routine brake inspection is recommended once a year, more often when the vehicle is driven an unusually high number of miles each year. If it is determined that braking is normal under all operating conditions, the inspection is only necessary to find conditions that may develop into a problem before the next regular brake inspection.

8-1 BRAKE SYSTEM INSPECTION

Brake inspection includes an inspection of the master cylinder, brake fluid, brake drums, brake rotors, linings, brake hardware, wheel cylinders, and brake lines, as well as the operation of the parking brake. Any brake inspection must be *a*

Brake Inspection

complete inspection. A failure in any one component can cause a complete failure of that portion of the brake system. This may result in a complete loss of braking effectiveness.

Master Cylinder Inspection. The brake inspection should begin by making sure that the master cylinder is operating properly. This can be accomplished by, first, lightly depressing the brake pedal and noting any "fading away" of the brake pedal. Pedal fading is a gradual decrease in the pedal reserve or pedal height as the pedal is lightly pressed for about 15 seconds. Next, the pedal should be applied with heavy force to see if there is any pedal fading during severe braking. If the pedal reserve decreases in both instances an external leak may be present. If the pedal reserve decreased when the pedal is applied lightly but holds firm under heavy application, the problem is often an internal leak in the master cylinder. The technician should next visually inspect the brake fluid level and contamination in the master cylinder reservoir. It is normal on vehicles equipped with disc brakes to have the fluid level lower in the large reservoir as lining wear occurs. With drum brakes, the fluid level should remain relatively unchanged throughout the lining life, unless a leak develops in the system.

The master cylinder is inspected for signs of external leakage. This kind of leakage is usually visible on the firewall or at the brake booster where the master cylinder is attached. External leakage may also occur at the brake line fittings leading to the brake system or from the master cylinder reservoir cover, if it has not been installed properly.

Brake Fluid Inspection. The brake fluid should be examined for any sign of dirt and/or mineral oil contamination. If the fluid is suspected of being contaminated, the fluid should be removed from the master cylinder and poured into a clean glass for inspection. If distinct layers of fluid form in the glass, contamination is present. Mineral oil attacks the rubber components in the system and causes them to swell and soften. If oil is present, the complete brake hydraulic system must be disassembled, flushed, and all rubber components replaced. If the

Figure 8-1 Contaminated brake fluid. Left container has water in the brake fluid and the container on the right has oil in the brake fluid.

fluid in the jar appears cloudy due to moisture and/or component wear, the system should be flushed *before* any new components are installed. Figure 8-1 shows a sample of contaminated brake fluid. Clean all the surfaces thoroughly with a clean shop cloth before removing any connection or cover that can expose the fluid to dirt.

8-2 BRAKE DRUM REMOVAL

Inspection of the service brake components and brake lines is accomplished with the vehicle raised from the floor surface. The vehicle may be raised with a jack and supported with safety stands or it may be raised with a hoist that will allow all four wheels to turn freely. The tire and wheel assemblies must be marked to identify their location on the vehicle as well as position on the axle or wheel hub. After the tire and wheel assemblies are removed, each brake drum should be marked to identify its location (Figure 8-2). The brake drums are now ready to be removed. Use the same procedure for removing the front brake drums as used for repacking wheel bearings, described in Chapter 2.

The rear brake drums on current domestic passenger vehicles are removed by first taking out any locating screws or speed nuts that hold the drum to the axle. Speed nuts are sheetmetal stampings that are placed over the wheel studs at the fac-

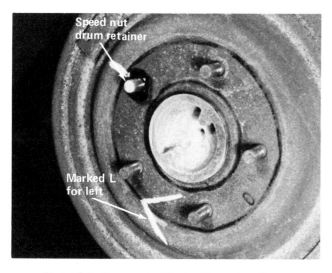

Figure 8-2 Drum marked for location and position.

8-3 LINING INSPECTION

The brake linings should be thoroughly inspected for wear, signs of contamination, heat cracking, and excessive glazing.

Wear of Linings. Wear shows as a minimal amount of lining material remaining on the brake shoe. It can be measured as shown in Figure 8-4. The lining may be worn to the point where the shoe has contacted the drum, causing damage to the drum. With riveted linings, the rivets themselves contact the drum before the steel brake shoe makes contact. Any signs of the rivets contacting the drum or nearing contact is cause for the replacement of the lining. Figure 8-5 and 8-6 show excessive lining wear. Loose riveted lining is also a reason for replacement. Whenever linings are replaced, they must always be replaced in axle sets. Never replace the lining on just one wheel because the friction characteristics of the old linings and the new linings may differ greatly. Different friction characteristics will cause wheel pull while braking.

Contamination of Linings. Contamination of linings occurs when a substance leaks onto the linings. The common sources of contamination are wheel

tory. These can be removed by grasping them with pliers and unscrewing them from the wheel studs. The drum can usually be lifted from the axle hub and wheel studs. Sometimes the drum and axle hub have become rusty, making removal difficult. Penetrating oil sprayed around the two mating surfaces and a slight tapping on the drum between the wheel studs will aid in loosening the drum from the axle hub. A puller of the type shown in Figure 8-3 can be used to remove a seized drum. This should be done with care because this may warp the drum and the brake surface. If it does, the drum will have to be replaced. Do not use a torch to heat the drum to loosen it. The heat will warp the drum so it will have to be replaced.

Figure 8-3 Removal of a seized drum can be accomplished with a special two-jaw puller as shown.

Figure 8-4 Measuring the thickness of the lining that remains on a brake shoe.

Brake Inspection

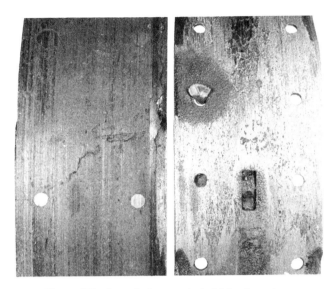

Figure 8-5 Excessively worn riveted lining from drum brakes. On the left the wear is into the rivets and on the right the lining is worn away completely so that there is only metal-to-metal contact.

Figure 8-6 Excessively worn bonded brake lining from drum brakes. The lining on the right is worn through the reinforcing mesh of the lining.

bearing grease, rear axle lubricant, and brake fluid. Whenever contamination is present, it is necessary to repair the malfunctioning component causing contamination before the linings are replaced. Wheel bearing grease can get on the lining if the wheel grease seal has become worn or damaged. The recommended practice is to repack front wheel bearings and replace front wheel seals whenever the brakes are relined. Rear axle lubricant will enter the brake mechanism if the axle seals or axle bearings have become damaged. An example of this is pictured in Figure 8-7. Brake fluid that leaks from the wheel cylinders will end up on the brake surface. When it shows signs of leakage, the wheel cylinder will require rebuilding or replacement prior to relining the brakes.

Heat Cracking of Linings. Heat cracking is a direct result of excess heat buildup on the lining surface. This appears as long radial cracks on the lining surface as shown in Figure 8-8. Anything that allows the lining to overhead repeatedly, such as excessive amounts of high-speed stopping, will cause this condition. Linings that appear

Figure 8-7 Brake assembly covered by rear axle lubricant coming out through a leaking axle seal.

8-4 Wheel Cylinder Inspection

Figure 8-8 Heat-cracked brake lining.

Figure 8-9 Glazed brake lining.

tion occurs when enough heat and pressure is put on the lining for fade to occur (see Chapter 7). A slight amount of glazing is normal and it will not adversely affect braking performance. Glazing of the lining usually precedes heat cracking when the linings are overheated.

8-4 WHEEL CYLINDER INSPECTION

An inspection to check wheel cylinder operation and leakage can be performed on some brakes with the linings and shoes remaining in position on the backing plate. The wheel cylinder boots are pulled away from the cylinder to permit visual inspection of the cylinder bore. In other brakes, as shown in Figure 8-10, the shoes must be removed. It is normal to have a slight fluid residue on the outside of the wheel cylinder cup. This residue lubricates the moving components and walls of the cylinder. This residue should not be confused with a cylinder leak. Excess fluid leakage is a reason to repair the brake and the wheel cylinder. The wheel cylinder will also have to be serviced when any corrosion ex-

Figure 8-10 Loosening a wheel cylinder boot to check for piston seal leakage.

overheated on all four wheels are normally the result of improper driving habits.

Glazing of Linings. Glazing of the brake linings is a result of overheating a portion or all of the lining. A glazed lining is shown in Figure 8-9. This condi-

119

Brake Inspection

ists in the wheel cylinder bore or if the piston sticks.

If the operation of one wheel brake is in question, the other three brake drums are placed over their respective shoes. While watching the brake in question, slight pedal pressure is applied. In this way the operation of the brake in question can be observed. This brake checking procedure should be used with care because excess pedal motion will force the piston from the wheel cylinder causing a loss of fluid. If there is any question about the proper operation of a wheel cylinder, the brake will have to be disassembled and the cylinder closely inspected.

8-5 BRAKE LINE INSPECTION

An often overlooked part of a brake inspection is a careful inspection of all the brake tubes and flexible hoses. Their inspection is of no less importance than the inspection of the master cylinder or wheel cylinders. If a brake tube or hose failure occurs, a sudden loss of hydraulic pressure, pedal reserve, and stopping ability will occur.

Brake Tube Inspection. The brake tubes should be inspected for proper fastening to the frame or underbody, corrosion that may lead to failure, cracks, and any damage that may cause the line to fail. An example is shown in Figure 8-11. Damage may result from road stones being tossed against the lines, careless attachment of tow hooks, or exhaust system components coming in contact with the tubing. Damaged tubing is most often found on the rear axle assembly. Corrosion may be found along the frame rails and in any place that holds road dirt and water.

Weak or damaged brake tubes must be replaced to assure safe brake performance. Only tubing that is designed for brake application should be used for repair of brake lines (see page 108). Brake tubes *must always* have double flared ends to provide the required strength. The double flare operation is shown in Figure 8-12. Copper tubing, as supplied by an automotive warehouse, should never be used for brake tubing because it is not strong enough and it has low fatigue, impact, and burst resistance. The use of copper tubing would soon lead to line failure and resulting brake failure.

Brake Hose Inspection. Flexible brake lines or hoses are designed to withstand the high pressures in the brake system. Because they are subjected to a great amount of flexing as well as temperature and climate extremes, the hoses may develop cracks over long periods of time (Figure 8-13). The hoses must be periodically inspected to detect any cracking that would soon lead to brake failure. They should also be inspected for routing that may cause chaffing to occur, as shown in Figure 8-14.

8-6 BRAKE DRUM INSPECTION

The brake drums are inspected for size, out-of-round, taper, cracking, and hard spots.

Brake Drum Wear. Brake drum size is measured with a drum micrometer (Figure 8-15). The measured size is compared with the discard dimension cast into the drum as shown in Figure 8-16.

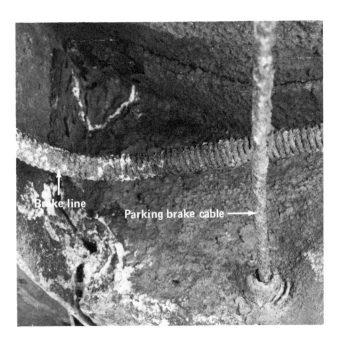

Figure 8-11 A rusted brake line.

Figure 8-12 Double flaring an end on metal tubing: (a) cutting the tubing; (b) clamping the tubing at the correct length in the flaring tool; (c) upsetting the tubing end in the first operation; (d) finishing the double flare in the second operation; (e) completed second flare operation; (f) first operation on the left and finished flare on the right.

Figure 8-13 A cracked brake hose.

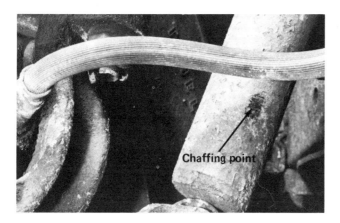

Figure 8-14 A brake hose that chaffed on the shock absorber.

Figure 8-15 Checking the amount of drum wear.

Figure 8-16 Typical maximum drum diameter marking.

The normal wear on the brake drum between brake relines is considered to be 0.030 inches (0.76 mm). This means that if a drum had a discard dimension of 10.090 inches (256.29 mm) the drum should only be resurfaced to a maximum of 10.060 inches (255.52 mm). This will provide a safe drum and an adequate service life for the new linings. Some manufacturers list the maximum resurface dimension as exactly 0.060 inches (1.5 mm) oversize. The strength and heat capacities of the drum are greatly reduced if the drum is resurfaced beyond these limits. Truck brake drums usually have a safe turn dimension that is 0.090 inch (2.3 mm) oversize. Drums must be *replaced* if they are worn too much for resurfacing. Always consult the proper shop manual if specifications are not given on the drum itself. Brake drums that have been manufactured since 1971 have a discard dimension imprinted on them.

Out-of-round. A brake drum may become out-of-round in service. This will result in pedal pulsations as the brakes are applied. An out-of-round drum can be detected with a brake drum micrometer by measuring the drum diameter in two or three different locations. If the measurements differ by more than 0.006 inch (0.152 mm), the drum will require resurfacing or replacement.

Taper. Brake drum taper, referred to as "bellmouthing," may be present in a drum. Bellmouthing can be detected with some brake drum micrometers or with inside calipers. A drum that is tapered more than 0.006 inch (0.152 mm) should be resurfaced. Figure 8-17 shows an exaggerated view of several types of drum wear.

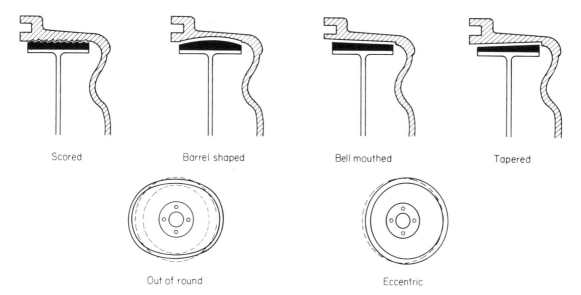

Figure 8-17 Types of drum wear.

Cracking and Hard Spots. Cracking and hard spots on the braking surface of the drum are usually indicated by discoloration of the drum braking surface. When the drum is machined, the hard spots will show up as bumps on the brake surface.

Figure 8-18 Hard spots in the surface of a reconditioned drum.

Figure 8-19 Cracks in the braking surface of a drum.

These can be seen in Figure 8-18. The hard spots may be ground for removal if the hardened area is not too large. It is accepted practice to *replace* a drum that has hard spots due to the metallurgical change that takes place in the hardened areas. Any cracks similar to those shown in Figure 8-19 that are visible on the braking surface are cause for replacement of the drum. No attempt should be made to repair a cracked brake drum.

8-7 DISC BRAKE INSPECTION

Disc brake service starts with a visual inspection of the wheel brake components. Because of the exposed nature of a disc brake, the lining material is visible when the wheel is removed as illustrated in Figure 8-20. This will allow assessment of the lining wear. Any fluid or grease leakage on the lining will also be visible. The rotor braking surface should be examined for excessive scoring and apparent wear. An example of the brake rotor can be seen in Figure 8-21. Several checks should be made before the brake is disassembled. The customer's complaint should be considered, the brake operation checked, and the disc runout, parallelism, and wheel bearing free play measured.

Disc Brake Runout. A dial indicator positioned against the surface of the rotor as illustrated in Figure 8-22 is used to detect runout of the disc brake surface. The dial indicator should be posi-

Figure 8-20 Brake lining material visible in an assembled disc brake.

Figure 8-22 Checking the rotor runout on an assembled disc brake.

Figure 8-21 The rotor braking surface can be checked on an assembled disc brake.

tioned approximately 0.5 inch (13 mm) from the outer diameter of the rotor and it should be perpendicular to the brake surface. Before measuring runout, the wheel bearing free play should be adjusted. For an accurate runout check, the wheel bearing is adjusted to zero free play and zero preload as discussed in Section 2-5. The runout can then be accurately checked. The total indicator reading (TIR) is the greatest dial indicator scale deflection and this reading is the rotor runout. This reading is used to help determine the cause of brake complaints. If it exceeds 0.003 inch (0.08 mm), it will cause brake pad knockback, resulting in a low brake pedal. A large runout can cause steering wheel shake. It may also cause disc brake noise, such as a rattle or periodic squeak. The high point, indicated by the dial indicator, is marked. The mark is used for a reference when setting the rotor in the disc brake lathe for resurfacing.

Parrallelism. The rotor thickness should be measured with a micrometer, as pictured in Figure 8-23. This will detect nonparallelism of the braking

Figure 8-23 Measuring the thickness of a rotor on an assembled disc brake.

Figure 8-24 Rust in the web area of a used rotor.

surfaces to determine if rotor resurfacing is necessary and to make sure the rotor is thick enough for resurfacing. Rotor parallelism is checked by measuring the thickness variation of the rotor at four or more points. Micrometer readings of more than 0.0005 inch (0.0130 mm) variation cause pedal pulsation while braking. If the variation is excessive the rotor will have to be resurfaced for smooth braking. The thickness measurement can be compared with manufacturer's specifications to determine if enough material remains for resurfacing. Consideration should be given to corrosion in the rotor web area, as shown in Figure 8-24. If excessively worn or corroded, the rotor must be replaced for satisfactory brake operation.

8-8 SUMMARY

The brake system is one of the major safety systems on vehicles. It should be given a thorough inspection when there seems to be a brake problem. This inspection includes an inspection of the master cylinder, lines and hoses, braking mechanisms, and braking surfaces. Any abnormal condition must be properly corrected before the vehicle is returned to service. Correction procedures are described in the following chapters.

REVIEW QUESTIONS

1. What conditions would cause a customer to have a technician inspect the brakes? [INTRODUCTION]
2. When should the brakes be routinely inspected? [INTRODUCTION]
3. What specific brake system components are checked on a complete brake inspection? [8-1]
4. What is the first thing to check in a brake inspection? [8-1]
5. Where could the brake system leak externally? [8-1]
6. What lining conditions are checked in a brake inspection? [8-3]
7. What should be done if the wheel cylinder is damp with fluid under the boot? [8-4]
8. What conditions are brake hoses and tubes checked for? [8-5]
9. What is the difference between the drum resurfacing dimension and the discard dimension? Where can the service technician find this information? [8-6]
10. What conditions indicate the need to replace the brake drum? [8-6]
11. What conditions indicate the need to resurface the brake rotor? [8-7]
12. Why is it important that no questionable condition exists in the brake system? [SUMMARY]

9
Drum Brake Service

Brakes affect vehicle control and safety as much as the steering system. The operational safety of the vehicle should never be compromised by poor workmanship or by a partial brake job. Driver complaints, a road test, and a brake inspection taken together will indicate when the brakes require service. Service for drum brakes may include replacing the brake linings, resurfacing the drums, and rebuilding or replacing the wheel cylinders. In some cases the master cylinder is also rebuilt or replaced along with replacement of damaged brake hoses and tubes. Proper brake operation can only be assured by completely servicing a faulty brake system.

Servicing procedures are similar for both front and rear dual-servo drum brakes. They may have different-size brakes and use different brake lining materials. Rear drum brakes have all of the components that are used in front drum brakes, plus a mechanically operated parking brake system mechanism.

9-1 BRAKE DRUM RESURFACING

The brake drum should be reconditioned first when doing a brake job so that the linings can be properly sized to fit the new arc of the reconditioned braking surface in the drum. It must be determined that the brake drums have enough material remaining to be safely reconditioned. This was discussed in the preceding chapter on brake inspection.

Resurfacing Front Brake Drums. The drums are mounted on a brake lathe for resurfacing. It is a good practice when reconditioning front drums to resurface them with the wheel mounted to the hub and properly torqued. This procedure will support the brake drum in the same way that it is held on the vehicle. In this way the refinished braking surface will be most accurate. When refacing front drums that can be removed from the hub, as opposed to those that are riveted on the hub, always reface the drum with the drum bolted to the hub. The wheel lug nuts can be inverted to hold the hub securely in the drum if the wheel is not readily available. If a drum is suspected of being loose on the hub, this method of bolting the drum to the hub must be employed to hold the drum in position. It is common practice to rely on rivets to hold the drum securely on the hub while resurfacing the drum. Relying on the rivets does not always produce a perfect concentric braking surface but it meets normal service standards.

Positioning the Drum. If the wheel bearings and seal have not previously been removed, they must be removed before the hub is mounted on the drum lathe. All grease is removed from the bearing races. Radii adapters of the proper size are selected to fit the bearing races, as shown in Figure 9-1. The bearing races should be inspected to ensure that they are tight in the hub. The hub and the selected radii adapters are slid over the arbor and the assembly is tightened. The anti-chatter belt (Figure 9-2) is then tightened around the drum. The drum should be checked to make sure it is concentric with the lathe arbor. This can be done by bringing the cutting tool into slight contact with the drum. The drum is then slowly rotated 180° by hand. The drum is concentric if the cutting tool maintains slight contact throughout the rotation. If this checking method shows a variation in the cutting tool contact, the arbor nut should be loosened, the drum rotated on

(a)

(b)

Figure 9-1 Drum lathe radii adapters for front hubs: (a) inner side; (b) outer side.

Figure 9-2 Installing an anti-chatter belt before turning the drum.

Drum Brake Service

the arbor, the arbor nut retightened, and the drum concentric alignment rechecked. If contact is the same, it is an indication that the drum is out-of-round. If it is out-of-round, resurfacing should correct the problem. After centering the drum, reconditioning of the braking surface can begin.

Resurfacing Procedure. The cutting tool is moved to the open portion of the drum and the stop positioner set. This will automatically shut the lathe off when a cut is completed. The cutting tool is then moved into the drum approximately one inch (25.4 mm) or to the deepest groove and the tool adjusted to just the contact the braking surface. The scale reference is set or a scale reading is taken, depending upon the lathe design, to indicate the diameter of the most worn surface. The tool is then backed away from the brake surface and moved toward the innermost portion of the drum. The machine is then turned on to rotate the arbor and drum. When resurfacing a drum, remove the least amount of material possible to clean the brake surface. Adjust the tool to make a rough cut depth of 0.006 inch (0.15 mm). If the drum is extremely scored, a deeper cut may be made. Figure 9-3 shows a drum being turned. A second cut may be required to remove all of the score marks, then a fine cut of 0.001 to 0.002 inch (0.03 to 0.05 mm) depth is made to smooth the surface. The drum is finished by breaking up the turning pattern using 80 grit sandpaper. This will establish a nondirectional pattern on the braking surface. The different finishes are shown in Figure 9-4. The reconditioned drum is removed from the lathe. The second drum from the same axle is reconditioned using the same setup procedure. It is generally recommended that the second drum be finished to approximately the same diameter as the first drum. This will help provide equal braking at both wheels on the same axle.

Some brake drum lathes can vary the drum rotation speed as well as the cutting tool feed across the brake surface. When using this type of machine, the depth of the finish cut is 0.001 or

Figure 9-3 Turning a drum from the inside toward the open side.

Figure 9-4 Different finishes during drum reconditioning: (a) rough first-turn finish; (b) smooth last-turn finish; (c) finish-turn surface broken up with sandpaper.

(a)

(b)

(c)

Figure 9-5 Typical mounting fixture used to install a rear drum on a drum lathe.

0.002 inch (0.03 or 0.05 mm), the feed is fine, and the rotating speed is high. This provides a smooth braking surface in the minimum amount of turning time.

Resurfacing Rear Brake Drums. The rear brake drums are reconditioned similarly to front brake drums. They require a slightly different mounting procedure. A spring-loaded tapered cone is used to center the drum on the arbor. Aligning cups are used on the inside and outside of the drum and these are tightened against the hub insert to support the drum. Typical mounting fixtures are shown in Figure 9-5. The drum can then be reconditioned using the same procedure that was used for the front drums.

9-2 BRAKE DISASSEMBLY

The brake drum was removed for inspection of the brake. Two special brake tools are required to remove the brake shoes from dual-servo type brakes. One is used to remove the return springs (Figure 9-6) and the other to remove the hold-down springs (Figure 9-7). It is a good practice for a beginner to finish the work on one brake shoe assembly before going to the brake on the opposite side of the vehicle. In this way, one brake is always assembled so it can be used as a reference for assembly of the other brake.

On dual-servo brakes, the return spring with the greatest amount of force is used on the secondary shoe. This is done to reduce the noise that would occur if the secondary shoe left the anchor pin before the primary lining made contact with the drum as the brakes are applied.

Disassembly and Servicing. The brake shoes are removed by first removing the return springs. Notice the order in which the springs are attached to the anchor pin; the color, size, and location of the springs; and which holes in the shoes are used for the springs. Carefully check the position of the

Figure 9-6 Tool being used to remove a brake return spring from the anchor end.

Figure 9-7 Tool being used to remove a hold-down spring.

Drum Brake Service

adjuster linkages and springs. They must be reassembled in the same order. Some of the automatic adjuster linkages may come off as the return springs are removed. The hold-down springs can be removed from the shoe after the return springs are disconnected from the anchor. Grasp one shoe in each hand and separate them enough to disengage the wheel cylinder push rods and, on rear brakes, the parking brake linkages as demonstrated in Figure 9-8. The secondary shoe of the rear brake will be held by the parking brake cable. The parking brake cable is disconnected by removing a clip or by simply twisting the lever and removing it from the pivot hole. The two shoes with the adjuster and spring can be lifted off when they are clear of the anchor. The brake shoes with the adjuster and spring may now be laid on the floor and twisted over one another to separate them as illustrated in Figure 9-9. All the wheel components (springs, adjuster, retainers, etc.) should now be inspected for signs of wear, rusting, or fatigue. Any spring that is stretched should be replaced. Experience has shown that if a wheel brake has been overheated because of a dragging brake, the springs have been heated, causing them to lose their strength. These springs must be replaced. The brake components are cleaned after the initial in-

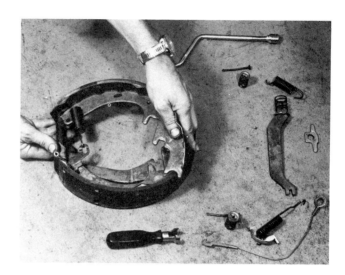

Figure 9-9 Overlapping shoes to remove the adjuster to separate them.

spection. A wire brush is used to remove rust and scale. Air should never be used to remove dust and dirt from the brake parts. It is best to remove the dust with a shop vacuum. Be certain that the work area is well ventilated and that safety goggles and a face mask are worn. A large portion of the dust on a brake assembly is asbestos particles. These particles are harmful if inhaled. After the brake shoes are removed and the initial cleaning of the parts is completed, the wheel cylinders, which are the only parts still attached to the backing plate, are serviced.

9-3 WHEEL CYLINDER SERVICE

Wheel cylinders should be serviced whenever a leak is present during inspection or whenever a complete brake relining job is being done. Wheel cylinders are serviced by either rebuilding the cylinder or replacing it with a new one. One reason for servicing is that new brake linings are thicker. This changes the position of the cylinder cups so they contact the contaminated inner portion of the cylinder. Although leakage may not be present at the time the brakes are relined, it will usually start shortly after. By polishing the wheel cylinder bore and replacing the cups, a positive long-lasting seal will be achieved.

Figure 9-8 Separating shoes to release them from the wheel cylinder push rods and parking brake linkage.

Preparation for Disassembly. Prior to any disassembly of the cylinder, it is a safe and time-

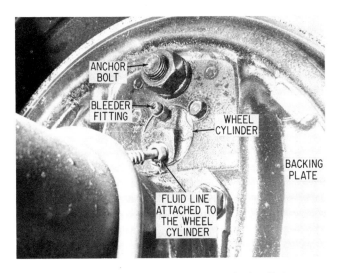

Figure 9-10 Typical location of the wheel cylinder bleeder screw fitting behind the backing plate.

saving step to loosen all of the wheel cylinder bleeder screw fittings, shown in Figure 9-10. Experience has shown that if a cylinder bleeder screw is frozen it may break. Time wasted in attempting to remove a broken bleeder screw will cost more than the cost of a replacement cylinder. If all the bleeder screws are free, the cylinders may be disassembled. It is a good practice to bleed the complete system *prior* to disassembly and again after reassembly. Bleeding *before* wheel cylinder disassembly will cause any rust or contamination in the lines and master cylinder to flow into the wheel cylinders. The contamination will be removed when the wheel cylinders are disassembled, instead of having it enter the rebuilt cylinder after it is reassembled.

Servicing the Wheel Cylinder. After disassembling the cylinder, the bore is polished or honed using a hone, as shown in Figure 9-11, driven by a drill motor. On some vehicles the cylinder must be separated from the backing plate to allow disassembly and honing to take place. The cylinder is measured as shown in Figure 9-12 to determine the original bore diameter. Only hone the minimum amount required to provide a smooth surface free from rust. Honing is done with the surface lubricated with brake fluid. Usually enough brake fluid from the brake line will seep into the cylinder during the honing operation. If additional fluid is needed while honing, use only brake fluid. The use of any petroleum-based fluid will destroy the new rubber cups in the rebuilt cylinder. The cylinder is remeasured after the honing operation to assure the bore size is not greater than 0.003 inch (0.08 mm) over the original bore diameter. If the cylinder requires more than the minimum honing to "clean up" the surface, the cylinder should be discarded. Attempting to "make do" with any part of a brake system is not a safe procedure.

The cylinder should be thoroughly cleaned with a shop cloth and brake fluid to ensure all the abrasive material from the honing process is removed. Forcing some fluid through the inlet

Figure 9-11 Honing a wheel cylinder.

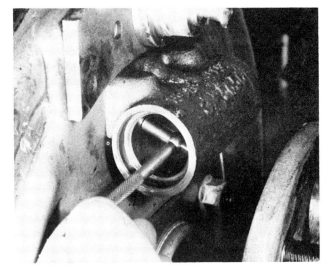

Figure 9-12 Measuring a wheel cylinder with a telescopic gauge. The gauge length is measured with a micrometer.

Drum Brake Service

opening will flush contaminants out. The cylinder is now ready to be reassembled.

Reassembly. Reassembly begins by coating the cylinder bore and rubber cups with either a special nonpetroleum-based brake assembly lubricant or a liberal amount of brake fluid. The bleeder screw is opened and the cup and piston components are carefully inserted into the cylinder. Be sure that the cup lip direction is correct and the sealing lips are not damaged as they are inserted into the bore. If the bleeder valve is allowed to stay open while the other cylinders are being serviced and the master cylinder reservoir kept over half full, often enough fluid from the master cylinder will slowly flow through the residual check valve into the assembled wheel cylinder to fill it. This will minimize the time required for brake bleeding. This process is called *gravity bleeding*.

Replacing the Wheel Cylinder. If the cylinder must be removed from the backing plate to facilitate replacement, the brake line is first loosened with a fitting wrench, as shown in Figure 9-13. All dirt and grease must be removed from the area of the connections. If a cylinder has a flexible hose threaded directly into the cylinder, the hose is disconnected at the opposite end. The hose can then be unthreaded from the cylinder. The bolts fastening the cylinder to the backing plate are removed to allow the cylinder to be freed from the backing plate. The replacement cylinder can then be installed.

Replacing Brake Hoses. Care must be used when reinstalling the type of brake hose that has one end screwed into the wheel cylinder to assure the copper washer is in place for a positive seal on assembly. The replacement hose is installed into the wheel cylinder or brake divider block (tee) and tightened. Always install and tighten the male end of the hose before connecting the other end to the line. The remaining end of the hose is then connected carefully to prevent any twisting of the hose when final assembly is complete. Some manufacturers paint a white stripe along the length of the hose so a quick visual check will show a twisted hose. A slightly twisted hose is shown in Figure 9-14. Final installation should include no more than a 15° twist of this painted line. After installation, the wheels should be turned to left and right extremes and the vehicle bounced to assure the hose does not contact the tire, wheel, or any underbody component.

Only hoses of the correct length should be installed as replacements. A hose that is only a slight amount too short could break due to stretching during certain wheel motions. A hose that is too long would result in failure due to chaffing on

Figure 9-13 Using a fitting wrench to remove the fluid line from the wheel cylinder.

Figure 9-14 White line shows that the brake hose is slightly twisted.

underbody components. The wheel cylinder repair is complete when all of the cylinders and lines are in perfect condition to accept the new linings.

9-4 LINING SELECTION

Lining selection might begin by choosing between new shoes with linings attached or relined brake shoes. This selection may not be possible if your supplier carries only relined brake shoes or only new brake shoes. Remanufactured or relined brake shoes are actually serviceable used brake shoes that have had the worn lining removed and a new lining bonded or riveted into place. Only the steel backing or shoe is reused in the remanufacturing process. Relined shoes, which are available at almost every automotive supplier, will provide good service as long as the remanufacturer uses only shoes that are in good serviceable condition prior to remanufac-turing. If the brake shoes have been scored, bent, twisted or if the anchor pin receiver is worn, their use will result in an unsatisfactory brake reline job. Although new brake shoes with new linings are top-quality replacements, they are generally more expensive than remanufactured shoes with new linings.

Brake Lining Characteristics. Brake linings are either riveted or bonded into place on the brake shoe. The riveted lining uses brass rivets passing through the lining and brake shoe to provide a tight and secure mounting to the shoe. The riveting process involves cleaning the shoe, clamping the lining securely to the shoe, inserting the rivet, and upsetting the rivet to further tighten the lining to the shoe. This sequence is illustrated in Figure 9-15. A bonded lining first has cement applied to the lining.

Figure 9-15 Riveting new lining on shoes: (a) remove old rivets; (b) cleaning the shoe; (c) countersink drill; (d) drilling the new lining; (e) new rivet upset in place.

(a)

(b)

(c)

(d)

(e)
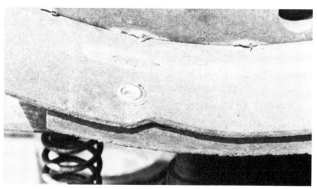

Drum Brake Service

Then special equipment is used to apply pressure and heat to the assembly, so that the cement is cured to provide a secure bond.

Bonded linings are generally less expensive to produce than riveted linings, with approximately equal life expectancy. The bonded lining has the tendency to be noisier because the bonded lining transfers more vibration into the brake shoe. The vibration of the larger brake parts causes undesirable noise during a stop.

Other lining choices that can be made are in the lining coefficient of friction. These are listed and described on page 85. If the lining coefficient of friction is too high, an overly sensitive brake pedal action may result. If it is too low, excessive pedal pressure will be needed to provide the required braking action. Special care should be taken to match the new linings to the coefficient of friction of the existing linings, especially when only one axle set of linings is being replaced. The manufacturer has taken great care to match the brake design to the vehicle braking requirements. Changes in the brake coefficient of friction will not necessarily improve brake response and it may even provide undesirable braking characteristics. It should be noted also that with dual-servo brakes, the primary and secondary linings provide different portions of the stopping power for the wheel. As a result, the coefficient of friction of the primary lining is usually different than that of the secondary lining.

Brake lining selection may also be made between a standard size lining or an oversized lining. Standard linings are used when brake drums have been resurfaced up to 0.060 inch (1.52 mm) oversize. If the drum is 0.060 inch (1.52 mm) oversize or more, while still remaining within the manufacturer's discard dimensions, oversize linings should be employed. Most passenger vehicles have a maximum turn dimension of 0.060 inch (1.52 mm) oversize and a discard dimension of 0.090 inch (2.29 mm) oversize. This means that oversize linings would only be needed if the drum were turned past the maximum dimension for standard size linings or if the drum were worn over 0.060 inch (1.52 mm) oversize.

9-5 BRAKE SHOE GRINDING

A critical and sometimes overlooked part of the shoe relining job is to accurately size the new brake lining to match the new turned diameter of the brake drum. A precise match of the diameters will give a better brake feel, prevent localized overheating of the lining, prevent noise that may occur, and give the operator maximum brake service life.

Brake shoe grinding or arcing is a procedure that allows the technician to remove high spots in the lining material so as to fit the lining to the arc of the drum. A refinished drum has a larger diameter (across the brake surface) than it had originally. The use of a standard size lining would cause the lining to contact the drum at the very center of the lining material only, as illustrated in Figure 9-16. This limited contact could cause

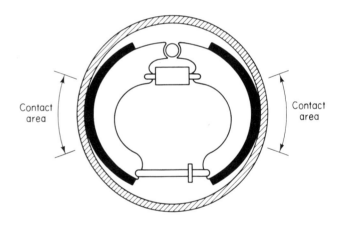

Figure 9-16 Standard-sized linings contact oversized drum at the middle of the lining.

localized overheating, damage to the new lining, and resulting poor braking performance. Brake shoe grinding corrects this problem.

Cam Grinding. As the brakes are applied, the brake shoe and brake drum try to change shape. These conditions are exaggerated in Figure 9-17. It can be seen that the shoe and lining tend to straighten out and the brake drum tries to become "egg" shaped. With high pedal effort like that which occurs in panic braking, the linings and drums do deflect slightly in this manner. To allow

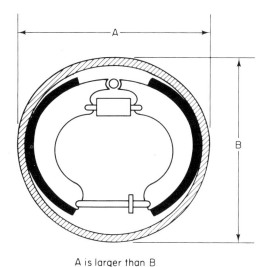

A is larger than B

Figure 9-17 Drum would tend to become egg-shaped if the linings are not properly matched to the drum.

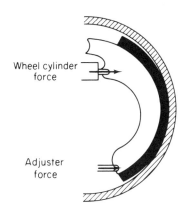

Figure 9-18 Cam grinding compensates for slight flexing so the lining matches the drum when the brakes are applied.

for this deflection, the linings are ground slightly cam-shaped so that the maximum lining to drum contact will exist during hard braking. *Cam grinding* provides a slight toe and heel clearance as the drum and lining are brought together during the fitting operation, where no force is exerted on the shoe. This is shown in Figure 9-18. The clearance is usually from about 0.008 inch (0.29 mm) to about 0.080 inch (2.03 mm), depending on the application. The larger clearances are necessary where a high pedal force is normal, such as the pedal force used in racing applications. This toe and heel clearance is established by grinding the shoe at a diameter approximately 0.030 inch (0.76 mm) less than the drum diameter. This type of cam grinding

9-5 Brake Shoe Grinding

will lead to more predictable braking and less noise during harsh stops.

Radius Grinding. An alternative type of grinding is *radius grinding,* where the lining is sized to the drum diameter. A small clearance should exist at the toe and heel after matching the lining to the reconditioned drum to prevent noise. The thought is that this close sizing will give a firm pedal during normal braking. This is due to the fit of lining to the drum when light braking occurs.

Various types of grinding machines are on the market. Each one has its own advantages. Differences exist in the way the shoe is mounted. A machine with a pin and selective sleeves, as pictured in Figure 9-19(a), is designed especially for fixed

Figure 9-19 Grinding the brake shoes: (a) cam grindding; (b) radius grinding.

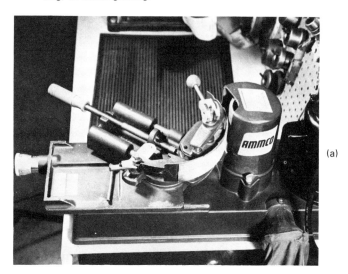

Drum Brake Service

anchor brakes (dual-servo). It allows the shoe to be fitted against and ground around the pin which acts as the anchor pin. This allows the lining to be ground around its natural pivot. Other types of grinders clamp solely on the shoe web and pivot on the machine center. The different types of equipment are not as important as the need for arcing shoes to provide maximum braking performance and maximum operator safety.

Brake shoe grinders differ in many ways, although the end result is the same. They make a better fit between the lining and drum. Because it is important for a correct fit, the exact diameter of the brake drums must be known. This dimension can be determined by using a brake drum micrometer, as shown in Figure 9-20. The dimension of the drum diameter is set on the brake shoe grinder.

Mounting the shoe into the holding vise is where brake shoe grinding machines differ. If the machine is specifically for the "fixed anchor" type of brake, sleeves and positioner fixtures are selected to properly position the shoe in the vice. If the machine is the general radius type grinder, as shown in Figure 9-21, no special preparation is necessary. The shoe is carefully clamped in the

Figure 9-21 A shoe clamped on a general radius-type grinder.

holding vise of the machine, making certain all necessary positioning points on the machine are in contact with the shoe. The machine is set to the diameter of the drum and the micrometer dial is set to the oversize dimension. The lining is now marked across its width with chalk to provide a visual aid in determining when the grinding disc has made contact with the full length and width of the shoe. This also provides a means of immediately seeing a twisted or distored shoe. After the lining has been properly ground, the shoe is placed inside the resurfaced drum and checked for correct toe and heel clearance.

9-6 BACKING PLATE SERVICE

The backing plate used on drum brakes serves many purposes. It must provide a means for holding the brake shoes in position; it must act as a guide to ensure the shoes and linings move out parallel to the brake surface without binding; and it must transfer the braking torque from the brake anchor pin to the spindle or rear axle housing.

Figure 9-20 Determining the diameter of the reconditioned brake drum.

Servicing Procedure. The backing plate for each wheel must be inspected for visible signs of damage or excessive wear. Damage occurs when an object that is struck physically twists or bends the backing plate. This type of damage is usually most obvious because it causes uneven lining wear.

Wear of the backing plate ledges is noted by careful inspection. The ledges are the guide surfaces for the shoes and linings. If wear is present as shown in Figure 9-22 and not corrected, it is possible the shoes will bind as they apply or release. This can cause erratic brake application, brake drag, or noise. Worn ledges are filed flat and lubricated with a high temperature brake lubricant.

Figure 9-22 Wear on the backing plate ledges.

The backing plate should be checked to make sure it is not loose on the axle housing or spindle, and then the attaching bolts should be torqued to specifications. The shoe can now be held in position against the backing plate and checked for equal contact on all the ledges. These procedures will assure the service technician that the brakes will provide maximum service life and brake safety.

9-7 PARKING BRAKE CABLE SERVICE

Prior to the installation of the brake shoes, the parking brake cables should be inspected. The cables may become frayed when used frequently and they often become frozen in their housings when used infrequently. A visual inspection will show any worn cables. A frozen or binding cable is found by checking for movement of the cable in the housing as the parking brake is applied and released. Lubricant applied to the area where the cable enters its housing will often help free the cable if some sticking is present. A cable that refuses to move will generally require replacement. An attempt should be made to free a frozen cable with penetrating oil although this takes valuable time and most often is unsuccessful.

Parking Brake Cable Design. The parking brake cables on most modern automobiles are in three or four separate sections. One example is shown in Figure 9-23. The front section is enclosed in a housing and connects the parking brake lever in the vehicle either to an equalizer or to a parking brake lever beneath the vehicle. From the equalizer, the cable runs either directly to the wheels or to an intermediate cable which connects the two rear-wheel cables. The most troublesome cables are those directly connected to the rear-wheel brakes, called the *rear cable sections*. These rear sections are subject to dirt, road salt, and water which tend to shorten their life.

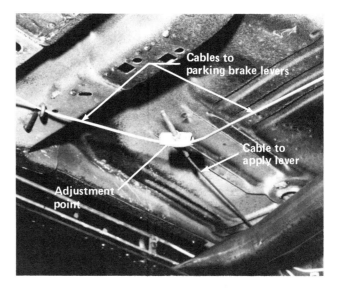

Figure 9-23 A typical parking brake cable equalizer adjustment point.

Cable Replacement. The rear cable sections are replaced by several different methods. They may be removed by loosening a clamp, compressing some tangs inside the backing plate, or loosening a clamp outside the backing plate. With the cable disconnected from the parking brake lever in the rear brake, the cable and housing are pulled from the backing plate. The forward end of the cable is either fitted into a bracket that attaches the two cables together or it is connected and secured with a retainer. Installation of the new cable is the reverse of the removal procedure.

It may be necessary to adjust the length of the parking brake cables prior to installation of the brake shoes. Adjustment is made by changing the length of a threaded rod on the rearward end of the front cable section or on the equalizer link. If the adjusting nut is rusted into position, it may be necessary to use a "nut cracker" to remove the old nut to avoid breaking the adjustment rod. If the rod breaks, an additional part will have to be replaced. To prevent the parking brake from interfering with the initial service brake adjustment, it is standard practice to have the parking brake cables adjusted as loosely as possible. They are properly adjusted *after* the service brakes have been adjusted.

9-8 SHOE INSTALLATION

Prior to shoe installation, all of the brake hardware (springs, adjuster mechanisms, hold-down pins, and retainers) must be inspected for wear, corrosion, and heat damage. All of these can be found through a visual inspection.

Springs are most often rejected because of heat damage or careless handling. If a wheel brake has been overheated, due to either a malfunction or an operator error, the possibility is great that all the springs on that wheel will require replacement. Overheated return springs appear stretched, as shown in Figure 9-24, and the hold-down springs appear collapsed. Mishandling may also stretch the hold-down springs or misshape the end loops so

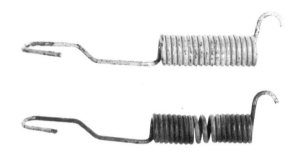

Figure 9-24 The upper brake spring is normal. The lower brake spring has been overheated and lost tension.

much that the spring is not usable. It is a good practice to replace the springs on both wheels of an axle set if one of the springs needs replacement. This practice will minimize the possibility of wheel pull during braking that would result from different return spring strengths.

Hold-down spring retainers and pins may suffer from wear that would warrant their replacement. Self-adjuster cables that have become stretched or frayed in service will require replacement. The self-adjuster screw assembly is inspected to ensure it moves freely and that the adjuster teeth are not worn or damaged. The adjuster should be checked to make sure it is on the correct brake (L or R) and that it is lubricated with high-temperature brake lubricant. It is then set aside for installation.

Every effort should be made to avoid getting grease, brake fluid, and/or other contaminants on the new brake linings. Figure 9-25 shows a new lining that was contaminated by careless handling.

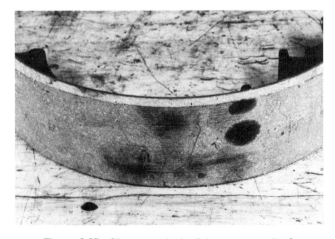

Figure 9-25 Oil on new brake linings as a result of careless handling.

After cleaning the brake assembly tools and your hands, the brake shoes are ready to be installed. The wheel cylinders are positioned on the backing plate and secured.

Installation Procedure. On the rear brakes the parking brake lever has to be connected to the secondary shoe of each rear wheel. The secondary shoe is put into position, the parking brake cable attached (Figure 9-26), and the hold-down spring and retainers put into place. The primary shoe is positioned properly and its hold-down spring and retainers locked in place (Figure 9-27). The self-adjuster screw, lever, and spring are connected between the shoes. The parking brake strut is positioned so the back of the strut engages the parking

9-8 Shoe Installation

Figure 9-28 Positioning the parking brake strut.

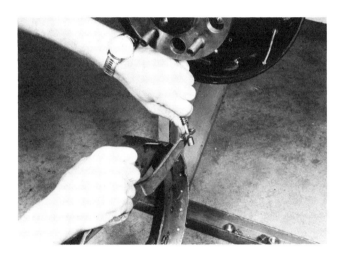

Figure 9-26 Typical parking brake cable attachment to the parking brake lever on a rear drum brake.

Figure 9-27 Attaching a shoe hold-down spring.

brake lever (Figure 9-28). The anti-rattle spring is placed on the front of the strut and then engaged in the slot on the front shoe. The anchor plate (if used) is positioned over the anchor pin and this is followed with the adjuster cable or link. The primary shoe spring is now properly positioned against the anchor. The cable guide (if used) is positioned in a hole in the secondary shoe along with the secondary return spring. The anchor pin end of the spring is then lifted into place with the spring tool as illustrated in Figure 9-29. The lower end of the adjuster cable is attached to the adjuster lever and the operation of the automatic adjuster checked. If the loops on the return spring ends have become opened due to removal and replacement, these can be closed with pliers. The assembly should be thoroughly inspected to make sure that everything is properly positioned. This brake assembly procedure will vary slightly from manufacturer to manufacturer and with types of self-adjuster mechanisms used, although the assembly sequence remains the same. After assembly, the rear brakes are ready for preliminary adjustment.

The front drum brakes follow the same assembly procedure as the rear drum brakes, with

Drum Brake Service

Figure 9-29 Lifting the end of the return spring onto the anchor.

the exception that they do not have the parking brake components.

9-9 BRAKE SHOE ADJUSTMENT

The brake shoes require a preliminary adjustment which can either be done prior to placing the drums in position or after they are installed.

Brake linings require a small operating clearance when they are applied. The preliminary adjustment will provide the required clearance so there is no possibility of continuous contact with the drum. At the same time they are close enough to the drum to provide satisfactory pedal reserve. Once the brakes are in service the normal clearance is between 0.005 and 0.010 inch (0.13 and 0.25 mm). This clearance prevents continuous lining contact on the drum which would reduce gas mileage and cause temperature extremes that could damage the new linings and reconditioned drums.

Adjustment Procedure. The first step in the brake adjustment procedure is to make certain that the parking brake cable is fully released. This can be

Figure 9-30 Setting the brake set gauge to the drum size.

checked before the rear brake drums are installed by checking to make sure the brake shoes are firmly seated against the anchor pins. If the shoes do not seat against the anchor pin and the shoes were properly installed, the parking brake linkage separates the shoes at the top. This can be caused by a parking brake cable adjustment that is either too tight or that is binding.

The preliminary brake adjustment may be obtained by using a brake set gauge, as shown in Figure 9-30. The gauge is adjusted to the drum diameter and locked. The opposite side of the gauge is then used to measure the setting of the shoes as shown in Figure 9-31. The brake adjuster

Figure 9-31 Using the brake set gauge to adjust the shoes to the drum size.

screw is turned until the brake set gauge just makes contact with the brake lining. This will provide the correct initial setting for the linings. The automatic adjusters will make the final adjustment when the brakes are used. The adjustment may be made with conventional brake-adjusting tools when a brake set gauge is not available, or when a manual brake adjustment is being made, and the drums are installed. On most domestic cars, the adjustment is made through an adjustment hole located either in the lower portion of the backing plate or in front of the brake drum. Adjustment is made by inserting a brake-adjustment tool through the access hole and making contact with the screw adjuster teeth. Moving the tool end up or down will force the adjuster screw to rotate and this will expand or contract the shoes. If it is necessary to loosen an adjustment, the automatic adjuster lever will have to be disengaged from the adjuster screw to allow the screw to be rotated backwards. The brake tool and a release tool (usually made from heavy gauge wire) are shown in Figure 9-32 and 9-33. The procedure to properly adjust the brakes is to tighten the adjuster until the wheel will not rotate. This will properly position the shoes. The adjuster is then loosened until only a slight brake shoe drag is felt. While performing rear brake shoe adjustment, parking brake interference will be noticeable as a locking wheel when the wheel and brake drum assembly is rotated in one direction but it can be rotated freely in the other direction. If this condition occurs while brake adjustment is being made, the parking brake cables must be loosened further. After all brake shoe adjustments have been made, the access hole covers must be inserted to prevent dirt and water from entering the brake assembly.

Reassembly. The rear drums and wheels must be installed prior to making the parking brake adjustment. The drums are installed by sliding the drums over the wheel studs and installing any holding devices (bolts or retainer nuts). The rear wheels can then be installed taking care to match the marks made during disassembly. The proper wheel nut torquing sequence was discussed in Section 2-5.

The front drums are most often riveted to the front hubs. This makes it necessary to remove the bearings for mounting the drum on the drum lathe. The wheel bearings and hubs must be thoroughly cleaned prior to reassembly and then the bearings are packed with the recommended wheel bearing grease. A thorough cleaning is necessary for three

(a)

(b)

Figure 9-32 Making the final brake shoe adjustment with a brake adjusting tool and a tool to release the automatic adjustment lever: (a) through the backing plate access slot; (b) through the drum access slot.

reasons: the machining operation may have left some metal chips in the bearing area; the manufacturer recommends packing the bearings whenever the drums are removed for inspection; various grease types do not mix and if all the old grease is not removed, bearing failure could result. *Never add grease to wheel bearings.* Follow the wheel bearing packing and adjustment procedures described in Section 2-5.

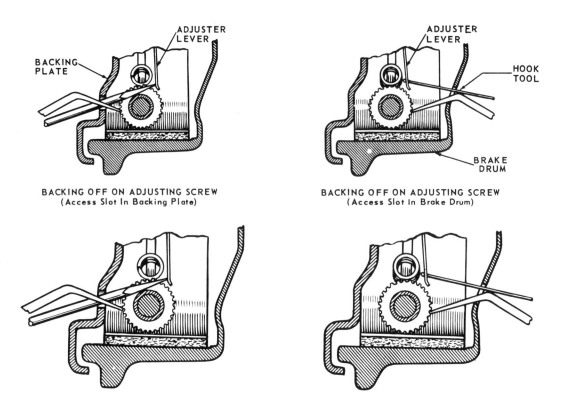

Figure 9-33 Section view of the use of the brake adjustment tools (Courtesy of the Bendix Corporation).

9-10 PARKING BRAKE ADJUSTMENT

After the rear service brakes have been adjusted and the wheels installed, the parking brake can be adjusted. The parking brake should never be adjusted without first adjusting the brake shoe to drum clearance. Each manufacturer has a specific adjustment procedure that applies to its own vehicles. A general adjustment procedure that applies to most brakes is described as follows.

Adjustment Procedure. The driver's parking brake lever is set to the second step from the fully released position (second click). The parking brake adjustment is tightened until a slight drag is evident at the rear wheels; the adjustment is then loosened until the wheels just rotate freely. The parking brake is now fully applied and then released. The wheels should rotate freely when the brake is released and the brake lever movement should not exceed 2/3 of its full travel when the brake is fully applied.

An overadjusted parking brake will not allow the brake shoes to come in contact with the anchor pin when the service brakes are released. This clearance will allow the shoes to "rock" each time the brakes are applied. This "rocking" will allow the self-adjuster to tighten the adjustment even though there is no clearance between the drum and lining. This overadjustment will result in tight dragging brakes and this will shorten the lining life.

Other Types of Parking Brakes. Our discussion of parking brake adjustment has focused on the parking brakes that are used primarily on dual-servo rear brakes. The parking brake on some heavy trucks and older passenger vehicles is an internal expanding drum brake, located on the rear of the transmission. This design prevents wheel rotation the same as parking brakes in the rear wheels. It does this by indirectly holding the rear wheels through the drive shaft.

Another type of parking brake is used on some vehicles equipped with rear-wheel disc brakes. This type of brake employs a special mechanism within the disc brake caliper, which forces the piston outward, clamping the shoes against the rotor.

REVIEW QUESTIONS

1. Describe the procedure used to resurface the brake drum. [9-1]
2. How does the brake lathe mounting of front and rear drums differ? [9-1]
3. How are brake components properly cleaned? [9-2]
4. When should a wheel cylinder be reconditioned and when should it be replaced? [9-3]
5. What is used to lubricate the brake hone while reconditioning the wheel cylinder? [9-3]
6. How are the wheel cylinders cleaned after honing? [9-3]
7. What standard procedures should be followed for lining selection and application to the brakes? [9-4]
8. What are the advantages of riveted linings and of bonded linings? [9-4]
9. Why are brake shoes ground after they have been fastened to the shoes? [9-5]
10. How does cam grinding and radius grinding of brake shoes differ? [9-5]
11. What service is done on parking brake cables during brake service? [9-7]
12. What precautions should be observed when assembling drum brakes? [9-8]
13. What must be done with the parking brake while making an adjustment to the service brakes? [9-9]
14. Describe the service brake adjustment procedure. [9-9]
15. How is the parking brake adjusted? [9-10]

10
Disc Brake Service

Before the disc brake is disassembled for service, the technician should review the customer's complaint, recheck to make sure the brakes still have the problem, and test the brakes to see if problems exist that the customer did not notice. When the operating condition of the brake system is known, the brakes are inspected, both visually and with measuring instruments as discussed in Chapter 8. After the technician has verified the brake condition, the brake can be disassembled.

It is generally quicker and easier to service disc brakes than drum brakes. This is true if the components are not rusted and seized together. If they are, freeing them can often take a considerable amount of time, thus reducing the advantage compared to the time required to service drum brakes.

10-1 DISASSEMBLY

After the wheel and tire assemblies are removed, the disc brake disassembly begins with removal of the caliper. If the hydraulic system is to be serviced, the technician should first open the bleeder valve, as shown in Figure 10-1, to be certain that it moves freely and then the brake hose or line is removed. If the hydraulic system is not going to be serviced, some brake fluid must be removed from the master cylinder reservoir. This will prevent fluid from spilling out of the master cylinder as the caliper pistons are pushed into the cylinders to provide clearance for pad removal and replacement.

10-1 Disassembly

Figure 10-1 Location of a typical caliper bleeder valve.

Figure 10-3 Lifting the caliper from the rotor.

Caliper Removal. The caliper is disconnected by first removing the attaching bolts, retainer pins, or guide pins. The caliper is then clamped with a C-clamp as shown in Figure 10-2. The clamp forces the piston inward to provide the clearance necessary for caliper removal. The clamp is removed and the caliper is lifted from the rotor (Figure 10-3). It must be supported as shown in Figure 10-4 so it does not hang on the brake hose while other parts of the brake system are being serviced.

Figure 10-4 Supporting the caliper so it does not hang on the brake hose.

Figure 10-2 Using a C-clamp to force the piston inward to provide the clearance necessary for caliper removal.

Brake Shoe Removal. The brake pads or shoes are usually retained in the caliper housing. These may be held in place with pins and retainer clips or by closely fitting tabs on the shoe that are bent over shoulders of the caliper as shown in Figure 10-5. Other types of calipers may retain the shoes in a portion of the knuckle assembly that is made as a guide and anchor for the shoes. In either type, the shoes are easily removed when the caliper has been lifted off the rotor. On some imported vehicles, the shoes are retained in the caliper with pins. Removal

Disc Brake Service

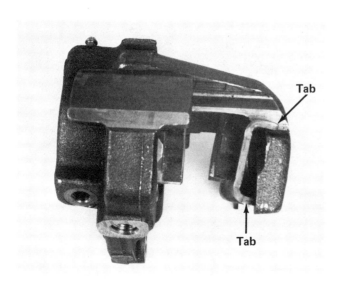

Figure 10-5 Tabs that hold the shoes in place.

of these pins allows the brake shoes to be lifted out without removing the caliper. The position of any anti-rattle or retainer clips should be noted to aid in the correct reassembly of these items.

Removal of Front Brake Rotors. After the caliper has been removed, the front rotors are removed from the spindles in the same way as the front brake drums are removed on a drum brake system. This includes removal of the grease cap, cotter pins, nut lock, bearing adjusting nut, and washer. Again, after removing the outer bearing, the nut may be threaded onto the spindle so it can be used to remove the inner bearing and the grease seal from the hub. The inner seal may also be removed by lifting it with a pry bar and then the bearing can readily be removed. Bearing assemblies are separated to prevent side for side mixing. The rotor itself should be marked with chalk to identify its correct location.

Removal of Rear Brake Rotors. On vehicles equipped with rear disc brakes, the rotors may be held with sheet metal nuts, bolts, or rivets, depending on design. Before removal, a lug and its corresponding lug hole should be marked to allow replacement in the same location. Sheet metal nuts or bolts holding the rotor are removed so the rotor can be lifted from the wheel lugs. Rotors that are riveted to the rear axle flange will require the rivets to be drilled out before the rotor can be removed. The rivets are only required for handling as the automobile is being built so replacing the rivets is not necessary when the brakes are reassembled. Some rear brake rotors have directional fins. These rotors must never be placed on the wrong side. When they require replacement, they must be ordered for the correct side of the vehicle.

Rotors need only be reconditioned if excess scoring, (as pictured in Figure 10-6) is present, if they have excess runout, or if the braking surfaces are not parallel. The only other time they will have to be removed is when the front-wheel bearings need to be repacked. Unlike brake drums, it is not

(a)

(b)

Figure 10-6 Typical rotor and lining wear: (a) normal; (b) excessive.

necessary to recondition the braking surface of the rotors each time the brake linings are replaced. Minor scoring does not affect the service life of the disc brake lining. Resurfacing will not improve the braking performance if the rotors are within service specification limits.

If the rotors are to be removed for service, they should again be measured to make sure they are within manufacturer's thickness specifications, which indicates the amount of material remaining for machining. The rotors should be carefully inspected for cracking on the brake surface.

10-2 ROTOR RESURFACING

Resurfacing disc brake rotors requires either a special disc brake lathe or a special rotor attachment for the drum brake lathe. It is of utmost importance that the rotor surfaces be *parallel*. This is accomplished by resurfacing both braking surfaces at the same time. Before putting the rotor on the machine arbor, the rotor is inspected for loose bearing cups. Excess grease is removed, and properly fitting bearing race radii adapters are selected. The adapters and rotor are positioned over the machine arbor, the required adapter spacers put in place (Figure 10-7), and the arbor nut tightened. The dial indicator is positioned against the rotor surface as illustrated in Figure 10-8, and the rotor is rotated to detect the amount of runout. If the point and amount of runout is the same as measured with the rotor on the vehicle, the rotor is ready to be resurfaced. If the point or amount of runout is not the same, loosen the arbor nut and turn the rotor while retightening the nut. This will minimize interference on the arbor and adapters. The runout is again checked. This procedure is repeated until the least amount of runout is obtained at the same location of maximum runout of the rotor when it was checked on the vehicle. An anti-chatter strap is positioned on the rotor as shown in Figure 10-9. The two cutting tools (Figure 10-10) are positioned to contact the smallest diameter of the rotor braking surfaces. The feed is engaged to move the cutting tools outward across the surface as the arbor turns the rotor. It is only necessary to remove the minimum amount of material required to smooth the rotor

(a)

(b)

Figure 10-7 Adapters used to mount the rotor on the reconditioning machine: (a) inner adapter; (b) outer adapter.

Figure 10-8 A dial gauge is used to detect runout on the mounted rotor.

Disc Brake Service

Figure 10-9 Installing an anti-chatter strap on the rotor.

Figure 10-10 The rotor is resurfaced with two tools cutting at the same time.

braking surface. The first cut is usually made by removing approximately 0.005 inch (0.13 mm) or less with each cutting tool. This should be repeated until both surfaces are smooth. The finish cut is usually made with a slow feed and reduced cutting depth, usually 0.002 inch (0.05 mm). Before removing the rotor from the machine, the machining pattern produced by the cutting tool should be broken up using medium-coarse sandpaper, about 80 grit. This is illustrated in Figure 10-11.

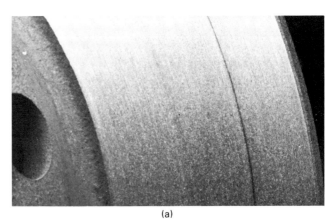

(a)

(b)

(c)

Figure 10-11 Rotor surface finishes: (a) first rough cut; (b) finish cut; (c) finish cut broken with sandpaper.

10-3 CALIPER SERVICE

Many of the problems associated with disc brake calipers result from pistons seizing in their bores. This can make piston removal quite difficult. If, during the inspection, it is found that one caliper piston is frozen on single piston caliper brakes, that piston can be forced from the caliper by lifting only that caliper from the rotor and spindle assembly. Then have an assistant depress the brake pedal hard enough to force the piston from its bore. If both calipers are to be disassembled, the line to the first caliper should be plugged and the remaining caliper disassembled in the same manner.

Some multipiston calipers are disassembled by first separating the two caliper halves and then removing the pistons from their bores. If the caliper is of a multipiston design, it is necessary to use clamps to hold all but the stuck piston while depressing the pedal when using this method to free the piston.

After the caliper is removed from the spindle or anchor plate and the brake hose is disconnected, the caliper can be serviced on the work bench. Various methods can be used to remove the pistons from their bores. Most methods use a tool that clamps inside the piston and with a rotating-pulling force the piston can be removed. This is shown in Figure 10-12. A special caliper service bench can be

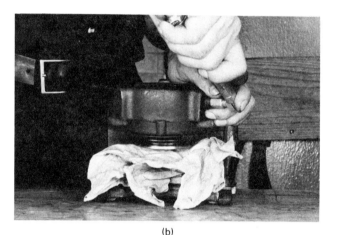

(b)

Figure 10-13 Removing the piston from the caliper: (a) with hydraulic pressure; (b) with air pressure.

used to force the pistons from their bores as shown in Figure 10-13. *Caution: Air pressure should be used with care to remove the caliper pistons. Their sudden release may cause personal injury as well as damage to the piston.*

After the pistons are removed, the dust boot can be pryed from the caliper housing. The hydraulic seals that are located either on the piston or in the caliper housing can be removed using a wooden or plastic pick to prevent damage to the seal grooves.

Servicing Procedure. A careful visual inspection should follow disassembly. The visual inspection includes determination of the condition of the cylinder bore and piston. The finish of the compo-

Figure 10-12 Removing a piston from a caliper with a special tool (Courtesy of Kelsey-Hays Company).

Disc Brake Service

nent that the hydraulic seal slides on is of primary importance. If the seals move with the piston (as in a drum brake wheel cylinder), the cylinder bore is the primary sealing surface which must be smooth. If the seal is mounted in the caliper housing and the piston slides through it, the piston surface is the primary sealing surface and it must be in perfect condition. The most common type of seal is the fixed seal mounted in the caliper.

Polishing of the cylinder bore can be done with a hone as illustrated in Figure 10-14. On the type with the moving seal mounted on the piston, the cylinder bore must be perfect. Honing is only done to polish the surface, not to remove metal. If the caliper is of the stationary seal type mounted in the caliper bore, honing is only done to remove rust that has accumulated near the outer edge of the cylinder bore. The bore finish does not have to be perfect. The piston, however, must be inspected. If it is pitted as shown in Figure 10-15, it must be discarded and replaced with a new one. The pistons are chromium-plated. Any pitting or other surface roughness on this surface will either destroy the seal or cause binding. Sanding or other methods of "improving" the roughness will not produce a satisfactory finish.

Before and after any honing operation, the bore sizes and piston clearances are measured and compared. Never remove over 0.002 inch (0.05 mm) from the bore diameter. This will normally

Figure 10-15 A pitted caliper piston.

result in a standard piston-to-bore clearance of less than 0.006 inch (0.15 mm).

Reassembly. The caliper assembly is flushed with brake fluid to prepare it for reassembly. Caliper reassembly begins by lubricating the new piston seal with disc brake assembly lubricant and carefully installing it into its groove. In most cases, the dust boot is now installed into the housing. Some manufacturers require the dust boot to be installed after the piston is started into the caliper bore. The dust boot inner lip is lubricated and the piston outer diameter coated with a light film of the assembly lubricant. The piston is guided into its bore, taking extreme care not to damage the seal in any way. A wood or plastic pick, as shown in Figure 10-16, may be helpful in positioning the seal. The piston should be able to be moved down-

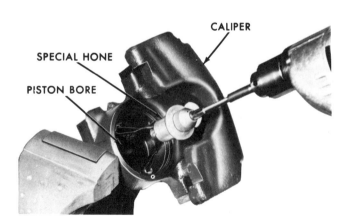

Figure 10-14 Polishing a caliper bore with fine stones on a hone (Courtesy of Chrysler Corporation).

Figure 10-16 Working the caliper piston through the seal.

ward in its bore with steady hand pressure. If it must be forced or pressed into its bore, a problem is present and it must be located and corrected. If the dust boot had not been installed previously, it is now installed. If the caliper halves were separated during disassembly, they must now be assembled using applicable fluid passage seals and their assembly bolts must be tightened to the correct torque specifications. The caliper is now ready to be mounted on the vehicle.

10-4 REASSEMBLY

Disc brake reassembly begins with a thorough cleaning of the bearings, hub, and spindle; removing *all* old grease. Following the procedure discussed in Section 2-5, the bearings are cleaned and thoroughly repacked with the recommended lubricant. The bearings and new seal are installed in the hub, the spindle coated with wheel bearing grease, and the rotor and hub assembly installed over the spindle. The spindle nut and washer are put into position and adjusted.

The brake shoes can now be positioned in the caliper or spindle, depending upon design, and any attaching pins put into position. Some shoes have tabs that are bent to securely hold the shoe in the caliper. The brake shoes may employ "anti-rattle" springs to prevent the pad from moving. Pad movement causes a rattling noise. Anti-rattle clips attach between the shoe and the caliper housing. An example of anti-rattle clips is shown in Figure 10-17.

Many manufacturers recommend using an insulating material that is applied to the back of the shoe, as illustrated in Figure 10-18, to prevent metal-to-metal contact between the shoes and caliper housing or piston. This insulation material may also be retrofitted to older-model brake shoes if an annoying squeal occurs during light brake application. The squeal is the result of a high-frequency vibration between the two metal surfaces that are in contact.

The caliper can now be positioned over the

(a)

(b)

Figure 10-17 Typical anti-rattle springs: (a) internal mount; (b) external mount.

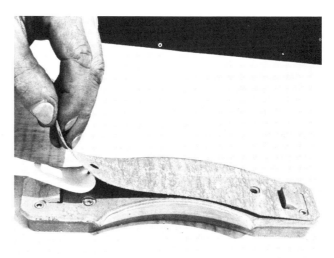

Figure 10-18 Insulation that fits between the shoe and caliper piston.

Disk Brake Service

rotor and its retaining bolts, pins, or clips installed. The guide pins or bolts should be lightly coated with brake lubricant just before they are installed. The retaining bolts must be torqued to specifications and then safetied if the bolts have provisions to do so.

The front brake hoses are attached to the caliper, making certain the copper sealing washers are in place. The washers provide a seal between the brake hoses and the caliper housing. The wheels are now ready to be installed after the bleeding operation is completed. Disc brakes are bled in the same manner as drum brakes.

REVIEW QUESTIONS

1. Why should some fluid be removed from the master cylinder reservoir before disassembling the disc brake? [10-1]
2. What measurements are taken on a rotor to determine that reconditioning is required? [10-1]
3. Describe the procedure to be followed to remove a caliper. [10-1]
4. Why is it important that the two faces of the rotor be parallel? [10-2]
5. How much material should be removed from a rotor to resurface it? [10-2]
6. Describe the rotor resurfacing procedure. [10-2]
7. Describe the procedures used to remove the piston from the caliper. [10-3]
8. When is it necessary to polish the caliper cylinder bore? [10-3]
9. How are pitted chromium-plated pistons serviced? [10-3]
10. Describe the procedure to be followed to assemble the caliper piston. [10-3]
11. Where is insulation used in disc brakes? [10-4]

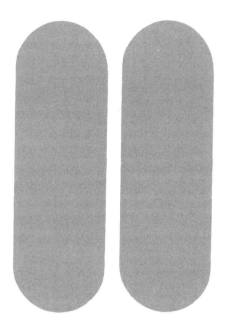

Hydraulic System Service

The brake hydraulic system transfers the brake pedal action through the master cylinder to the wheel cylinders and brake shoes. It provides pressure to all four wheel cylinders so the brakes can work together to produce a smooth, straight stop. Disc-drum brake systems require auxilliary hydraulic valves to modify the brake pressure going to the different types of brakes to produce the intended retarding and stopping action. Two of the most common auxilliary valves used to modify the brake hydraulic pressure are the metering valve and the proportioning valve.

The hydraulic system is checked as a complete system. If it does not function properly, the individual parts are tested. The faulty part is either overhauled or replaced.

11-1 HYDRAULIC CONTROL VALVE TESTING

It is important on vehicles equipped with metering and proportioning valves to have these valves operating properly. A faulty metering valve could prevent front brake operation although it is more common for the metering valve to cause premature front brake application. This occurs when the metering valve sticks in the open position. Premature application would be most noticeable on a slippery road when the front wheels would lock easily on light brake application, inhibiting steering control. A malfunction of the proportioning valve would be most noticeable as the rear wheels tending to lockup

Hydraulic System Service

under hard braking. This also could result in loss of control by allowing the rear end of the vehicle to spin out. The operation of the proportioning valve and metering valve can both be checked using two high-pressure gauges.

Checking the Metering Valve. One gauge is connected into one of the master cylinder lines entering the metering valve with a "T" fitting. The other gauge is connected into the metering valve port for the front brakes. A typical gauge installation on a demonstration board is shown in Figure 11-1. The pedal is slowly depressed and gauge readings are observed. The metering valve inlet pressure should be approximately 110 psi (750 kPa) to 160 psi (1100 kPa) depending upon application before the metering valve opens to allow pressure to rise in the front brake system as indicated on the second gauge. This can also be visually inspected by observing the metering valve manual override that is used when pressure-bleeding the brakes. The override button,

(a)

(b)

(c)

Figure 11-1 Testing a metering valve on a demonstration board. Pressure from the master cylinder is on the lower gauge and front brake pressure is on the upper gauge: (a) low master cylinder pressure; (b) moderate master cylinder pressure; (c) high master cylinder pressure.

Hydraulic Control Valve Testing

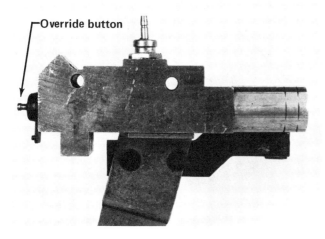

Figure 11-2 Metering valve override button on a brake system combination valve.

as shown in Figure 11-2, should move when the pressure at the front wheels begins to rise.

Checking the Proportioning Valve. To check the proportioning valve, the gauge is removed from the front brake outlet port and the front brake line is reconnected. The rear brake line is disconnected at the valve and the gauge is installed in this port as pictured on the demonstration board in Figure 11-3. The inlet and outlet pressure gauges should increase pressure equally until approximately 300 to 500 psi (2070 to 3450 kPa) is reached, depending on application. After that point the pressure should rise in the inlet line at a greater rate than the rear brake pressure rises. For example, assume the valve was set for proportioning at 300 psi (2070 kPa). As the pressure on the inlet side reached 500 psi (3450 kPa), the outlet pressure going to the rear brakes would only be approaching 400 psi (2750 kPa). This is shown as a graph in Figure 11-4. The manufacturers specifications should be checked for correct values because valve characteristics vary from model-to-model as well as between manufacturers. If either the metering or the proportioning valve is not functioning properly, the valve must be replaced. There is no repair procedure for either valve.

Figure 11-3 Testing a proportioning valve on a demonstration board. Pressure from the master cylinder is on the lower gauge and pressure going to the rear drum brakes is on the upper gauge: (a) low master cylinder pressure; (b) moderate master cylinder pressure, (c) high master cylinder pressure.

(a)　　　　　(b)　　　　　(c)

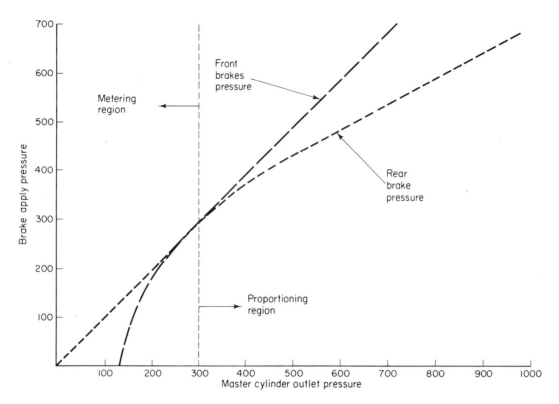

Figure 11-4 Graph of typical master cylinder outlet pressures and brake application pressures after going through the metering and proportioning valves.

11-2 MASTER CYLINDER SERVICE

The master cylinder is serviced whenever it is found to be the source of a problem. It is also serviced during a complete hydraulic system servicing job. The master cylinder is *not* commonly rebuilt each time the brakes are serviced. Master cylinder problems that require service or replacement fall into two basic areas: external leakage and internal leakage

External Cylinder Leakage. External leakage usually results when the rearmost piston seal (primary piston secondary seal) does not effectively retain the fluid in the cylinder. This leakage will show as fluid traces on the engine side of the cowl panel or on the front side of the power booster. It may also cause a brake fluid odor in the passenger compartment.

Internal Cylinder Leakage. Internal leakage may occur past any of the seals within the master cylinder. The location of possible internal leaks are identified in Figure 11-5. If leakage occurs at either primary cup, pedal fade will be noted during light brake applications. If the primary seal fails, one half of the master cylinder will not operate. Leakage past the secondary cup of the secondary piston may allow some fluid from the front master cylinder reservoir chamber to leak into the rear reservoir chamber of the master cylinder. This transfer of fluid will result in a low fluid level in the front reservoir and overflowing of the rear reservoir. Leakage past the inverted secondary cup or O-ring on the rearmost end of the secondary piston can allow a transfer of fluid from the rear reservoir to the front reservoir of the master cylinder. The purpose of the inverted secondary cup is to seal the forward end of the primary section. Whenever any of the above conditions exist, the master cylinder will have to be repaired or replaced.

Adjustment of Master Cylinder Push Rod. Maladjustment of the master cylinder push rod is another source of trouble that can easily be corrected. Incorrect adjustment either leads to a low brake pedal, when there is excess free play, or brakes that

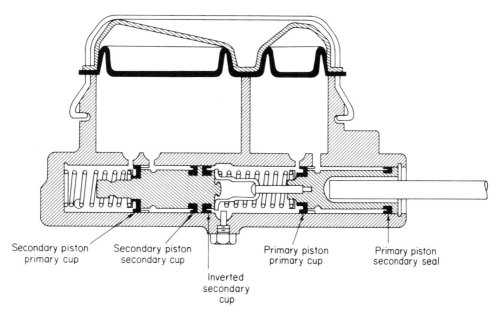

Figure 11-5 Location of possible master cylinder leaks.

will not fully release, if there is no free play. The length of the push rod is adjusted to provide a specified clearance for proper operation. With a power booster, a gauge is used to establish the correct push rod length. With manual brakes the push rod is adjusted to provide approximately 1/16 inch (1.5 mm) of free play (Figure 11-6). In either situation, the adjustment is made to provide full piston and pedal return under all brake release conditions.

Removal and Disassembly. The master cylinder must be removed from the vehicle for servicing. The brake tubes are first disconnected using a flare nut wrench, and plugs are installed in the master cylinder outlets to seal the outlets during removal. The brake pedal is disconnected if the vehicle is equipped with manual brakes. The cylinder can now be unbolted from the power booster or cowl panel and removed from the vehicle.

Disassembly of the master cylinder should begin with a thorough cleaning of the outside of the cylinder. The reservoir cover is removed and the fluid emptied from the reservoirs into a waste container. A secondary piston retainer screw must be removed. It is located either internally in the bottom of the reservoir or in the outside housing. A typical retainer screw is shown in Figure 11-7. At the push rod end of the cylinder, either a piston stop or snap ring is removed. This is done by depressing the primary piston and removing the ring or stop. The primary piston assembly should now slide from its bore. The secondary piston can be removed by tapping the open end of the cylinder on a wooden block to bring it to the end of the bore where it can be grasped and lifted out. If the secondary piston will not move when tapped, low-pressure air may be applied to the secondary outlet fitting to force it from its bore, *taking extreme care* to avoid damage. The outlet tube seats can be removed by threading a small screw into the seat and prying it out, by tapping the hole and then tightening a screw against a washer to pull it out, or by using an easyout tool to pull it out (Figure 11-8).

Figure 11-6 Measurement of brake pedal free play.

Hydraulic System Service

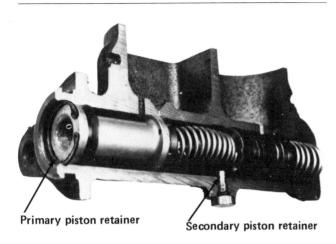

Figure 11-7 Typical location of master cylinder piston retainers.

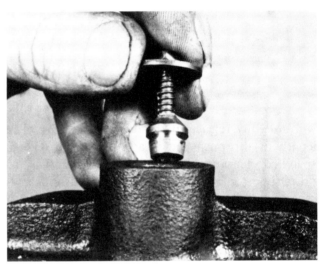

Figure 11-8 Removing the outlet tube seat.

An exploded view of a master cylinder is shown in Figure 11-9.

Cleaning and Honing. The cylinder bore is cleaned with brake fluid and inspected. If only a light film of rust or slight scratches exist, the bore can be cleaned using a very fine abrasive called *crocus cloth*. If the scoring is too deep for crocus cloth, the cylinder can be honed like a wheel cylinder in an attempt to clean it. A typical bore condition is shown in Figure 11-10. Never increase the bore size over 0.002 inch (0.05 mm) unless a greater amount is allowed by the manufacturer's specifications. If the cylinder bore is not cleaned and all pits and scratches removed when this limit is reached, the cylinder must be replaced. After honing, clean the cylinder thoroughly with brake fluid and wipe with clean shop towels until *all* signs of the grit have been removed.

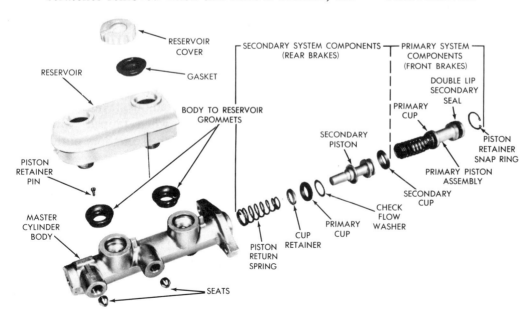

Figure 11-9 An exploded view of a master cylinder that has a transparent plastic reservoir (Courtesy of Chrysler Corporation).

Master Cylinder Service

Figure 11-10 A typical corroded master cylinder bore.

All component parts of the master cylinder are cleaned and all rubber parts replaced with new ones. Most cylinder repair kits, as shown in Figure 11-11, include the primary piston assembly and secondary piston cups and washers. If the secondary piston is damaged, it must also be replaced. After the piston parts are completely assembled, the components are dipped in brake fluid or coated with assembly lubricant so they will move smoothly and not damage the seals. The piston assemblies, seals, and springs are carefully worked into the master cylinder bore until the primary piston stop or snap ring is positioned. The primary piston is depressed with a dowel and the secondary piston retainer screw tightened into place. The residual check valves and tube seats are positioned in the outlet bores (Figure 11-12). A flare nut can be threaded into the outlets to force the seats firmly into their bores. If the vehicle uses disc and drum brakes, the residual check valve is only used in the drum brake portion of the cylinder. This is on the small reservoir part of the master cylinder.

Bench Bleeding. After the master cylinder is assembled, the air should be removed from the hydraulic chambers before installing it on the vehicle. The process of removing the air from the master cylinder is called *bench bleeding*. It is possible for the cylinder to be air-locked so no fluid is displaced as the pedal is depressed. Bench bleeding eliminates this problem. It will also save the technician valuable time while bleeding the rest of the brake system.

Bench bleeding is accomplished by installing two fluid lines that are connected in the outlets and routed to the reservoir as illustrated in Figure 11-13. The cylinder is mounted in a holding fixture or held firmly in a vise. With the cylinder filled

Figure 11-12 A flare nut is used to force the outlet tube seat firmly in the bore.

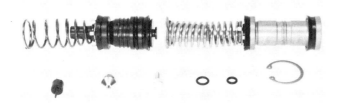

Figure 11-11 New parts that are included in a typical master cylinder repair kit.

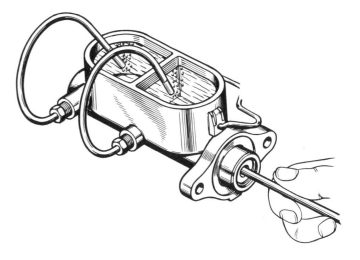

Figure 11-13 Bench bleeding a reconditioned master cylinder (Courtesy of the Bendix Corporation).

with fluid, the pistons are slowly moved through their full stroke until the discharge going into each reservoir is air-free. The bleeding lines can then be removed, the openings plugged and the cover installed. The cylinder is then ready for installation on the vehicle.

Power Brake Push Rod. If the master cylinder is from a power brake-equipped vehicle the push rod length may be checked to make sure that the correct clearance will exist between the piston and push rod when the master cylinder is installed. It is usually unnecessary to check this clearance unless the technician is aware that a problem exists that could be related to the adjustment. Brake dragging or excessive pedal free play are two of these problems.

Power brake push rod adjustment is performed by lengthening or shortening the push rod until the proper length is obtained. A reference measurement is usually given by the brake manufacturer. A quick check may be made by comparing the distance from the push rod end to the power brake housing with the distance from the push rod seat in the cylinder piston to the master cylinder flange. A depth gauge, as shown in Figure 11-14 can be used to make this measurement.

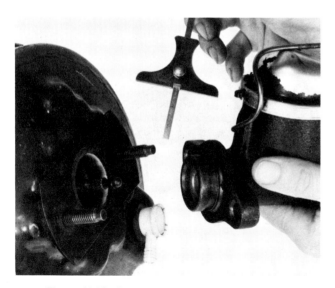

Figure 11-14 A quick check of the correct push rod length can be made with a depth gauge.

Installation. The master cylinder is now ready to be bolted into place on the power brake unit or on the cowling. The brake lines are connected and tightened securely with a fitting wrench and the linkage is connected. If the vehicle is equipped with manual brakes, an adjustment of the cylinder push rod may be necessary to obtain the required pedal free play. On some vehicles, it is not possible to make this adjustment because of the pedal linkage arrangement. If adjustment can be done and is required, the locknut on the push rod is loosened and the rod is threaded in or out until the proper amount of free play is obtained. The free play is usually between 1/16 inch and 1/8 inch (1.5 and 3 mm), measured at the brake pedal. This was shown in Figure 11-6. After adjustment is made, the locknut is tightened.

Stoplight operation should now be inspected if a mechanical switch is used. The mechanical type stoplight switch is adjustable and should be adjusted at this time. If a hydraulic switch is used, the operation of the stoplight will have to be checked after the brakes have been bled.

11-3 BRAKE BLEEDING

Brake bleeding is the process used to remove all of the air trapped in the hydraulic brake system. Any air left in the system will compress when the brake pedal is applied. Braking will feel like stepping on a spring. This pedal feel provides very little braking. This is due to the compressibility characteristic of the liquid/air mixture, which will not transmit the required force to the wheel brake.

The brakes will need to be bled whenever the hydraulic system has been opened to the atmosphere for service or whenever a spongy brake pedal action indicates that air is present in the brake hydraulic system. If air continually enters the system, the system is in need of repair and it should be serviced immediately.

Flushing the Brake System. It may be necessary to flush the brake system to remove contaminants that have accumulated in the brake fluid. Flushing can take place with the wheel cylinders disassembled or removed. If flushing were to be done with the wheel cylinders in place and assembled, any sludge, abrasive, or dirt would be carried into the wheel cylinder where it would lodge. This can result in scoring of the walls and piston cups of the wheel cylinder. It may eventually cause failure of the wheel cylinders. This is why bleeding before disassembly is recommended.

Periodic bleeding of the brake hydraulic system for purposes of replacing the fluid in the system is not recommended. This could result in failure of the cylinders from contaminants as discussed.

Never use alcohols or flushing fluids to flush the brake system hydraulic lines. Complete removal of such fluids is impossible. They will remain in the system to contaminate the new fluid. Use only brake fluid of the type specified for replacement by the manufacturer to flush the hydraulic brake system.

It is extremely important that no contamination is allowed to enter the master cylinder reservoir while checking the brake fluid level. Contamination can be prevented by using the precautions listed as follows:

1. Thoroughly clean around the master cylinder area prior to opening the cylinder cover.
2. Do not open the master cylinder reservoir cover out-of-doors while precipitation is present in the air.
3. Do not open the master cylinder reservoir cover while dust is present in the surrounding air.
4. Do not allow the reservoir to remain open any longer than required to add fluid or, if pressure bleeding, to connect the apparatus.
5. Be certain when closing the system that an airtight seal is made.
6. Use only fresh brake fluid approved by the manufacturer and make sure that it has been stored in an airtight container.

11-4 METHODS OF BLEEDING

There are three basic methods used for bleeding brakes. The methods are pressure bleeding, pedal bleeding, and gravity bleeding. Of the three, pressure bleeding is the most effective and efficient method, while gravity bleeding is the least effective. Gravity bleeding is usually effective on disc brakes, but it will not work as effectively, if at all, on drum brakes. In the following sections, each method will be discussed with an explanation of the equipment required and a comment will be made on the advantages and disadvantages of each method.

General Procedure. It is recommended that all brakes be properly adjusted before starting the bleeding operation. This will be helpful because minimum travel will be present in the wheel brakes. Brakes having little clearance are easier to bleed.

Several pieces of equipment are necessary, regardless of which method of bleeding is used. These include the correct-sized bleeder wrench for the vehicle, a small rubber hose approximately two feet in length, and a jar that is partially filled with brake fluid. These are shown in Figure 11-15 with a pressure bleeder. One end of the hose is connected on the bleeder fitting and the opposite end is placed in the jar beneath the level of the fluid. The bleeder valve is opened. The bleeding operation is completed when no more air bubbles are discharged into the fluid. The bleeder valve is then closed. When pedal bleeding, placing the hose end under the fluid will prevent air from being drawn into the system when the pedal is slowly released. The fluid that is discharged into the jar must never be used to refill the system. It would contaminate the new fluid. It should be thrown away when the bleeding operation is completed.

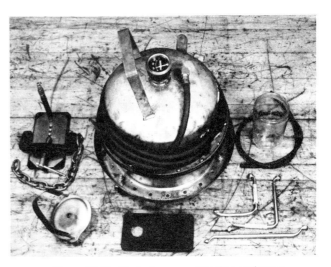

Figure 11-15 Typical pressure-bleeding equipment.

Bleeder valves use a tapered seat to prevent leakage. It is, therefore, unnecessary to tighten them so tight that they break. The technician should use the torque specification that is given in the shop manual.

Pressure-bleeding. The technician can pressure-bleed brakes without help. The pressure bleeder is designed to hold pressure in the hydraulic system while keeping the master cylinder reservoir full.

The pressure bleeder consists of a tank which contains brake fluid on one side of a diaphragm and air pressure on the oppostie side. A float in the fluid side closes the outlet when the fluid gets low to prevent air from entering the brake hydraulic system. The air side of the tank is pressurized to approximately 30 psi (200 kPa). This, in turn, places the brake fluid under the same pressure. An adapter is clamped on the master cylinder in place of the master cylinder cover as pictured in Figure 11-16, and this provides a cover connection from the pressure bleeder to the master cylinder. When a valve to the master cylinder is opened, the hydraulic pressure in the bleeder equipment pressurizes the brake hydraulic system. The valve is closed to prevent fluid loss when the pressure bleeder is to be disconnected.

On vehicles equipped with disc-drum brakes, a metering or hold-off valve may be used to prevent premature front brake application as explained in Section 7-5. Since this valve prevents fluid from entering the front brakes at any pressure below approximately 100 psi (690 kPa), it is necessary to override the metering valve when pressure-bleeding at 30 psi (200 kPa). The valve can be bypassed manually using a special tool, as shown in Figure 11-17, or by hand. The tool must allow the valve to move slightly while the brakes are being pressurized. The valve can fail if it is held open rigidly.

The bleeding sequence or the order in which each wheel is bled, depends upon the design of the system. It is generally considered acceptable to bleed the wheel cylinder farthest from the master cylinder first and the cylinder closest to the master cylinder last. On vehicles using special power boosters, such as those used on heavy-duty vehicles, it is imperative to follow the correct bleeding sequence. The technician should consult the appropriate shop manual for the correct sequence for the vehicle being serviced. With the pressure bleeder attached and the bleeder hose valve open, follow the proper brake bleeding sequence. Clean the bleeder fitting, then place one end of the bleeder hose on the fitting and the other end in the jar of clean brake fluid. With a fitting wrench open the bleeder screw on the wheel cylinder three-quarters of a turn. The pressurized bleeder will force new brake fluid into the system as the air is forced out through the hose attached to the bleeder screw. When the bubbles stop forming in the fluid jar, close the bleeder screw. Repeat this bleeding procedure on each brake following the proper brake bleeding sequence.

It may be helpful on disc brake calipers to lightly tap the caliper with a plastic mallet once or

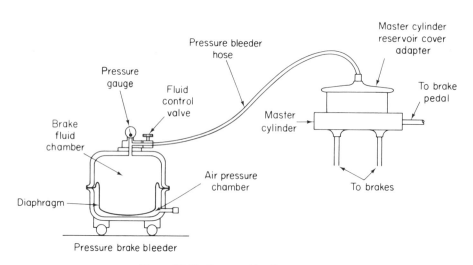

Figure 11-16 Pressure-bleeding connections.

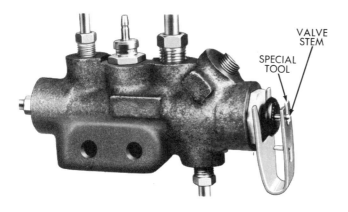

Figure 11-17 Use of a special tool to hold the override button to bypass the metering valve for brake bleeding (Courtesy of Chrysler Corporation).

twice to free any air bubbles that may be trapped within the caliper. Remember that the bleeder passage is always located at the highest part of the wheel cylinder bore. If this were not the case, not all of the air could be removed. If the bleeder valves are not at the top of the cylinder bores, the problem is caused by cylinders or calipers that have been mixed, side for side.

When all air has been removed from all of the cylinders and bubbles no longer appear in the jar, the bleeding operation is completed. The metering valve tool is removed and the brake pedal inspected to make sure it is high and firm. If any sponginess is present, some air still remains in the system. The bleeding operation will have to be repeated to clear the air from the system. The brake warning light should be checked to make sure the valve is centered. This procedure is discussed in Section 11-5.

Pedal-bleeding. Pedal-bleeding requires two people. The vehicle master cylinder is used as the pressure source for this bleeding procedure. One person pushes the service pedal while the other performs the bleeding operation at each wheel. After each wheel is bled, the master cylinder is refilled to make sure the reservoir outlet remains covered with brake fluid. If it is accidentally uncovered, air will be forced into the system. Extra bleeding will be required to clear this air from the system.

It is not necessary to use a tool to hold the metering valve open when pedal-bleeding. The pressure usually exceeds the 100 psi (690 kPa) that is necessary to open the metering valve so that normal bleeding will occur.

The pedal-bleeding operation requires cleaning of the bleeder fitting, connecting the hose to the bleeder valve, and submerging the opposite end of the hose in the jar of fluid. The bleeder valve is opened one-quarter of a turn and the pedal is *slowly* depressed and released repeatedly until an air-free stream of fluid is discharged into the jar. The bleeder valve is then closed and the technician moves to the next brake in the bleeding sequence. After the sequence is completed, the pedal is checked to assure firmness. When all the air is removed from the brakes and the master cylinder is filled with brake fluid, the bleeding job is complete.

Gravity-bleeding. Gravity-bleeding is the least efficient means of bleeding brakes. The method relies on the natural gravity flow of fluid from the master cylinder to the wheel cylinders to force the air from the system. This method is usually slower than other bleeding methods discussed, because it relies solely on gravity. An advantage of this method is that the pressure is so low that the metering valve will not restrict the flow. Another advantage of gravity bleeding is the pressure differential switch will not be disturbed on brake systems that are not self-centering. Gravity bleeding is usually unsuccessful on drum brakes because the residual check valves, as mentioned in Section 7-5, often restrict the normal movement of fluid at these low pressures.

Gravity-bleeding at the brake is done in the same manner as previously discussed while keeping the master cylinder reservoir full of fresh fluid. This method can be used while the brakes are being serviced. The technician can allow one wheel to bleed while he assembles the opposite wheel. When the brake service is complete, the air is usually out of the system. The brakes are checked for pedal firmness and readjusted as necessary.

11-5 BRAKE WARNING LIGHT

After the brakes have been adjusted and bled, it is necessary to center the pressure differential switch. This switch is described in Section 7-5. If the warn-

Hydraulic System Service

ing light is on and will not go out when the service brake pedal is pushed firmly, it will be necessary to center the switch by bleeding. This is accomplished with the pedal depressed by opening a wheel cylinder bleeder valve either on the front brake system or the rear brake system to create a drop in pressure. The pressure drop on one side of the valve will force the warning light switch valve to move. Most brake warning light switches are spring-loaded to automatically recenter the switch. They should not require additional centering operations.

The AMC brake warning light switch, on many models, is a delicate switch. These may require removal before the bleeding operation takes place. Others can remain installed while bleeding. If the delicate switch is not removed the switch control pin will shear, and the brake warning light will remain illuminated. After the brakes are bled, the switch should be reinstalled and a check made to insure that the warning light remains off. If the light is on, the valve is centered by bleeding. Some valves are centered by depressing the service pedal while one rear bleeder screw is loosened to lower the pressure in that part of the system, then retightening the screw. If the light is out, the job is complete. If the light remains on, one front bleeder screw is loosened to lower the pressure in that part of the system and then that screw is retightened. The light operation is again checked. If the light is out, the operation is complete. If the light remains on, the process is repeated until the light remains off or the valve is replaced.

The Ford Motor Company brake warning light switch may or may not be self-centering. When both parts of the system function normally, the switch will center as the pedal is depressed. If the light remains on, the switch will have to be centered by bleeding the brakes, one part at a time, as previously described. As soon as the light goes out, the technician will close the bleeder valve preventing further movement of the valves in the pressure differential switch. The brake warning light operation is then rechecked.

The General Motors and Chrysler-type switches are self-centering and should present no problems. If the switch does stick at one end of its travel, it can be centered by pedal-bleeding the brakes alternately, from front to rear, until the light remains off.

If the valve cannot be centered, it may be necessary to replace the valve assembly. The shop manual should be consulted to determine if new centering procedures are recommended.

After the warning light is operating properly and all the brakes are adjusted, the wheels that have not previously been installed are mounted on their respective hubs. The wheel lug nuts are torqued in the proper sequence to specifications and the hub caps replaced. The master cylinder is completely filled and properly sealed. The vehicle is lowered to the floor surface and the brakes tested. The vehicle is then taken for a test drive. The test drive is important and it should be done with care. Any wheel pull or noise that is traceable to the operation of the brake system should be noted. If any noise or pull exists, the troublesome wheel should be removed and the brake inspected to positively identify the problem.

REVIEW QUESTIONS

1. Under what conditions would the operator notice that the metering valve was malfunctioning? [11-1]
2. Under what conditions would the operator notice that the proportioning valve was malfunctioning? [11-1]
3. How does the output pressure of the metering valve differ from the output pressure of the proportioning valve? [11-1]
4. How are the metering valves and proportioning valves serviced? [11-1]
5. What problems indicate the need to rebuild the master cylinder? [11-2]

6. List the steps in the procedure used to remove the master cylinder. [11-2]
7. List the steps (in the proper order) to disassemble a master cylinder. [11-2]
8. How is the master cylinder bore reconditioned? [11-2]
9. What is used to clean the master cylinder before assembly? [11-2]
10. Why is the master cylinder bled on the bench? [11-2]
11. Why is it necessary to have pedal free play after installing a master cylinder? [11-2]
12. Why is brake bleeding necessary? [11-3]
13. Name three brake bleeding methods. How do they differ? [11-4]
14. Look up the bleeding sequence of a specific vehicle. [11-4]
15. How can the technician tell that bleeding is complete? [11-4]
16. How is the brake warning light centered? [11-5]

12 Power Brake Systems

The size and weight of vehicles increased during the 1950's and 60's and into the 70's. As the weight increased so did the required braking force. This led to brake pedal forces that were unacceptably high for a large portion of the driving public. Power-assist brakes were introduced during this period to relieve the operator from a large portion of the required braking effort. With power-assisted braking, the same amount of braking could be accomplished with less pedal effort and less pedal travel. Several different approaches are taken to provide power-assist for the brakes. All units used on automobiles and light trucks employ either vacuum or hydraulic pressure as a source of assist power for braking.

12-1 THEORY OF OPERATION

With the use of our present drum, disc, or drum-disc brakes, a high mechanical advantage is required between the operator's foot and the brake shoes. This mechanical advantage is achieved through the use of a lever system at the brake pedal and hydraulically through the ratio of wheel cylinder area to master cylinder area as described on page 102. To achieve greater braking at least one of the following is used: the force applied to the brake pedal must be increased, the ratio of wheel cylinder area to master cylinder area must be increased, or the brake pedal mechanical advantage must be increased.

An increased pedal force would be objectionable to the operator. An increase in the wheel cylinder to master cylinder area ratio or an increase in the brake pedal mechanical advantage requires a longer brake pedal travel. This too would make braking difficult. The demand was thus created for an assist mechanism that could sense the operator's brake apply force. It would have to either assist the operator's mechanical effort at the pedal or directly increase the brake line pressure. In either case there would be no increase in the pedal travel distance.

The use of the disc brakes has also increased the need for power-assisted braking. This is necessary because the disc brakes are not self-energizing. Dual-servo drum brakes are self-energizing. Without the self-energizing effect, an increase in the force on the brake lining is required. This makes power-assisted braking necessary on all large vehicles equipped with disc brakes. Because disc brakes provide better braking, as discussed in Section 6-2, their continued use will require the continued use of power-assisted brakes, even on many lightweight vehicles.

When power brakes are used, the master cylinder bore diameter is usually increased and the brake pedal linkage ratio is usually decreased. This with a booster will provide braking with less pedal effort and less pedal travel.

Either vacuum or hydraulic pressure provides the power for the booster. Vacuum systems utilize engine manifold vacuum and atmospheric pressure. The pressure difference between these two is applied across a diaphragm(s). During application, the atmospheric pressure on one side of the diaphragm and the vacuum on the other side will add force to the driver's effort that is being used to apply the brakes.

Vacuum-assist Power Brakes. Brake boosters that use vacuum as the power-assist force on passenger vehicles are from 8 to 12 inches (200 to 300 mm) in diameter. Some of the smaller diameter units employ two diaphragms, one in front of the other. They are called *tandem* boosters. Figure 12-1 shows a typical vacuum-assist power brake booster. The various size boosters have a reaction

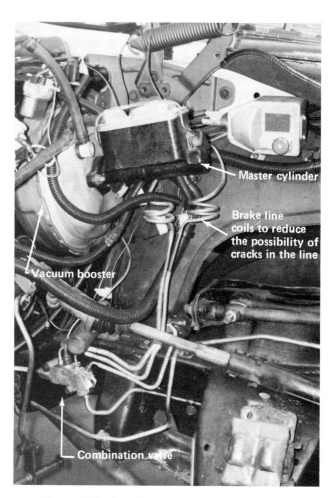

Figure 12-1 A typical vacuum-assist power brake system.

area of between 75 and 100 square inches (484 to 646 cm^2). The area depends on the size of the diaphragm. If a manifold vacuum of approximately 16 inches (406 mm) of mercury is used (which is equal to about 7 psi or 48 kPa absolute) and atmospheric pressure is approximately 14.7 psi (101 kPa), there will be a pressure difference across the diaphragm during application of approximately 7.7 psi (53 kPa). Using customary units in Figure 12-2, this pressure will increase the force on a 100-square-inch booster by a maximum of 770 pounds (F = 100 sq. in. × 7.7 lb/sq. in.). This force is achieved with a very small force on the brake pedal, usually about 40 pounds (178 N). After the maximum amount of assist or boost has

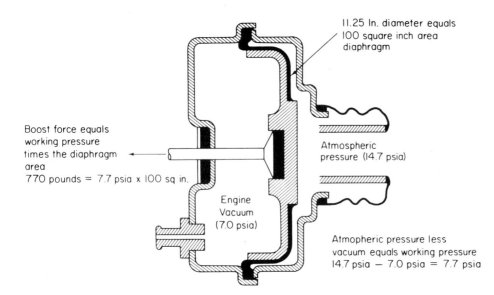

Figure 12-2 A section view of a simplified vacuum booster showing the principle of force increase.

been achieved from the power brake booster (called *runout*) the operator can continue to apply additional force to the brake pedal to increase the brake line pressure. This force will be in addition to the force provided by the booster. It will increase the force on the master cylinder push rod as the input force is increased.

The preceding calculation for booster force has not considered pedal force. If the average person can generate between 100 and 150 pounds (450 and 675 N) of force on the brake pedal, the total force on the master cylinder push rod would be as high as 600 pounds (2670 N) when the mechanical advantage of the brake pedal linkage is 4. This could increase the force after runout up to 400 pounds (1950 N). The force on the master cylinder would be increased from 770 pounds (3425 N) at runout to over 1200 pounds (5350 N). This provides an approximate 30 percent increase in line pressure after runout has been reached, as shown in Figure 12-3.

The amount of assist that is possible from a vacuum booster is dependent on the active area of the diaphragms and the pressure difference across them. The only means to increase boost in this type of unit is to increase the physical dimensions of the booster, increase the diaphragm area or increase the pressure difference across the diaphragm. With the addition of emission controls and changes in the combustion process, the amount of engine vacuum has decreased. The size of the boosters is limited by the space available under the hood. The space is being utilized to a greater extent by other required

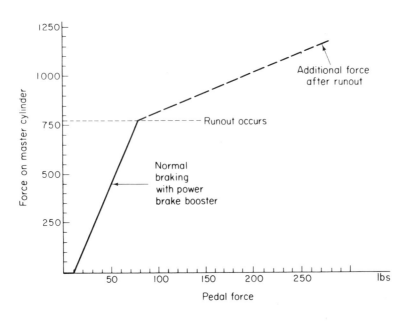

Figure 12-3 Using a vacuum-assist power brake, the pedal force is plotted against the force produced on the master cylinder push rod.

equipment in small-sized vehicles. The limited space and decreased engine manifold vacuum has reduced the amount of boost effort available from the power brake boosters. This has brought about the need for utilizing an alternative power source for the power-assisted brakes. The natural power source in most vehicles is the hydraulic pressure in the power steering system. Power steering is standard on many vehicles and it is optional on others.

Hydraulic-Assisted Power Brakes. Hydraulic-assisted braking is an alternative to vacuum-assisted power braking. Because a large percentage of vehicles are manufactured with power steering, this source is often used. A typical hydraulic power brake unit is pictured in Figure 12-4. The high-capacity, high-pressure hydraulic pump presently used for power steering is ideally suited for the requirements of the hydraulic brake booster. The power steering pump requires little change to adapt it to provide hydraulic pressure for the brake booster.

The use of hydraulic pressure, up to 1500 psi (10,350 kPa), allows the booster to be small while still providing more assist than is presently available from vacuum power. The hydraulic pump output is not affected by engine load or by manifold vacuum. It provides adequate output for the brakes at any engine rpm.

The hydraulic brake booster provides the capacity to have power-assisted braking on diesel engines as well as on other engine types that have low or positive manifold pressure. For example, it can be used on turbocharged engines and turbine engines. It is adaptable to both light and medium-sized trucks and offers them the same benefits as on the passenger vehicles.

A dual system, employing both a vacuum system and a hydraulic system, is presently being used on some heavy trucks. This system offers additional boost over either the hydraulic or the vacuum systems when used alone. The two boosters are interconnected to create the total boost required for stopping the heavy vehicle.

12-2 VACUUM POWER BRAKE OPERATION

Vacuum power boosters that are presently being used on passenger vehicles and trucks fall into two basic categories. These are the *intergral type* and the *multiplier type* which is sometimes called a Hydra-Vac unit. Both of these are *vacuum-suspended*. The integral unit is used on current passenger vehicles. The multiplier type is generally used on medium-sized trucks and motor homes. The operational differences of each type are discussed in the following paragraphs.

Integral Type. A typical control valve assembly and associated linkage components are shown in Figure 12-5. It should be examined to identify the parts while studying the following description. The brake pedal in the passenger compartment is connected to the valve *operating rod*. The other end of the operating rod attaches to the *valve plunger*. A rubber *poppet assembly* is situated in the *valve housing* on the operating rod side of the valve plunger. A resilient rubber disc, called the *reaction disc,* is positioned between the valve plunger and the end of the master cylinder push rod that is located in the *diaphragm hub*. Springs on the reaction disc side of the valve plunger and on the operating side of the poppet assembly hold each in their respective positions when the brake pedal is in the released position.

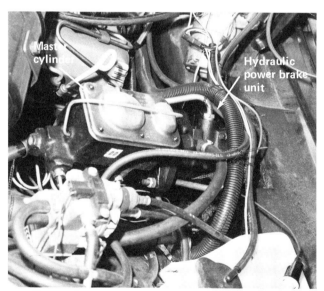

Figure 12-4 A typical hydraulic-assisted power brake installation.

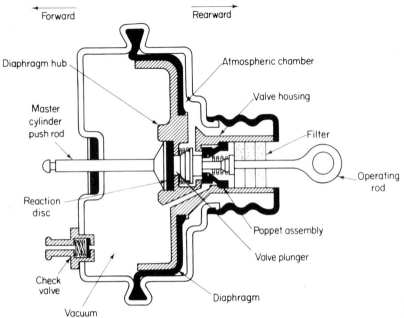

Figure 12-5 A section view of a typical vacuum booster and control valve.

The chambers on both sides of the diaphragm are open to manifold vacuum when the brake pedal is released. This balances the force on both sides of the diaphragm so it provides no output force. This is why this type of power brake unit is called *vacuum-suspended*.

The power brake booster unit is usually mounted on the engine side of the firewall. Therefore, in the following description *rearward* means toward the brake pedal and *forward* means toward the master cylinder push rod. In the release position the valve plunger is moved rearward in the housing by spring force (not shown in the illustration). This rearward position holds the rear face of the plunger in contact with the poppet face. At the same time it moves the poppet face rearward, opening the vacuum port between the chambers. In this position the plunger and poppet will seal out atmospheric pressure from the rear chamber. In this valve position the rear chamber is evacuated because the poppet has opened the passage between the front and rear chambers. The front chamber is always subjected to engine vacuum. It therefore serves as a vacuum reserve for the brake booster. This *release position* of the power brake control is shown in Figure 12-6.

System Operation. As the operator depresses the brake pedal, the plunger moves forward. Spring force causes the valve poppet to move with the plunger. As long as the valve poppet face is in contact with the rear plunger face, no atmospheric pressure will enter the unit. Manifold vacuum continues to be applied to the rear of the diaphragm until the poppet face comes into contact with the vacuum seat in the diaphragm housing, closing it. At this time, the atmospheric port is also closed because the plunger rear face still maintans contact with the poppet face. This condition is called the *hold* or *lap position* as illustrated in Figure 12-7. Vacuum and atmosphere are both prevented from entering the rear chamber and the valve plunger has just made

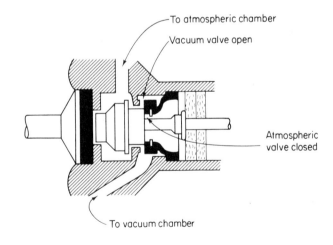

Figure 12-6 The position of the control valve in the release position.

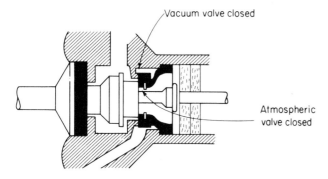

Figure 12-7 The position of the control valve in the hold or lap position.

Vacuum Power Brake Operation

contact with the reaction disc. In the hold position nothing is changing within the booster.

As the operator depresses the brake pedal further, the valve plunger begins to penetrate the reaction disc. This forward movement causes the plunger to move away from the poppet face and opens the atmospheric port. The pressure in the rear chamber will rise as atmospheric pressure enters. The pressure difference on each side of the diaphragm produces a forward force. This force moves against the master cylinder push rod to help push the piston into the master cylinder to apply the brakes. This position of the poppet assembly is called the *apply position*. It is shown in Figure 12-8. It should be noted that the control valve housing is carried forward by the diaphragm. The control valve housing moves forward until the poppet once more comes into contact with the rear face of the plunger. This shuts off atmospheric pressure to balance the forces on the diaphragm assembly. This will again bring the control valve assembly back to the hold position.

Reaction Disc. If a greater amount of braking is demanded by the driver, the plunger is again moved forward away from the atmospheric port and into the reaction disc. The open atmospheric port allows additional atmospheric pressure to enter the rear chamber. This increases the pressure differential across the diaphragm, which in turn increases the force being applied on the master cylinder push rod. The control valve mechanism always tries to seek the hold position. This allows the operator to apply the exact amount of braking desired. The power brake continues to operate in this way until the booster has achieved runout. Runout occurs when full atmospheric pressure is on the rear of the diaphragm and full engine vacuum is on the forward side. At runout, the atmospheric port remains open in the apply position. Any additional input force from the operator would apply mechanically to the master cylinder, with no additional assist from the booster.

The reaction disc has another important function during brake application. All of the force going to the master cylinder goes through the reaction disc. The plunger is forced further into the disc as the brake pedal force is increased. The forward force on the master cylinder is resisted by the reaction disc as it produces force toward the rear. This makes it necessary to increase the amount of plunger force in order to penetrate the disc far enough to open the atmospheric port. The force increase required to penetrate the disc is what gives the operator the required *brake feel* that allows him to apply just the right amount of braking needed. The reaction disc size is selected by the manufacturer to achieve the required feel for each specific vehicle model application.

Vacuum Reserve System. A vacuum reserve system plus a mechanical backup is present in all vacuum power-assist brakes to provide braking if the booster should fail. The vacuum reserve system is required in the event that the engine stops running and manifold vacuum is lost. The reserve

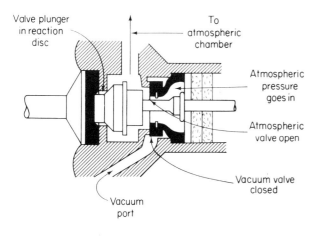

Figure 12-8 The position of the control valve in the apply position.

gives the operator normal assist to apply the brakes for one or two brake applications when the engine is not running. The reserve system is incorporated in the large vacuum reservoir in the front chamber of the booster. A separate tank is used if the booster is atmospheric-suspended. If the reserve is depleted due to brake applications, the straight-through mechanical design of the booster carries the brake pedal force directly to the master cylinder push rod. Of course, it will require much greater pedal effort to achieve the same amount of braking.

The vacuum is held in reserve in the front chamber of the booster by a vacuum check valve. This valve will open whenever engine manifold vacuum is greater than the vacuum in the booster. When the engine stops, the check valve remains closed to hold vacuum in the unit. An additional function of the vacuum check valve is to allow air to flow only from the booster to the engine. If the check valve were not used, the vacuum in the booster would tend to draw some of the air-fuel charge out of the manifold and into the booster front chamber as manifold vacuum decreased during acceleration. Gasoline will contaminate and attack the booster parts. This will cause sluggish action and poor durability. In some cases a charcoal filter is used in the vacuum line to trap gasoline vapors.

Multiplier Type. The multiplier type of power booster is a self-contained brake-assist unit. It is used primarily on medium-sized trucks and motor homes. This type of unit, shown in Figure 12-9, is installed in the brake line between the master cylinder and the wheel brake. The units are often installed in tandem with one unit operating each portion of a split brake system. The unit location is flexible and can be varied. It is usually mounted along the frame member on the underside of the body where there is enough free space.

The multiplier type of unit is vacuum-suspended and employs line pressure *from* the master cylinder to move the operating valve that allows atmospheric pressure to enter and apply the booster. A pressure differential is created as atmospheric pressure enters on one side of the diaphragm with a vacuum present on the other side. This pressure differential on the surface area of the diaphragm creates the force to apply a piston in a booster slave cylinder. The slave

Figure 12-9 A typical multiplier type power brake booster.

cylinder is similar to a master cylinder. The movement of the slave piston forces brake fluid to the brake wheel cylinders. The piston within the booster has master cylinder pressure on one side and wheel cylinder pressure on the opposite side.

The valving is moved by the master cylinder pressure. It consists of a control valve and two poppet valves: a vacuum poppet and an atmospheric poppet. The spring-loaded control valve piston moves along its axis as master cylinder pressure is applied. This first closes the vacuum poppet. An increase in master cylinder pressure will move the control piston further and this opens the atmospheric poppet. The resulting pressure increase on the slave piston side of the booster counters the application force on the small control valve piston. This closes the atmospheric poppet to place the booster in the hold position. When the master cylinder pressure is reduced the slave cylinder pressure opens the vacuum valve to the atmospheric side and thus lowers the amount of brake application pressure. Brake modulation is achieved by the balance of the vacuum and hydraulic forces.

12-3 HYDRAULIC POWER BOOSTER

Hydraulic brake boosters work in series with the power steering system. The pressurized oil from the power steering pump is directed to the power brake booster. The oil is pumped through a portion of the power booster to the high-pressure line going to the power steering gear. A third line is used with the hydraulic brake booster. It returns the high-pressure fluid directly to the power steering pump after brake application. The hoses interconnecting these two systems are shown in Figure 12-10.

Power Steering Pump Operation. A thorough understanding of the power steering pump operation is essential to a knowledge of the hydraulic power-assist unit. The details of the power steering pump operation can be reviewed on pages 285 to 288. In brief, the power steering pump has two valves which

Figure 12-10 Schematic drawing of a typical hydraulic brake booster system (Courtesy of the Bendix Corporation). (a) With an open centered valve and a spring accumulator, (b) with a gas pressure accumulator.

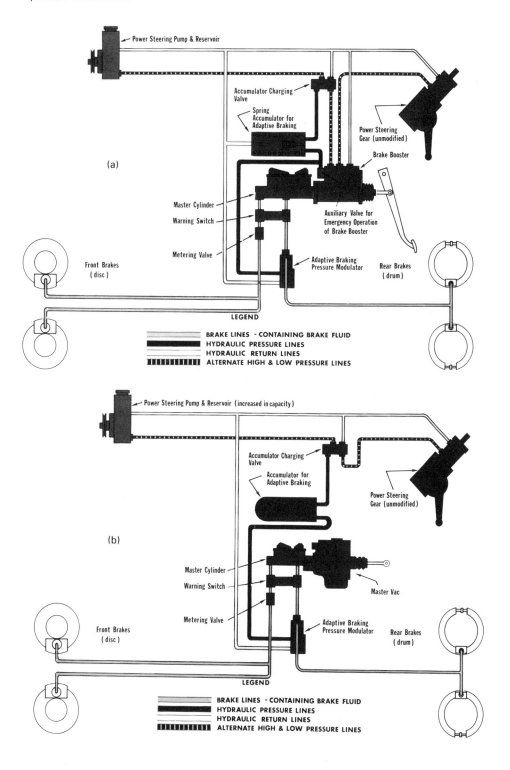

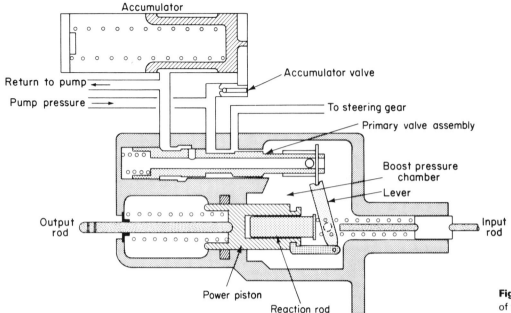

Figure 12-11 A section drawing of the essential parts of a hydraulic booster.

function to control its output: a *flow control valve* and a *pressure control valve*. Their function depends upon the demand by the power steering system. The flow control valve limits the amount of oil that can flow to the steering gear when the demand is low and the pump is running at high speed. This is done by routing the oil within the pump from the pump outlet to the pump inlet. The pressure control valve limits the maximum pressure in the system to approximately 50 psi (345 kPa) when there is no demand from either the power steering gear or the hydraulic brake booster. This is called an *open center system*. With maximum demand, the control pressure will rise to a maximum usable pressure of approximately 1500 psi (10,350 kPa). The pressure will rise when the free flow is restricted at either the steering gear or the booster. This occurs anytime there is a need to assist either steering or braking.

Hydraulic Brake Booster. The hydraulic brake booster is made up of a power piston, a hollow primary valve, an accumulator, input and output rods, a reaction rod, some internal linkage, and a body as shown in the section view in Figure 12-11. The key to the basic operation of this unit is the lever which connects the power piston to the primary valve with its intermediate connection to the reaction rod. This linkage is shown in Figure 12-12.

When brake boost is required, the movement of the input rod moves the reaction rod directly. Since the power piston does not move immediately, the reaction rod linkage pivots and moves the primary valve, as shown in Figure 12-13. The amount of movement of the primary valve is controlled by the linkage pivot ratio. Its movement closes off flow to the steering gear and, at the same time, it opens a passage to the hollow primary valve. The restricted flow to the power steering gear is reflected back to the pump and the pressure begins to rise. This pressure increase is sensed in the cavity at the rear of the power piston through the hollow primary valve. The increased pressure acts on the power piston and against the forward end of the reaction rod piston. The increased pressure causes the power piston to move but the input rod is held in its position. This causes the linkage to pivot and to open

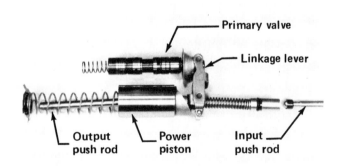

Figure 12-12 Linkage and levers removed from a hydraulic booster.

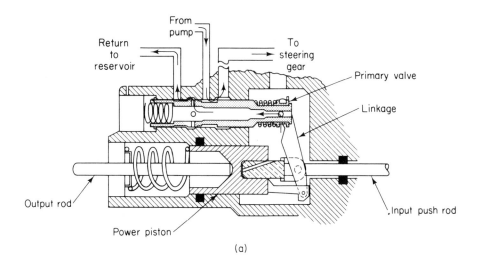

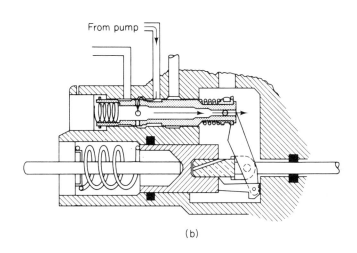

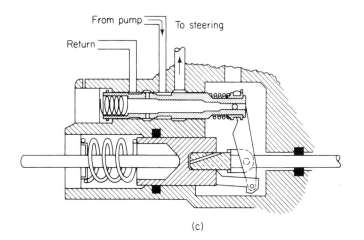

Figure 12-13 Schematic drawing of a hydraulic booster operation shown without the accumulator assembly. (a) release, (b) apply, and (c) hold.

the passage going to the steering gear. At the same time it will maintain the pressure demanded by the operator for braking. The pressure of the fluid on the reaction rod face acts to provide operator feel as the demand for braking is increased.

On release, the reaction rod moves rearward, opening a passage that allows the fluid used for boost to flow back through the primary valve to the pump reservoir. The fluid pressure demand by the power brake booster does not affect the power

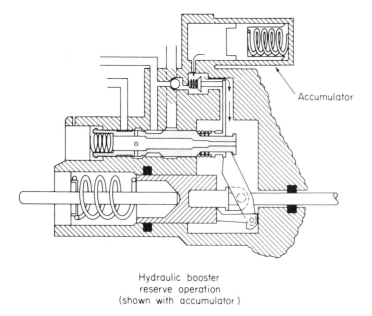

Figure 12-14 Hydraulic booster operation from the accumulator reserve when the engine is not running.

assist for steering during normal stops. When the booster is in a runout mode of operation, which occurs when the piston is moved to its maximum limit, steering can be more difficult because the flow of fluid to the steering gear is limited.

Reserve Power Backup System. A reserve power system and mechanical backup system is incorporated into the hydraulic power brake booster unit. An accumulator holds reserve oil under pressure for use in the event the power steering pressure system fails or the engine stalls. The reserve system will provide at least one full power-assisted stop. With no pump pressure, the input force on the reaction rod shifts the position of a valve sleeve that surrounds a journal of the primary valve. This movement releases the accumulator charging check valve. The check valve opening allows high-pressure oil from the accumulator to force its way into the area at the rear of the power piston to provide assist. This position is shown in Figure 12-14. The runout of this unit occurs in the reserve system at about 500 psi 3450 kPa).

Manual backup is provided by the assist unit's straight through design. When no pressure is available, the input rod will bottom on the reaction piston. This will apply mechanical force directly against the power piston. The power piston movement will move the master cylinder piston to apply the brakes.

12-4 DUAL POWER SYSTEM

The dual power system employs a vacuum-assist unit operating through a hydraulic-assist unit. The two systems work together to improve braking.

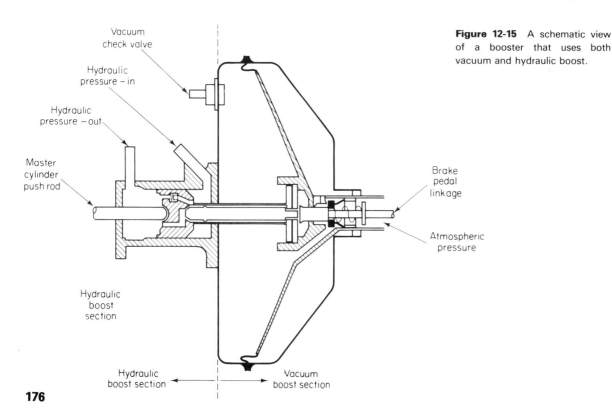

Figure 12-15 A schematic view of a booster that uses both vacuum and hydraulic boost.

The vacuum portion is much the same as the vacuum boosters previously discussed. The vacuum chamber ahead of the diaphragm provides vacuum for reserve braking as shown in Figure 12-15. The output push rod, which operates the master cylinder on the standard vacuum booster operates the control valve of the hydraulic unit in this system.

The hydraulic unit is simplified considerably when using vacuum power to apply it. Reserve braking and driver feedback feel are achieved by the vacuum portion of the system. Movement of the hydraulic control valve restricts oil flowing to the steering gear. This raises the pressure on the back side of the power piston. Hydraulic pressure acting on the piston adds to the vacuum assist to provide additional braking capacity. The hydraulic unit employs a bypass valve to allow pressure to enter the steering gear unit when the hydraulic unit reaches runout.

12-5 POWER BRAKE SERVICE

Some power brake service may be done while the power assembly remains on the vehicle. On-the-vehicle repair is limited to a few external checks and tests. Some malfunctioning components can be serviced while the booster is installed. Prior to any disassembly or removal, a thorough test should be done while the unit is on the vehicle. This should follow a careful analysis of the operator's complaint. Replacement of the entire assembly, with either a new or a rebuilt replacement unit, is usually done if the booster problem can't be corrected while it is installed on the vehicle.

On-the-Vehicle Testing. It is important to be aware of braking problems that are not the fault of the booster assembly. One of the important points to remember about all power-assist mechanisms, with the exception of the multiplier type, is that each is provided with a mechanical backup system. Problems, such as the brake pedal going clear to the floor or a brake pedal that moves slowly towards the floor on application, are not the fault of the booster. Wheel pull is not booster-related. A booster affects all wheels alike. A soft, spongy, or low pedal would not be the result of a malfunctioning brake-assist unit. The technician should keep in mind that the master cylinder is always operated when the pedal is depressed, whether or not the booster malfunctions.

Vacuum Power. Vacuum-assist problems can be diagnosed by following a logical sequence of steps. The first step is to make sure that the unit is receiving the proper vacuum. This is first checked by visually inspecting the hoses from the engine to the booster. A typical hose installation is pictured in Figure 12-16. This will assure that no restrictions are present in the vacuum passages to the booster. A restricted hose would result in little or no assist and after repeated brake application, the pedal would become more difficult to depress. A vacuum supply problem can be repaired while the booster remains mounted on the vehicle.

Another test is to check for vacuum leaks on the booster itself and for faulty operation of the check valve. The engine is turned off and the technician listens for a vacuum leak. If no vacuum leaks are heard, wait two or three minutes, then depress the pedal to see if vacuum is still present in the reserve system. A leak is present if there is no assist on the first brake application. The leak will be through either the check valve or an external leak. To test for a malfunction of the check valve,

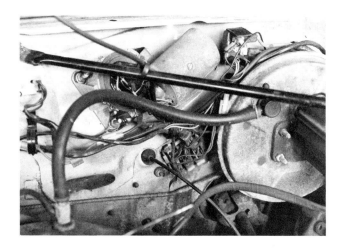

Figure 12-16 A typical hose installation between the intake manifold and vacuum booster.

Power Brake Systems

the vacuum hose is closed off after the engine has idled to make sure full vacuum is in the system. The above test steps are then repeated. If vacuum is held in the unit with the hose closed off, the check valve is faulty.

A check for assist can be made first by depressing the pedal several times with the engine off to make sure that no vacuum is in the system. The brake pedal is then held down and the engine started. This should cause the pedal to move down slightly.

If the unit does not leak during the reserve check and yet boost is not achieved, it is possible that a problem exists inside the booster unit. With the brake pedal depressed and the engine running, listen for a hissing sound caused by vacuum leakage. If it cannot be heard, stop the engine and continue listening. A steady hissing sound with the pedal depressed that goes away with the pedal released indicates a diaphragm or poppet valve malfunction. Repair or replacement of the booster unit is required to correct this problem. Vacuum brake-assist units are only repaired at shops specializing in this type of service.

Hydraulic Assist. The hydraulic power-assist unit must be accurately diagnosed by the technician before any attempt is made to remove it from the vehicle. Most of the problems encountered with this type of unit will be oil supply problems. For this reason, an accurate check of the pressure system is necessary while the unit is installed on the vehicle.

Since the same supply oil is used in both the power steering system and the power brake system, many of the oil-pressure related problems should affect the operation of both systems. A slipping drive belt will cause booster chatter, pedal vibration, and steering gear chatter as they are operated. A similar condition may be found if the power steering fluid level is low. Lack of power-assist in either the steering or the brake can be caused by a malfunctioning power steering pump, by loss of fluid, or by a broken drive belt. If the pump is suspected of being faulty, a pump pressure check should be performed as outlined on page 289.

A slow or no pedal return after application would result from a restricted return line or from a broken return spring within the assist unit. A unit that self-applies is the result of a plugged return line or faulty valving in the accumulator circuit.

Excessive pedal effort, while still having normal steering gear operation, could be the result of internal leakage within the booster. This is corrected by removing the booster and replacing all of the internal seals.

Assist can be checked by pumping the pedal with the engine off to deplete all of the accumulator reserve pressure. Then, while holding the pedal depressed firmly, start the engine. The pedal should move slightly downward as the engine starts. It should be noted that the accumulator reserve pressure should be discharged before the hydraulic booster type power brake unit is removed from the vehicle.

Dual Power System. On-the-vehicle testing of this power assist unit requires test procedures incorporating both vacuum and hydraulic tests. The vacuum section functions similarly to the typical vacuum booster previously discussed. When checking assist, it is difficult to inspect each system unless one is disabled by removing a belt or disconnecting the vacuum supply. This allows each system to be tested separately. Each part of the system is checked in the manner previously discussed for vacuum and for hydraulic-assist booster units.

12-6 POWER BRAKE REMOVAL AND REPLACEMENT

The basic removal and replacement procedures of vacuum power-assist and hydraulic power-assist units are covered in the following discussion. Unit disassembly is not discussed here because this is generally not considered normal shop service but rather a specialty repair. In all instances, it is recommended that an appropriate service manual be used to find out the specific tools required and service steps for the unit being repaired when the unit is to be disassembled.

If it is determined that the unit is faulty after the power-assist unit has been properly diagnosed, it will be necessary to remove the unit from the vehicle. For ease in disassembly, as well as for safety on some units, the reserve system is exhausted before the unit is removed. This can be accomplished when the engine is not running by repeatedly depressing the brake pedal until there is no more assist.

On the integral and hydraulic-assist units the control linkage must be disconnected by either removal of a shouldered bolt or by the removal of a lock pin. These are usually located on the brake pedal linkage under the dash. On all units, the brake tubes are disconnected from the master cylinder or, in the case of the multiplier type, from the slave cylinder. The hydraulic lines or vacuum hoses are disconnected from the booster unit. The booster is then unbolted from the firewall or bulkhead attaching bracket. It can then be removed from the vehicle.

Integral Vacuum Booster Replacement. In this type of repair, the master cylinder should be serviced. It is removed from the booster, carefully inspected, and rebuilt or replaced as necessary. Prior to bolting the master cylinder back on the booster it is necessary to check the length of the master cylinder push rod. A typical checking method is shown in Figure 12-17. The push rod length is very important. If it is too long the master cylinder piston will not clear the compensation port. This will not allow the brakes to fully release. If the push rod is too short excessive pedal free play will occur. The push rod length should be adjusted to the manufacturer's specifications as found in the applicable shop manual. After attaching the master cylinder, the unit is ready to be installed in the vehicle. Brake bleeding procedures as discussed on page 161 are completed and the system is retested as described in Section 12-5.

Multiplier Type. The multiplier unit is a self-contained unit mounted remotely from the brake pedal. The unit contains the booster, the slave cylinder, and the control valve assembly. As in any special unit repair, the applicable shop manual should be consulted for specific procedures. The slave cylinder in this type of unit is always rebuilt or replaced along with the booster assembly.

The unit is reinstalled in the vehicle and all trapped air is properly bled from the system. The operation of the unit is then retested as described in Section 12-5.

Hydraulic Booster Replacement. As the hydraulic booster has gone through various changes since it was first marketed, the applicable shop manual should be used for the particular unit being serviced.

Installing the unit in the vehicle requires that both the power steering and the brake system be refilled with the specified fluids. The power steering system can be purged of air by rotating the steering wheel from lock to lock with the engine running and the front wheels off the ground. The fluid reservoir must be kept full of fluid while purging the system. The brakes are bled using standard bleeding procedures discussed in Section 11-4. After bleeding and stabilizing the steering fluid level, the wheel should be fully rotated against one lock position and held approximately five seconds with the engine running at fast idle. This will recharge the brake reserve system with fluid under pressure. The fluid level in both the master cylinder and the power steering pump should be rechecked and refilled to the normal level, if necessary. The pump belt is inspected and the tension properly adjusted. The unit is given a final operational test using on-the-vehicle procedures previously described in this chapter.

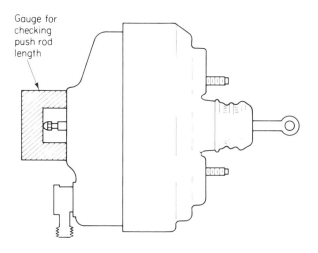

Figure 12-17 A gauge can be used to check the booster push rod length.

REVIEW QUESTIONS

1. What is the advantage of power brakes? [12-1]
2. What forces are used to provide power assist in power brakes? [12-1]
3. What does the term "vacuum suspended" mean as it applies to power brakes? [12-2]
4. Describe the operation of a typical vacuum-operated power brake during release, hold, and apply. [12-2]
5. How does the reaction disc help the driver control braking? [12-2]
6. What is the purpose of the vacuum reserve system? [12-2]
7. Describe the operation of the hydraulically operated power brake. [12-3]
8. What is the purpose of the accumulator on a hydraulically operated power brake? [12-3]
9. How can the automotive service technician determine that the power brake is functioning? [12-5]
10. What service can an automotive service technician do on power brakes with the booster installed on the vehicle? [12-5]
11. What power steering system problems will affect a hydraulically operated power brake? [12-5]
12. Describe the procedure used to remove a vacuum power brake booster from a vehicle. [12-6]

13
Suspension Systems

The suspension systems on modern vehicles are designed to combine many important variables in a compromise to achieve suitable handling and ride characteristics. Factors that influence the design of the suspension system are often more subjective in nature than objective. In suspension systems, subjective evaluations are made for: ride quality, handling ease, harshness, and passenger comfort. Objective evaluations include tire wear, wheel pull, and tire squeal.

Suspension systems have been changed and refined considerably since the advent of the family passenger automobile. Design objectives differ between large luxury sedans, smaller compact vehicles, and light trucks.

Development of High-Ratio Steering. Prior to the use of power steering, the domestic vehicles traditionally grew heavier and larger. This required the manual steering ratios to be increased. The *steering ratio* is the angular rotation of the steering wheel divided by the angular steering change of the wheels. Due to the increased vehicle weight, the front wheels were more difficult to steer. This fact coupled with the increased number of drivers who did not have the strength to park a vehicle, forced manufacturers to design high-ratio steering. The high-ratio steering provides maximum mechanical advantage for the driver, resulting in less steering effort. It also results in an increased number of turns of the steering wheel and a slower

Suspension Systems

response to make the same turn. As a result of high-ratio steering, soft springing, and tires designed to give a smooth, soft ride, American drivers have been poorly conditioned for responsive steering. Vehicles in the past responded slowly and gave early warning of an impending loss of control. Tire squeal and body roll developed early in a turn so the driver could sense the approach of control loss long before a skid developed. The driver could, therefore, take corrective action to prevent the skid.

Power Steering. Power steering is now used on many domestic vehicles. With power steering doing most of the steering work, steering ratios have been decreased so that the vehicle again has responsive steering. The size and weight of present day vehicles are decreasing to the point where high-ratio manual steering is no longer needed. As a person becomes accustomed to the driving characteristics of a domestic passenger sedan, it is often difficult to smoothly control a precise-handling, responsive vehicle. Manufacturers have gradually improved handling and response characteristics so that there is no inherent danger when a driver changes to a different model year vehicle having slightly different handling characteristics.

Other Developments. Tire improvements, along with improvements in shock absorbers, steering systems, and improved suspension control devices, have continually upgraded vehicle handling characteristics. Drivers are becoming conditioned to improved vehicle handling and responsiveness. More of the same type of changes can be anticipated in future vehicle designs.

Radial ply tires have caused a problem for the vehicle design engineer because they have different operating characteristics than bias or bias belted tires (see Chapter 3). Suspension systems are being designed for radial tires on most vehicles to take advantage of the superior operating characteristics of these tires.

The main priority in suspension design is to maintain maximum tire-to-road contact to produce safe, positive control under all operating conditions. All four tires must remain in contact with the road at all times so that maximum vehicle control will exist. Compromises in handling response, tire wear, driver comfort, and ride harshness are made to achieve positive vehicle control.

13-1 SUSPENSION TYPES

Dependent-type front suspension systems were used on early domestic vehicles. Dependent-type solid axle suspension is in use today on some heavy-duty trucks and farm vehicles. This suspension (Figure 13-1) consists of a solid member, called an I-beam or solid axle, that connects the left and right front wheel knuckles to the front springs. The knuckles pivot on a shaft, called a *kingpin,* which fits into the end of the axle. The wheel hub and bearings are mounted on the outer portion of the knuckle which is called the *spindle*. The steering linkage is connected to an arm attached to the knuckle for controlled steering. When required, link and lever control devices attached between the axle and frame are used to absorb braking and turning forces as well as improve ride quality.

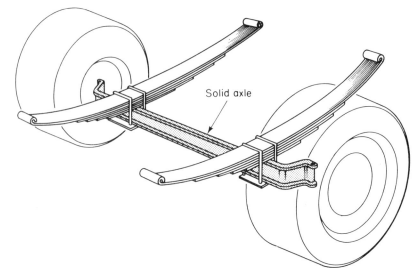

Figure 13-1 A typical dependent type front suspension with a solid axle.

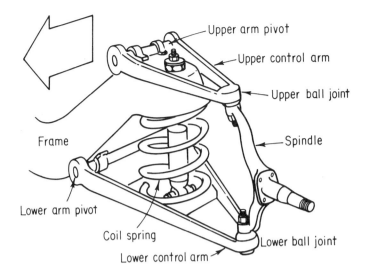

Figure 13-2 A typical long-short control arm independent front suspension.

The solid axle is durable for slow speeds on rough roads. The solid axle is presently used extensively on large trucks and off-the-road vehicles because it is the strongest type of front suspension system.

The need for an improved suspension system grew from the demand for improved ride quality and improved handling at increased highway speeds. The independent front suspension was developed to meet these requirements.

Modern independent suspension systems (Figure 13-2) consist of control arms, A-frames, steering knuckles, ball joints, torsion bars or coil springs, stabilizer bars, brake struts, and shock absorbers. Selected combinations of these parts provide the required support for braking and turning forces, steering, and vertical wheel movement while maintaining vehicle control.

A variation of the independent design, referred to as twin I-beam suspension, is encountered on some trucks (Figure 13-3). This design uses two long members, called I-beams, with kingpins on the outer ends to allow the wheels to pivot for turning. Struts are used to absorb braking forces, and most often, coil springs are used to support vehicle weight. A pivot connects the I-beam to the vehicle frame at the end opposite from the kingpin. The twin I-beam design provides a compromise between the independent suspension ride quality and the ruggedness of the solid axle.

13-2 SPRUNG AND UNSPRUNG WEIGHT

In the design of suspension and suspension control devices, the difference between sprung and unsprung weight must be considered. Sprung weight is all of the weight that is supported by the vehicle springs. Unsprung weight is the weight of parts between the springs and the road surface. These can be seen in Figure 13-4. Sprung weight includes the body frame, engine, transmission components, and everything contained therein. Unsprung weight on most vehicles includes tires, wheels, wheel brake mechanisms, steering knuckles, and rear axle assemblies. Suspension control devices, such as shock absorbers, control arms, brake struts, some stabilizer bars, tie rods, and springs, have one end sprung and the other end unsprung. They do not clearly tie into either category. Their weight is usually considered to be sprung weight. The quality of the vehicle ride diminishes as a higher percentage

Figure 13-3 A twin I-beam front suspension.

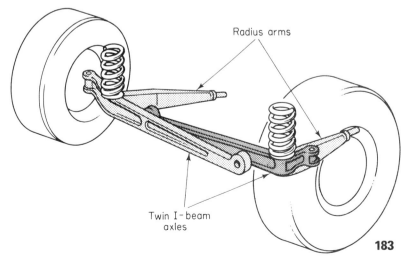

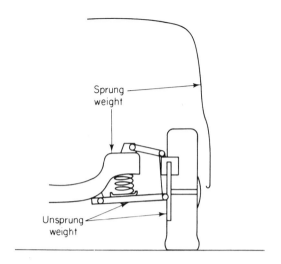

Figure 13-4 Parts included in the sprung and unsprung vehicle weight.

of the vehicle weight is unsprung. This is especially noticeable while traveling over rough roads. On light vehicles the ratio of unsprung weight to sprung weight is kept low to provide the best possible ride.

13-3 SPRING REQUIREMENTS

Some type of spring is required to provide the necessary support for the vehicle body weight and the vehicle load. At the same time the springs provide flexibility to minimize the shock transmitted to the passengers and vehicle load by irregularities in the road surface. The deflections of the suspension system while the vehicle is in motion must be absorbed to provide passenger comfort and to maintain tire-to-road contact. Part of the irregularities in the road surface are dampened within the tire itself as the tire flexes. Large deflections are transmitted to and are absorbed by the vehicle springs. Vehicle springs are designed to provide a compromise between handling characteristics and riding comfort demanded by the customer. *Spring rate,* as illustrated in Figure 13-5, is the load required to deflect a spring one inch. A vehicle designed with a lower spring rate will allow more

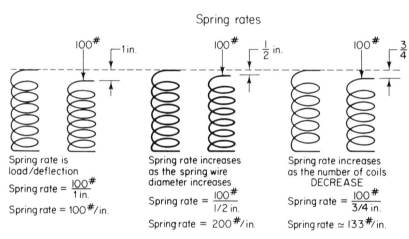

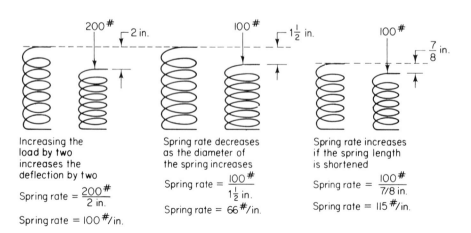

Figure 13-5 Examples of coil spring design on the spring rate.

suspension movement giving a softer ride for passenger comfort. That same vehicle designed with a greater spring rate would have improved handling characteristics but it would provide less passenger comfort.

Vehicle Requirements. Small subcompact vehicles, vans, pickups, and station wagons place an even greater demand on vehicle springs. With less vehicle weight, a lower spring rate is required to support the weight at each wheel. A low spring rate provides satisfactory ride and handling characteristics during normal day-to-day use. This same vehicle may be required to transport four more persons and cargo for an occasional trip thus carrying an additional weight of up to 1,000 lb (454 kg). This extra weight would lower the ride height of the vehicle to the point where the suspension would bottom as the vehicle rolls over a small bump. If the designer put a higher rate spring on the vehicle to carry an additional load, it would produce a harsh ride in normal day-to-day use under lightly loaded conditions. What is required is a spring or mechanism that can sense vehicle load and provide additional support for the added load. This would provide the necessary support, handling, and ride quality under all operating conditions. Air suspension, air-assist suspension, mechanical assist, and hydraulic assist are methods that either have been employed, are currently being used, or are being studied to provide the required variable support in future vehicles.

13-4 SPRING TYPES

Spring types that are currently being used in vehicles include leaf springs, torsion bars, and coil springs. *Leaf springs* may be separated into two categories, multiple-leaf, the most popular, and single or monoleaf design. The leaf spring has a long history of proven performance, from the days of the horse drawn buggy to today. This spring design provides all the necessary restraint for the wheel assembly during braking, acceleration, cornering, and suspension flexing that results from road surface irregularities.

Multiple-leaf Springs. The spring consists of a number of slightly arched sections of varying lengths of steel. Figure 13-6 shows a typical leaf spring. The spring sections or leaves, are fastened together by a through bolt placed slightly ahead of the center of the spring. This bolt is called the *center bolt*. The use of multiple leaves provides the necessary suspension control and compliance while maintaining high fatigue strength. As the spring flexes, the individual leaves slide over one another. This sliding can be a source of noise and it also produces friction. These problems are helped by interleaves of zinc and plastic placed between the spring leaves. Friction as the leaves slide produces ride harshness as the spring flexes but they dampen the oscillation produced by large spring deflections.

Monoleaf Spring. The monoleaf spring design provides the same restraints as the multiple-leaf spring in controlling movement. It does not have the noise and static friction of the multiple-leaf spring. It does have less fatigue resistance. Monoleaf springs are made of chrome carbon steel. They vary in thickness from a maximum at the center to a minimum at the spring eyes on each end. After being rolled to provide the correct taper, they are shot-peened to provide maximum fatigue resistance. Fatigue results from continued flexing of the spring. As the thickness of the spring section is increased, the resistance to fatigue is decreased. This inverse relationship results from the change in length that must occur on the upper and lower leaf surface as the spring flexes. It is proportional to the distance from the center line of the leaf to its surface as illustrated in

Figure 13-6 A typical rear leaf spring installation.

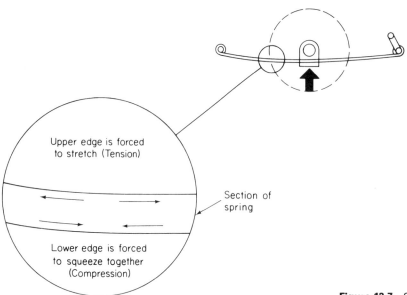

Figure 13-7 Stresses within a spring leaf.

Figure 13-7. The multiple-leaf spring incorporates more leaves of thinner sections which will minimize the stress buildup in the leaf that occurs as the spring flexes. Additional leaves, however, increase the weight of the spring.

Leaf Spring Shape. Leaf springs are designed to function with positive arch (camber), flat, and reverse arch. A leaf spring designed to operate relatively flat gives greater control of lateral forces applied to the suspension and thus minimizes side sway.

Leaf Spring Failure. Leaf spring failure often occurs at the center bolt because of the reduced spring section width at the bolt hole. Failure may also occur near the spring mount as a result of localized stress buildup caused by overtightening the spring pads. Overloading the vehicle is another major cause of leaf spring failure. Premature failure may also be caused by worn shock absorbers which result in little or no dampening. The suspension will then have excess travel and the number of oscillations will increase for each road surface irregularity encountered. As springs are used they will fatigue. This leads first to sag and eventually to broken leaves.

Torsion Bars. Torsion bars are another type of spring used to support vehicle weight and provide the necessary suspension movement. Torsion bars are made of heat-treated alloy spring steel. They provide for suspension deflection by twisting. One end of the torsion bar is attached to the vehicle frame while the opposite end is connected to the lower control arm. This is shown in Figure 13-8. The bar is usually formed with a round cross-section having the ends formed into a hexagonal shape to fit into a socket in the lower control arm (moving end) and the vehicle frame (stationary end). Torsion bars are most commonly found on the front suspension, with coil or leaf springs providing the rear suspension control. The independent front suspension is readily adaptable to the use of torsion bars. The torsion bar uses very little

Figure 13-8 A typical longitudinal front torsion bar installation.

space under the vehicle. With dependent rear suspension, the use of a torsion bar using a conventional mounting is difficult. During manufacturing, torsion bars are prestressed to provide fatigue strength. Because of directional prestressing, the torsion bars are directional. They must only be used on designated sides of the vehicle. The torsion bar is marked either *right* or *left* to identify the correct installation position. Surface imperfections on the bar will produce an area of high localized stress which may result in premature failure. This type of failure can be prevented by using care during removal and installation of the torsion bar so that undue marking and resultant stress concentration does not occur.

Torsion bars are mounted longitudinally (parallel to the center line of the vehicle) or transversely (perpendicular to the vehicle center line). With the bars mounted in a longitudinal orientation, the front suspension noise is transmitted along the bar to its fixed end. The longitudinal bar has its fixed-end socket in the cross-member that is located beneath the front seat of the vehicle. This location will transmit the noise and vibration to the cross-member. It is then difficult to isolate the noise from the passenger compartment. Mounting the torsion bars in a transverse location as shown in Figure 13-9 allows easier noise isolation at the front of the vehicle. When the fixed end of the transverse torsion bar is attached to the front cross-member, the noise can be isolated on the front suspension and cross-member. The use of torsion bars provide the manufacturer with good underhood space by not requiring the space that is needed for coil spring towers.

Coil Springs. Coil springs are the most common vehicle spring. They support the vehicle load while allowing the vehicle wheels to deflect as they encounter road surface irregularities. Coil springs, like monoleaf and torsion bar springs, do not have inherent noise problems nor do they develop the static friction that causes ride harshness as in multiple-leaf springs. Coil springs have load-carrying lower spring seats on the vehicle A-frames (control arms) and reinforced upper spring seats in the vehicle frame or underbody as illustrated in Figure 13-10. This type of mounting requires reinforcement not required with other types of spring.

A coil spring develops its force in a manner similar to the twisting of a torsion bar. If a short section of the coil were examined while the spring was compressed, it could be seen that a twist was present in that section of the coil. Twist is il-

Figure 13-9 A typical transverse front torsion bar installation.

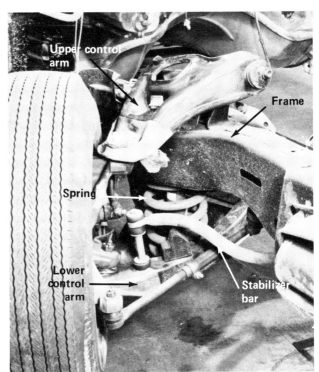

Figure 13-10 A typical front suspension with the coil spring mounted between the frame and lower control arm.

Suspension Systems

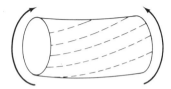

Figure 13-11 Twist occurring in a short section of a coil spring or a torsion bar.

lustrated in Figure 13-11. It can now be seen that a coil spring is actually a torsion bar wound in a coil to make a compact suspension component design that will absorb vertical loads.

Coil springs can be designed with a variable spring rate. This means that they become stiffer as they are forced to deflect as illustrated in Figure 13-12. The variable-rate design provides ride smoothness while operating under light loads and, at the same time, it has the ability to carry heavy loads. The variable-rate springs can be easily identified by the presence of more closely wound coils in one part of the spring than at other parts or they may be taper wound. As the vehicle is loaded with more weight, the closely wound coils come in contact with each other and cease to function. The more widely spaced coils will then provide the spring deflection. Because the widely spaced part of the coil has a higher spring rate, the spring has the ability to carry a heavy load. This design has been used in the rear of light trucks which are required to carry varying loads. Variable-rate coil springs are more costly to make because the manufacturing uses two separate coil spring winding processes. Coil springs only have the capacity to absorb loads applied axially (vertical). They provide no control of lateral (sidewise) or longitudinal (front and rear) forces that develop in the suspension system. For this reason a complex linkage system is necessary when coil springs are used in the rear to restrain the torque loads that occur during acceleration and braking, and the side loads that occur during cornering.

13-5 SUSPENSION CONTROL DEVICES

Due to the inability of some spring designs to control forces developed during braking, acceleration, and cornering, additional control linkages are needed. A spring provides a very small amount of natural dampening of body oscillations after encountering road irregularities. For this reason devices must be installed to provide the required damping.

A vehicle manuevering around turns produces side loads on the vehicle suspension. These side loads must be transferred to the tires of the vehicle to maintain control. In a turn a side load force is developed as the sprung weight tends to remain moving in a straight line that is tangent to the turning arc, as illustrated in Figure 13-13. This force is called *inertia* or the force that tends to keep a moving body in motion in a straight line. The suspension transfers this force through the vehicle tires

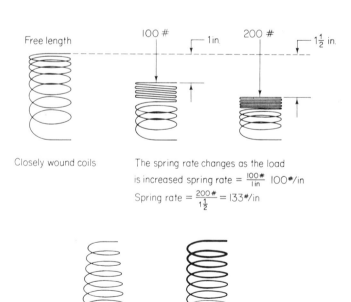

Figure 13-12 Types of variable-rate coil springs.

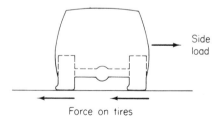

Figure 13-13 Side loads produced as a vehicle moves through a turn.

to the road. The tire-to-road contact produces the reaction required to turn the vehicle. In the solid axle front suspension system design using leaf springs, the springs restrain the body movement during cornering. Leaf springs provide the required restraint to transmit the inertia side loads to the road surface. A vehicle equipped with torsion bar springing in the front will transfer the inertia side load forces to the control arms or A-frames and then to the road surface through the wheel and tire. The control arms provide the restraint for side load forces.

Coil springs use a variety of linkages to transmit side load forces to the road surface. In front suspensions using upper and lower control arms, the side loads are restrained by the control arms. With twin I-beam suspension, the front axles (I-beams) transfer these loads to the wheels. Rear suspension, depending on manufacturer, may transfer the side loads through angular linkages, A-frames, a Panhard rod (conventionally called a track bar), or any combination of the above.

Angular Linkages. The most common arrangement is angular linkages connecting the vehicle body and frame to the rear axle as shown in Figure 13-14. By angling the links to the center line of the vehicle, the inertia side loads and the torque loads from braking and acceleration can be transmitted to the frame. When an independent rear suspension is employed and A-frames are used as shown in Figure 13-15, the A-frames restrain the side loads. A variation of the conventional A-frame design may be used with dependent type rear suspension where the pivot point of the A-frame on the axle is behind the fixed pivot. This is illustrated in Figure 13-16. When the pivot point trails the fixed pivot axis, the link is called a *trailing arm*. In these suspension systems the side loads are transferred to the road surface through the arms. Their load capacity is limited by the spacing between the two legs of the A-frame. Many arrangements of angles and design of the A-frames have been used. The most important considerations when determining their angle is the weight of the vehicle and the amount of torque the engine will produce.

Track Bars. A vehicle may contain parallel linkages for torque control and a track bar to transfer the side loads to the tires and road surface. A *track bar or Panhard rod* is a link connecting the vehicle frame on one side to the rear axle on the opposite side. This can be seen in FIgure 13-27. The track bar transfers all the side loads to the axle. This design frees the trailing arm so it will only have to control torque. Control of the side loads is through the track bar. Small vehicles sometimes use a short torque tube (a tube connected to the nose of the differential) approximately 2.5 feet (0.76 m) long. The front end of the torque tube is

Figure 13-14 A typical dependent rear suspension with a Hotchkiss drive using coil springs and control arms to retain side loads and torque loads (Courtesy of Buick Motor Division, General Motors Corporation).

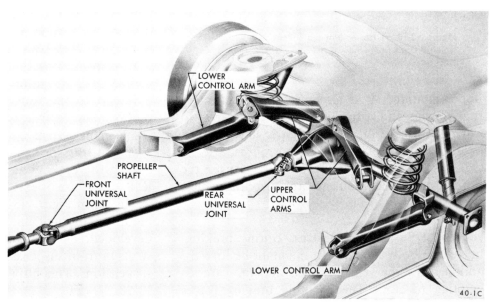

Suspension Systems

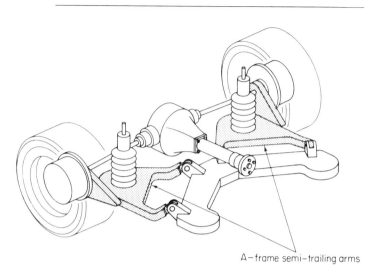

Figure 13-15 An independent rear suspension using A-frame semitrailing arms to retain side loads and torque loads.

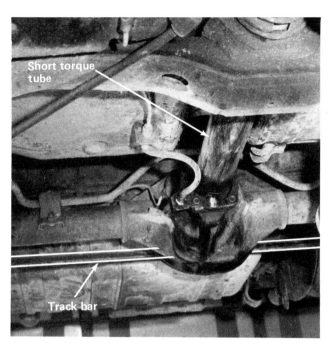

Figure 13-17 A rear suspension using a short torque tube to retain braking and acceleration torque loads and a track bar to retain side loads.

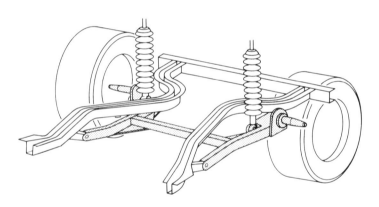

Figure 13-16 A trailing-arm dependent rear suspension used on a front wheel drive vehicle.

rubber mounted. It is also necessary to use two trailing arms and a track bar or Panhard rod to provide full control. This design is illustrated in Figure 13-17. The track bar controls side loads while the trailing arms and torque tube transfer the torque from acceleration and braking to the vehicle frame.

Additional Suspension Devices. During braking, the vehicle body attempts to maintain its forward motion while the vehicle wheels and tires are attempting to stop. The resulting braking forces (Figure 13-18) develop a torque or twisting force that is transmitted to the suspension system. If these forces, as well as the inertia force of the vehicle, were not restrained, the vehicle body would attempt to continue in straight ahead motion, even though the vehicle brakes were attempting to stop its motion.

During acceleration the differential housing attempts to rotate in a direction opposite to the direction the rotation of the tires (Figure 13-19). The forces on the suspension of the driving wheels during acceleration must restrain the axle housing to prevent rotation. Vehicles employing independent suspension at the driving wheels transmit the torque of acceleration directly to the differential mounted to the vehicle frame or to the vehicle power plant which is mounted on the vehicle frame. This system can be seen in Figure 13-20.

The leaf springs provide enough strength to absorb the braking and acceleration forces if the leaf springs are oriented in a longitudinal direction as shown in Figure 13-21. Transverse leaf springs require additional linkages to restrain the forces developed. These linkages take the form of brake struts in front suspensions and trailing arms in rear suspensions. This suspension is illustrated in Figure 13-22.

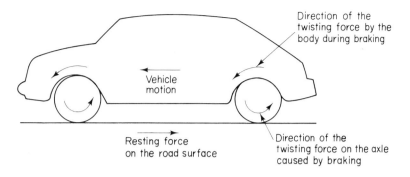

Figure 13-18 Torque loads during braking.

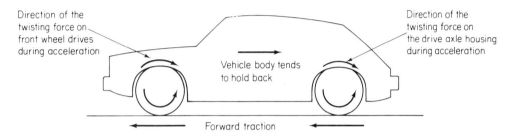

Figure 13-19 Torque loads during acceleration.

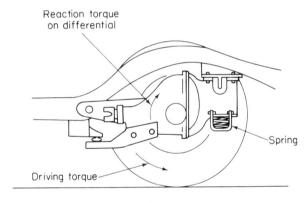

Figure 13-20 Frame-mounted differential retains acceleration torque loads. Swing axles are used with this arrangement.

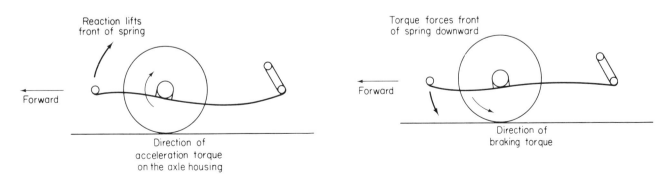

Figure 13-21 Torque and brake loads absorbed with a leaf spring.

191

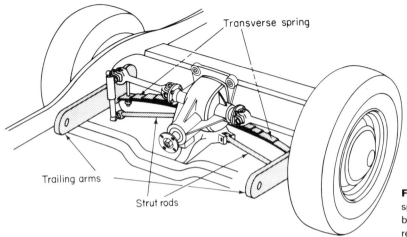

Figure 13-22 A rear suspension using a transverse leaf spring to retain side loads and trailing arms to support braking torque loads. The acceleration torque loads are retained by the frame mounted differential.

Coil springs by themselves provide no lateral or longitudinal restraint. With coil spring front suspensions (Figure 13-2) A-frames are generally employed to properly locate and restrain both upper and lower wheel pivots. Their design uses widespread fixed pivot points which enables the A-frame to transmit the braking forces. If the coil springs are mounted between the upper A-frame and the body or spring tower as shown in Figure 13-23, a lower control arm of reduced section can be used with a brake strut to absorb braking forces.

Brake struts usually consist of a heavy rod or a stamped steel strut, one mounted on each side of the vehicle. One end of the strut is attached to the frame and the other end is connected to the axle or the lower control arm near the wheel pivot. The strut can extend either forward or rearward. The brake strut transmits all rearward forces placed on the lower control arm during braking to the vehicle body as well as the rearward forces on the suspension that result from road surface irregularities.

A radius arm may be used in the front suspension for the same purpose as a brake strut. The name radius arm implies one more important function, that of controlling the arc or radius in which the suspension travels, as viewed from the side. This can be seen in Figure 13-24. The radius arm has its centerline parallel to the vehicle centerline while a brake strut has the fixed strut pivot (Figure 13-25) and the fixed lower control arm pivot approximately the same distance from the centerline of the vehicle. Rear suspension trailing arms are canted from the centerline of the vehicle to absorb the forces from acceleration and braking. These are illustrated in Figure 13-26.

A torque tube is another method of absorbing these forces. The torque tube is a large tube connecting the nose of the differential to either the vehicle underbody or the transmission extension housing as illustrated in Figure 13-27. The torque tube absorbs rearward motion and torque forces due to braking, as well as forward movement and torque forces due to acceleration. A Hotchkiss drive, which has one universal joint on the differential companion flange and another at the rear of the transmission can incorporate some of the features of a torque tube drive. This is accomplished by using a steel strut fastened rigidly to the differential and pivoting at the transmission. This design, shown in Figure 13-28, is used on vehicles where space for rear suspension linkages is limited.

Longitudinally mounted torsion bars use the same method for absorbing forces on the suspension

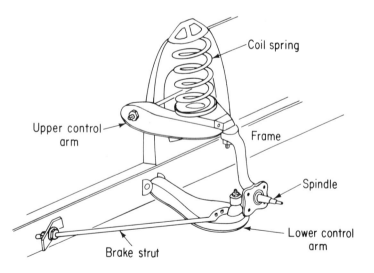

Figure 13-23 A front suspension having the spring mounted above the upper control arm.

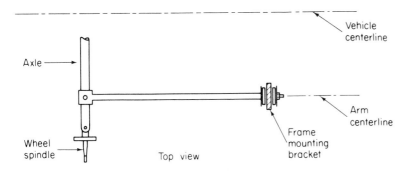

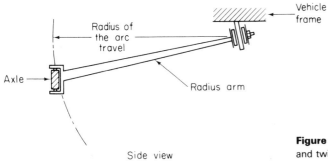

Figure 13-24 Radius arm used with dependent solid and twin I-beam front suspensions.

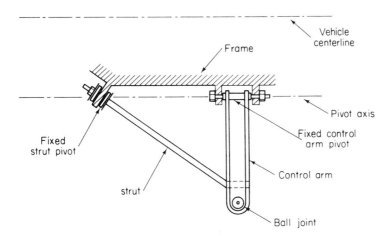
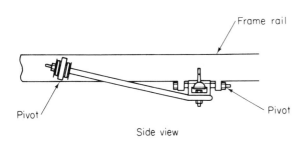

Figure 13-25 Brake strut used on an independent front suspension.

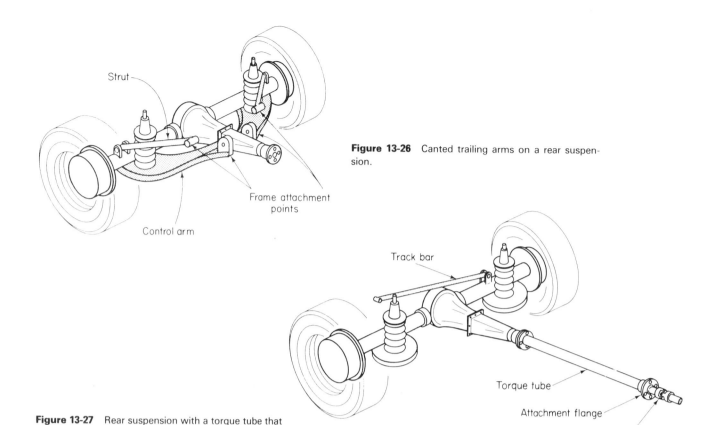

Figure 13-26 Canted trailing arms on a rear suspension.

Figure 13-27 Rear suspension with a torque tube that retains both braking and acceleration torque. A track bar is used to retain side loads.

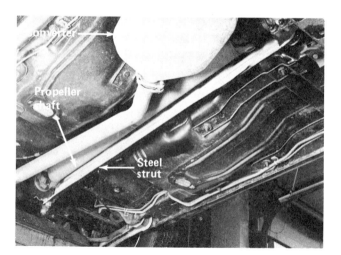

Figure 13-28 A steel strut adjacent to the propeller shaft retains braking and acceleration torque.

arms as coil springs. When the torsion bars are mounted transversely, the torsion bars themselves may be used to absorb the forces. This is accomplished by positioning the rotational end of the torsion bar near the wheel pivot and using the torsion bar for restraint.

Control arms on the rear suspension, used in conjunction with a beam-type axle, perform one more important function, that of controlling differential nose angle. Most modern vehicles use the Hotchkiss drive rear axle as shown in Figure 13-14. Due to the characteristics of universal joints operating through angles (see Chapter 17), the front and rear universal joints must operate near the same angles. These angles will remain nearly equal as the vehicle travels through jounce and rebound only when the proper control arm length, pivot point location, and operating angle have been selected by the suspension engineer.

Trailing arms and radius arms can be used on either front or rear suspension. A leading arm suspension shown in Figure 13-29 is another design

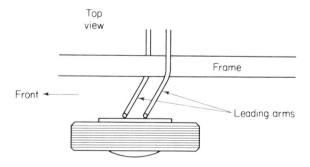

Figure 13-29 Leading arms hold the wheel ahead of the frame mountings.

alternative where the movable pivot leads the fixed pivot point. The leading arm suspension, as well as the trailing arm type, is adaptable to transverse torsion bar springing and it is used on some imported vehicles.

MacPherson Strut Suspension. On the MacPherson strut front suspension, a lower control arm provides lateral support and the stabilizer bar often doubles as a strut bar. MacPherson strut suspension consists of a coil spring mounted over a telescoping shock absorber. The upper end of the shock absorber is connected to the vehicle underbody. A lower control arm and a stabilizer bar complete the assembly as shown in Figure 13-30. A modified design places the coil spring between the control arm and frame, inboard of the strut. This can be seen in Figure 16-10. Light weight, simplicity, and the use of minimum space are the major benefits of the MacPherson suspension design. During braking, the stabilizer bar acts as a brake strut to absorb rearward force while the upper end of the shock column absorbs the brake torque.

Roll Center. The *roll center* is the point about which the body rolls while making a turn. The roll center is the only point on the vehicle centerline that does not move up or down during a turn. The front and rear suspensions each have their own roll centers. A line connecting the front and rear roll centers is called the *roll axis*.

The *instant center* has to be determined before the roll center can be located. The instant center is determined by extending an imaginary line through both pivots of each control arm to the point where the lines intersect.

The roll center is located at the vehicle center on an extended line connecting the instant center and the center of the tire patch. This can be seen in Figure 13-31.

The roll center is an important handling consideration. Body roll increases as the roll center is moved further from the vehicle center of gravity. The amount of roll is controlled by the stabilizer bar. A larger stabilizer bar is required when the distance between the roll center and center of gravity is great.

Stabilizer Bar. A stabilizer bar, anti-roll bar, or anti-sway bar usually has a solid cross-section. It is connected from the lower suspension arm on one side of the vehicle to the lower suspension arm on the other side. The bar may be connected to the suspension arms by short links, allowing the bar to pivot on mounts attached to the vehicle frame. This can be seen in Figure 13-32.

Connecting the suspension control arms together with the stabilizer bar produces many benefits. When the vehicle body comes closer to the road surface the suspension is in *jounce* or *compression*. When jounce affects both left and right wheels equally, the stabilizer bar follows the suspension and so it has no effect on the vehicle. As the vehicle encounters a rise in road surface with one wheel (the wheel goes into jounce), the upward force is transferred across the stabilizer bar producing an upward force on the suspension on

Figure 13-30 A typical MacPherson strut front suspension.

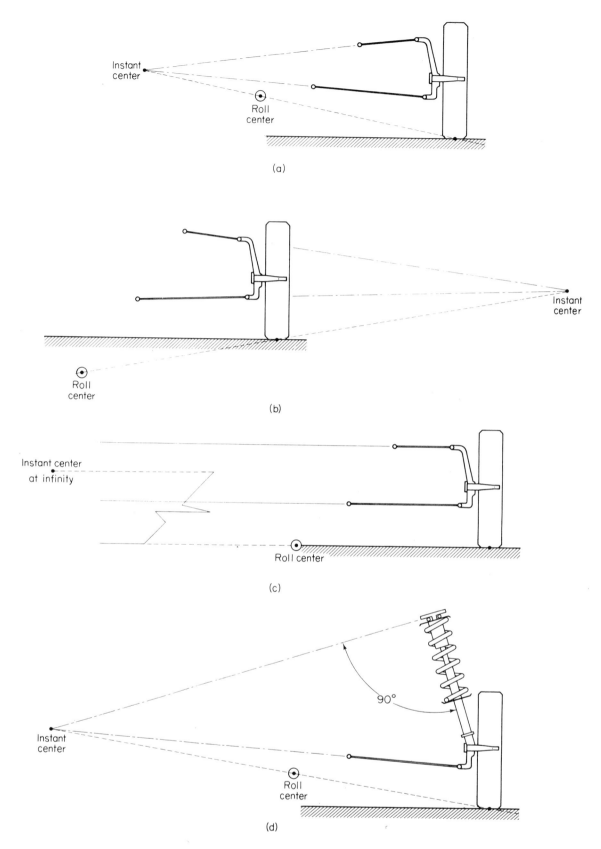

Figure 13-31 Illustrations showing how the roll center is located: (a) long-short control arm suspension converging *toward* the vehicle center; (b) long-short control arm suspension converging away from the vehicle center; (c) control arms parallel makes the instant center at infinity and the roll center at the road surface; (d) the instant center on a MacPherson strut is extended at 90° from the top of the strut and through the lower control arm pivot points.

Suspension Control Devices

Figure 13-32 A typical front stabilizer bar installation.

the opposite side of the vehicle. This is illustrated in Figure 13-33. This transfers a portion of the jounce load to the opposite suspension. The action twists the stabilizer bar much like a torsion bar. This twisting action increases the effective spring rate of the suspension in jounce.

The stabilizer bar can be considered to be a spring that will resist the independent action of the suspension. It will transfer some of the load from one suspension to the other to help prevent excessive vehicle body roll. As one wheel drops, or *rebounds,* due to a drop in the road surface, the stabilizer bar exerts an upward force on the suspension of the wheel encountering the low spot. This will reduce the effective spring rate of that wheel. As a vehicle enters a turn, the vehicle body leans towards the outside of the turn and this forces the outside suspension into jounce (compression) while the suspension on the inside will go into rebound. The stabilizer bar opposes both of these motions by increasing the effective spring rate of the outside wheel and decreasing the effective spring rate of the inside wheel. In a turn, the stabilizer bar reduces the degree of body roll and hence it has acquired the name *anti-roll bar*. The stabilizer bar dampens body movements due to wind gusts and rolling road surfaces that may be noticeable at highway speeds by applying forces on the suspension that oppose vehicle input forces (one wheel going into jounce, while the other goes into rebound).

The stabilizer bar came into being on vehicles designed for handling, such as road race cars. They have been incorporated, first on the front suspension of passenger vehicles and then on the rear suspension (Figure 13-34). Many vehicles now have both front and rear stabilizer bars to provide the handling and ride quality demanded by the operator. A reduced vehicle spring rate can be used when sturdy stabilizer bars are used. This will provide the vehicle with soft ride springing while at the same time it has the handling characerstics of stiffer springing. The diameter and shape of stabilizer bars vary according to the manufacturers design objectives. Due to the heavy weight bias on front-engine vehicles, the front stabilizer bar will usually have a heavier cross-section than the rear stabilizer bar. If stabilizer bars have a cross-section that is too large, the vehicle will be difficult to control on rough road surfaces. Again, a compromise must be made between rough road and high-speed smooth road handling.

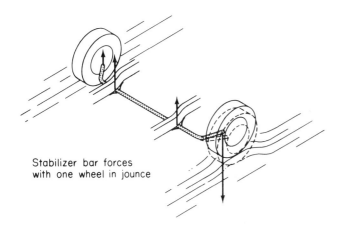

Figure 13-33 The action of a stabilizer bar.

Figure 13-34 A typical rear stabilizer bar installation.

Suspension Systems

The suspension control devices such as control arms, stabilizer bars, A-frames, track bars, brake struts, radius arms, and vehicle springs are isolated from the vehicle body and/or frame. Isolation of the suspension components from the frame or underbody reduces the transmission of noise from the tires and road surface into the driving compartment. Bushings of a rubber or elastomer material are placed in the pivots of control arms, spring eyes, and idler arm of the steering linkage. These bushings usually have an inner and outer liner of steel with the rubber bonded to the liners as illustrated in Figure 13-35. Rotation can take place between the liners due to the flexing of the rubber. Bushings of this design allow relative movement of the suspension, help absorb some of the road shock, and prevent the transmission of noise into the vehicle body. The relative movement allowed is referred to as *compliance*. Due to the bonding of the liners to the rubber or insulating material, no lubricant should ever be used on them. Lubricant will soften the rubber-type elastomer material and it will usually separate the bonding that is required to control the tolerances in the components. Elastomer bushings will tolerate a slight amount of misalignment. They, therefore, reduce the need for closely controlled tolerances in the manufacturing of the suspension components. This, in turn, reduces the manufacturing cost.

Bushings are also used between suspension components and cross-members at the ends of the brake struts and at stabilizer bar supports. These bushings may or may not have steel liners, although their function is similar to those requiring steel liners. In some cases wide tolerances are satisfactory, which reduce the cost of manufacturing. Examples of these wide tolerances are the typical stabilizer bar mounting brackets shown in Figure 13-36 where carefully maintained tolerances are not required.

Other types of insulators may be found between the coil springs and the spring seats, between the leaf springs and their mounting pads, and between cross-members that carry all of the suspension components for the front or rear of the vehicle. The main purpose of this type of insulator is to prevent the transmission of irritating vibrations and noise into the passenger compartment. Radial ply tires have compounded the problems of noise isolation due to their inherent low-speed harshness. Suspension components are isolated and tuned so that radial tires may be used with the same passenger comfort as obtained with other types of tires. Large-diameter spring eye bushings with voids will allow suspension compliance for extra

Figure 13-35 Typical steel encased elastomer suspension bushings.

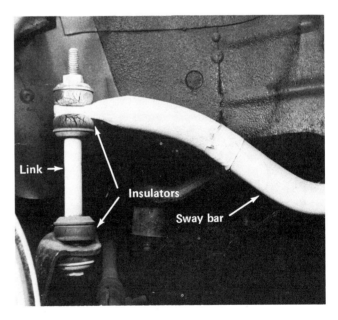

Figure 13-36 Stabilizer bar insulator bushings. The bushings pictured show aging cracks.

Shock Absorbers

Figure 13-37 Bushing designed with voids that allow greater suspension compliance than solid bushings.

13-6 SHOCK ABSORBERS

The shock absorbers used in todays' vehicles are direct, double-acting shock absorbers. Most vehicles use four shock absorbers, one near each wheel. Additional shock absorbers have been used to produce special ride qualities. The shock absorbers are called direct acting because of their direct connection between the vehicle body or frame and the suspension member near the wheel. The double action describes the direction of motion control, both jounce (compression) and rebound (extension).

Shock absorbers are placed near each wheel as pictured in Figure 13-39 to help dampen uncontrolled up and down wheel motion as well as to control the rolling motion of the body that results from rises and dips in the road surface. The *standard shock absorber* differs from an air-lift or spring-assist shock absorber because it does not support vehicle weight. The spring supports the vehicle weight. The shock absorber's only function is to dampen body oscillations and wheel

damping and vibration control. An example of this is pictured in Figure 13-37. Torsion bars are isolated at front cross-members. In some cases weight is added to suspension components to change the frequency of vibration. An example of this is shown in Figure 13-38. These are some of the ways the chassis engineer will reduce or change the frequency of the noise and vibration reaching the vehicle occupants.

Figure 13-38 A weight added to the rear suspension control arm to change its vibration frequency to reduce noise and vibration.

Figure 13-39 A typical rear shock absorber mounting location.

Suspension Systems

Figure 13-40 A typical spring-assist shock absorber.

movements that could affect the contrallability of the vehicle. This control will improve ride quality and, as a result, it will lengthen suspension component life. Air- and spring-*assist shock absorbers* increase the effective spring rate to aid the vehicle springs when the vehicle is carrying heavy loads. The assist shock absorbers are easily identified by a coil spring mounted over the conventional shock absorber on the spring-assist, or an air chamber with air connection on the upper portion of the air-assist type. A spring-assist shock absorber is illustrated in Figure 13-40. Figure 13-44 shows an air-assist shock absorber.

Shock Absorber Operation. The shock absorbers are hydraulic devices that convert the kinetic energy of the vehicle body motion (in and up and down direction) to thermal energy (energy of heat). You should recall that energy cannot be created or destroyed, but that it can be converted to other forms. The shock absorber dampens the motion by forcing the hydraulic fluid through small openings between chambers within the shock absorber. For all practical purposes, the hydraulic oil contained within the shock absorber is incompressible. If the incompressible fluid is forced through small-sized openings called orifices, it requires a period of time for the fluid to flow from one side of the orifice to the other. Fluid movement in the shock absorber is created by a piston connected to a rod, which, in turn, is connected to the vehicle body. Each time the wheel of the vehicle is forced upward (jounce) the shock absorber piston displaces the fluid. The fluid beneath the piston is forced through the orifices into the chamber above. As the suspension returns to its normal position, the piston moves upward and displaces the fluid above it by forcing the fluid to return to the lower chamber through a separate set of orifices. It also draws some fluid from a reservoir where some was displaced when the wheel went into jounce. The shock absorber operation is illustrated in Figure 13-41.

The shock absorber requires a fluid reservoir to assure that enough fluid is available as the shock absorber travels through its range. Reserve fluid will prevent air from being drawn in on either side of the piston. When air is drawn in, it is called *aeration*. Air is a gas and it is compressible. When air is mixed with the fluid, the mixture will produce erratic shock absorber operation. Aeration normally occurs during rapid cycling. It is minimized in standard shock absorbers through the use of properly placed baffles in the shock absorber. The baffling on standard shock absorbers is usually visible as spiral grooves around the shock absorber body. A small air chamber at the top of the fluid reservoir column is present to allow for expansion of the fluid as it warms and for contraction of the fluid as it cools. The size of the fluid reservoir may vary from manufacturer to manufacturer, even though the pistons are the same diameter. A large fluid reservoir will improve the life expectancy of the shock absorber and it will operate better under heavy duty operation because it has the ability to reject more heat. The heavy duty shock absorbers are usually identifiable because they have a larger diameter than standard shock absorbers. The fluid in a shock absorber is subjected to high heat loads when the vehicle is driven over rough roads for an extended period of time. The shock absorber may reach temperatures that are too hot to touch. Because of these heat loads, the fluid used in shock absorbers must be able to withstand high temperatures.

Gas-Filled Shock Absorbers. Vehicles may be equipped with a gas-filled shock absorber. The inert gas, usually freon, is placed in a flexible container that

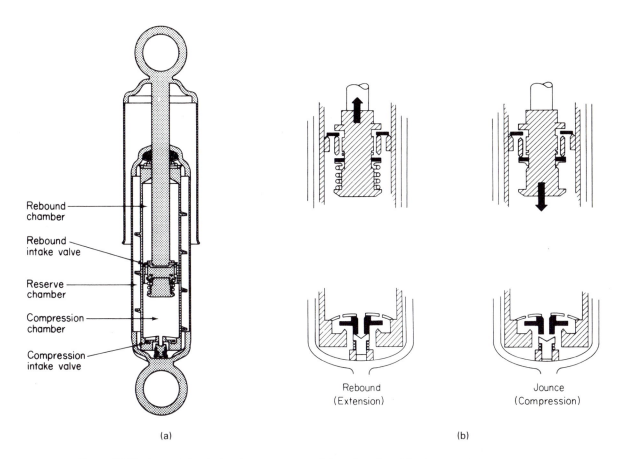

Figure 13-41 Internal shock absorber components: (a) section view; (b) valve operation.

takes the place of the air chamber in standard shock absorbers. This gas-filled chamber eliminates all of the air in the shock absorber. This will prevent fluid aeration even when a shock absorber is cycled rapidly as the vehicle is driven over rough roads. The inert gas expands with increasing temperature. Thus, if the temperature rises during a high-demand situation, it exerts more pressure on the hydraulic oil. This increase in pressure eliminates any possibility of fluid aeration.

Damping Control. Shock absorbers vary in the damping control placed on the suspension in jounce versus rebound. Shock absorbers designed to improve handling usually approach an equal amount of control in both jounce and rebound, such as 40–60 (40 percent jounce and 60 percent rebound) or 50–50. Shock absorbers used on drag racing vehicles generally are 90–10 on front and 50–50 on the rear. The 10 percent of the control on extension allows for weight transfer to the rear wheels while accelerating and the 90 percent retards front-end drop during the gear change. The 50–50 rear shock absorbers control rear wheel motion for maximum traction. The public demand for more responsive vehicles has led to shock absorbers approaching the 50–50 range. Compact vehicles, due to their light weight and soft springing, require a stiffer (more control in both jounce and rebound) shock absorber. This will allow the frequency of the vehicle suspension to stay in its natural 60–80 cycle/minute range. This frequency range provides the best ride characteristics. Damping rates within the shock absorbers are controlled by the size of the piston and the size of the orifices.

Mounting Shock Absorbers. Shock absorbers are mounted at varying angles to achieve the desired ride control. The shock absorbers may be angled inward at the top to provide lateral and body roll damping or angled forward at the top to provide more damping of the road shock that produces rearward motion of the wheels. The shock absorbers are mounted vertically when only the up and down motion of the vehicle wheels is to be controlled.

Care should be used in chassis design or when adapting shock absorbers to different vehicles. If a standard shock absorber is mounted upside down from its intended position, after a few cycles it will

Suspension Systems

not provide any damping action whatsoever. This condition is occasionally found on home-built vehicles and occasionally even on racing vehicles.

13-7 STEERING PIVOTS

Some type of wheel pivot is required to allow for front suspension movement and front-wheel steering at the same time. Vehicles using independent front suspension of the twin I-beam and dependent I-beam design suspensions use a pin fitted through the steering knuckle and into the end of the axle as shown in Figure 13-42 to allow the steering knuckle to turn. This pin is referred to as a *kingpin*. The kingpin is accompanied by a thrust bearing fitted between the lower portion of the axle and the steering knuckle to support vehicle weight through the bearing. The kingpin is well suited for use in I-beam type axles where the vertical wheel angle does not change in relation to the axle. Kingpins have the advantage of greater strength over other types of wheel pivots. Kingpins were used on early independent suspension systems prior to 1954. The kingpin is at present limited to use on I-beam type suspension and on straight axles.

Ball Joint Wheel Pivot. Independent front suspension of the long–short-arm design (upper and lower control arms) and MacPherson strut all use spherical ball type pivot points to allow for vertical suspension movement while turning the wheel. The ball and socket allows freedom of movement of the suspension. This type of wheel pivot is called a *ball joint*. The ball joint is by far the most popular type of wheel pivot used on passenger cars. Depending upon their function in the particular suspension system, a ball joint may be a load-carrying joint or a nonload-carrying joint. These are shown in Figure 13-43.

Vehicle-Loaded Ball Joint. It is known that one of the control arms on the suspension has the vehicle suspension spring forcing it down and in the case of a coil spring, the opposite end forcing up on the vehicle frame or body. The ball joint attached to this load-supporting arm is called a *load-carrying* or *vehicle-loaded ball joint*. The load may be in tension, attempting to pull the ball from the socket, or in compression, attempting to force the ball through the socket.

Friction Ball Joint. One ball joint on each suspension acts as a pivot point for the suspension system and controls the direction of one end of the steering knuckle. It does not support vehicle weight and so it is called a *friction ball joint*. This unloaded ball joint has, within it, a means of taking up wear that will normally occur. Suspension looseness would occur without this feature. A preload spring or elastomer disc is used to maintain zero clearance in the friction-loaded ball joint. A vehicle-loaded ball joint will maintain zero clearance as long as vehicle weight is on the ball joint. Some vehicle-loaded ball joints have a preload spring or disc to remove clearances that may occur as the weight is

Figure 13-42 Typical dependent I-beam front axle assembly with a kingpin pivot.

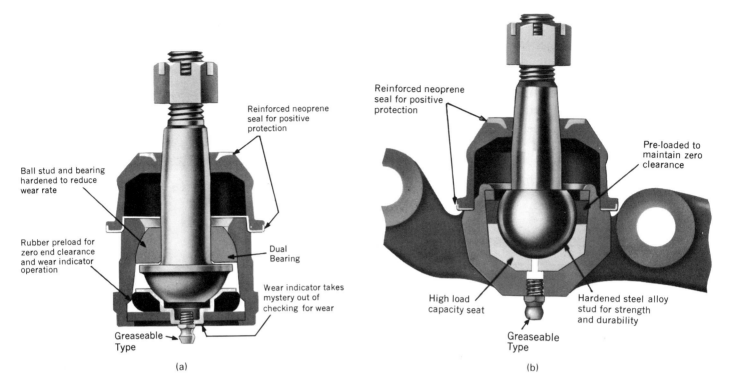
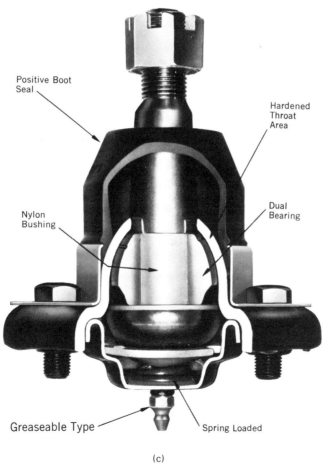

Figure 13-43 Typical ball joint types: (a) tension-type ball joint; (b) compression-type ball joint; (c) preloaded ball joint (Courtesy of TRW Inc.).

Suspension Systems

periodically removed from the ball joint. This situation can occur when traveling over extremely rough roads where the wheel cannot continually follow the road surface. End play in the loaded ball joint may be a source of the additional suspension noise and wear due to this impact loading.

13-8 LEVEL CONTROL SYSTEMS

Air-assist shock absorbers are intended for use as leveling devices when the vehicle is carrying loads at or near maximum capacity. Conventional shock absorbers do not have the ability to level the vehicle. Air-assist shock absorbers are offered as standard or as optional equipment on many passenger vehicles. When required, they may also be fitted to vehicles that were not originally so equipped.

Level-control air shocks are not intended to be used for replacement of worn or damaged springs. Neither are they intended to be used to support vehicle overloads. Their use is to keep the ride height near normal to maintain the suspension ride clearance. This is called *trim height*. It will prevent the suspension from bottoming against the stops as the vehicle goes over bumps. Maintaining the proper rear suspension height will prevent contact of the underside of the rear of the vehicle while driving over sharply inclined surfaces. When the vehicle stays at the correct ride height the suspension geometry angles will operate near normal. This will give the vehicle better handling than it would have if the rear end sagged. The effect of rear suspension height is discussed in Section 14-3.

Leveling-Type Shock Absorbers. The air-assist shock absorbers that are designed to level the vehicle are very similar to conventional dual-acting shock absorbers. They are mounted in place of the standard shock absorber. The hydraulic valving and operation is identical to the conventional shock absorbers discussed in Section 13-6. The difference is that these shock absorbers have an air chamber connected between the upper and the lower portions of the assembly. The air chamber is

Figure 13-44 Typical air-assist shock absorber.

fitted to the upper housing. It takes the place of the protection shield on conventional shock absorbers. This is pictured in Figure 13-44. The adapter provides a place for high-pressure air to be introduced into the air chamber. The high-pressure air forces the upper and the lower shock absorber sections apart. This helps to level the body by increasing the distance between the centerline of the axle and the body. Since air is compressible, it has no real effect on the overall dampening action of the shock absorber. The air only supports some of the load, similar to an additional spring. Failure of the air circuit or the assist portion of the shock absorber will cause the system to operate as a standard suspension without leveling control. The air pressure in the shock absorber may vary from as low as 15 psi (103 kPa) to nearly 200 psi (1380 kPa), depending on the type of shock absorber and the amount of leveling assist required. Maximum pressure will normally support up to about 250 pounds (113 kg) per shock absorber, or about 500 pounds (227 kg) for the entire rear suspension. The maximum assist is designed to support the weight of three average adults plus some luggage without appreciably lowering the rear.

Three types of air level control systems are used: air-assist shock absorbers that are manually adjusted by the operator, air-assist shock absorbers that are adjusted by an operator control in the vehicle, and automatically adjusting air-assist shock absorbers.

Manual Adjust Systems. The manual adjust level control systems can be ordered on new vehicles or they can be installed on almost any type of vehicle that was produced without them. This system is the least expensive and the least complicated of the leveling systems. The only active part of the system is the shock absorber itself. Air lines are routed from the shock absorber to a tire-type inflation valve located in a convenient and accessible place on the vehicle. Factory-installed systems and most retrofit systems use one valve to provide equal assist on both shock absorbers. Separate valves can be installed for each shock absorber if unequal vehicle loading is encountered or if side-to-side trim balancing is needed.

The shock absorbers are inflated after the load is in the vehicle when using the manual adjust system. After the load is removed the air must be released to once again level the vehicle.

Operator-Controlled Air Leveling Systems. Various shock absorber manufacturers produce level control systems that can be fitted to vehicles that do not have level controls. The shock absorbers are the same as those used on the manual adjust systems. These retrofit systems will allow the operator to control leveling by using a switch located within easy reach of the driver.

The operator-controlled air leveling system consists of a pair of air-assist shock absorbers, an air compressor, a control switch, and connecting lines. By positioning the control switch, air can be admitted to or exhausted from the shock absorbers. The compressor is connected to the shock absorbers through small-diameter high-pressure nylon or plastic tubing. When the vehicle is to be leveled to compensate for loading, the control switch is moved to the raise position. This starts a small compressor to increase the pressure in the shock absorber system. The vehicle is lowered by venting some of the air to the atmosphere to reduce the pressure in the system.

Air Leveling Compressors. The air compressors used on air leveling suspensions all perform the same function, although their basic designs differ.

Air leveling compressor types include engine vacuum-driven diaphragm compressors, electric motor-driven diaphragm compressors, and electric motor-driven piston compressors.

The working force in the engine vacuum-driven compressor, as shown in Figure 13-45, utilizes engine vacuum on one side of a diaphragm and atmospheric pressure on the opposite side. Diaphragm movement from side to side moves a small two-headed piston. The piston pumps up high air pressure to control the shock absorber assist. As in the power brakes shown in Figure 12-2, the difference in vacuum and atmospheric pressure on opposite sides of the diaphragm moves the diaphragm. The diaphragm moves the piston in one direction, compressing the air ahead of the piston. When the piston reaches the end of its travel, a valve mechanism reverses the vacuum and atmospheric pressures to opposite sides of the diaphragm. This drives the piston in the opposite direction to build up pressure ahead of the opposite piston. This valve mechanism reversal keeps the piston in motion, compressing air alternately on each end of the two-headed piston. This pumping fills the reserve tank with air pressure. The vacuum-operated compressor will only function when the difference between vacuum and atmospheric pressure is great enough to overcome the air pressure on the piston head. This limits the maximum system air pressure and maximum assist.

An electric motor is used on some compressors to drive the diaphragm. The diaphragm operates similarly to the vacuum-driven compressor. Either a spring between the diaphragm and it actuator, or a relief valve is used to regulate maximum compressor pressure.

The electric motor-driven piston compressor is similar to air compressors found in a service facility for shop air, only they are smaller. An electric motor drives the compressor crankshaft. This, in turn, moves a piston which draws in outside air and discharges compressed air to the system. A relief valve on the system limits maximum pressure while a timer keeps the compressor from running continuously. The timer will allow the compressor to run long enough to adjust trim height. If trim

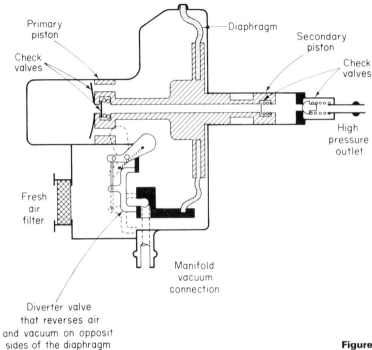

Figure 13-45 A section view of a vacuum-driven diaphragm compressor.

height cannot be attained in the time set, the compressor will automatically turn off. This prevents continual operation of the compressor in case of a system malfunction. Resetting the timer is accomplished when the ignition is turned off and then turned on again.

A reservoir is used on some systems to provide relatively rapid adjustment when the system is equipped with a small compressor. The reserve air will allow the system to adjust to level and then the compressor will slowly replenish the pressure in the reserve tank. The vacuum-diaphragm compressor uses a reservoir to provide some adjustment capacity when the engine is not operating. Some piston type compressors do not use a reserve tank because the compressor has an adequate displacement capacity.

Automatic Leveling Control Systems. Automatic leveling control is available from the manufacturer on most full-size domestic passenger vehicles. The fully automatic height-control leveling systems do not require operator control. The system consists of two air-adjustable shock absorbers, an air compressor, a reservoir or reserve tank, and a height-control sensor assembly.

The height-control sensor or leveling valve assembly is the component that automatically adjusts the pressure in the air shock absorbers. It senses the vehicle body position in relation to its trim height position and determines whether air must be added to or released from the shock absorbers. The height-control sensor is connected between the vehicle body and a suspension member to enable it to sense changes in ride height. When a change of the load level is noted by the ride height sensor, air is forced into the shock absorber until the vehicle has returned to the trim height. When the load is removed the vehicle body rises. This causes the suspension to extend beyond its normal ride height so air is exhausted from the shock absorber to once again set the trim height.

The sensors for the electric motor-driven compressor are an electrical switch and a relay that starts the compressor to raise the vehicle. The sensor will open a solenoid, allowing air to escape; this, in turn, will lower the vehicle to trim height. Another type of sensor is completely mechanical in operation. In this system a valve is connected to the suspension through various linkage components. The movement of the linkage and valve directs air from the reserve tank to the shock absorbers to increase height. It exhausts air from the shock absorbers to the atmosphere if the vehicle is to be lowered. This type of system is shown in Figure 13-46.

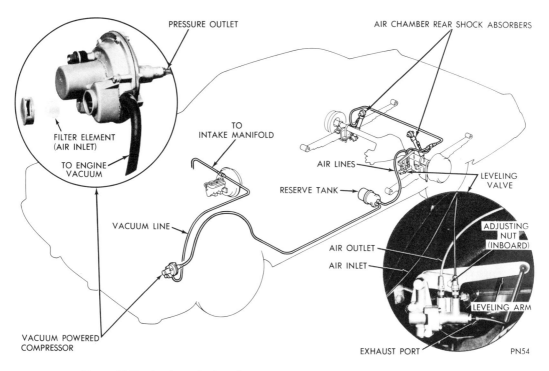

Figure 13-46 A schematic view of a typical automatic leveling control system. This system uses a vacuum-powered air compressor (Courtesy of Chrysler Corporation).

Each type of sensor is equipped with a delay mechanism. The delay time is usually 10 to 15 seconds. This delay prevents the system from continually trying to adjust as the vehicle encounters bumps in the road. If the position of the vehicle body remains in the new position longer than the delay time, the sensor detects this as an actual change in loading, not a momentary change. The automatic adjustment system will then begin to make level adjustments. As adjustments occur the switch or valve will gradually move toward the neutral position. When it reaches neutral it will stop further adjustment at the trim height.

The only periodic maintenance required on the adjustment systems is air filter cleaning. Air leakage can occur but it is easily detected. Occasionally problems result from improper positioning of the ride height sensor. This will have to be adjusted. Failure of the shock absorbers can also occur. All air shock absorbers require a minimum air pressure of 15 psi (103 kPa) to provide a normal service life. If this is not maintained, early shock absorber failure will result. Failure to maintain minimum pressure could be due to tubing failure, height sensors improperly adjusted, a compressor failure, or a failure in the valving mechanism.

REVIEW QUESTIONS

1. What is the main priority in suspension design? [INTRODUCTION]
2. How does an independent front suspension differ from a dependent front suspension? [13-1]
3. On a typical vehicle, list the parts that are considered unsprung weight. [13-2].
4. What is spring rate? [13-3]
5. In what way are the three basic types of springs the same? [13-4]
6. What is the advantage of each type of spring? [13-4]
7. What type of springs can have a variable spring rate? [13-4]
8. What type of springs require suspension control devices? [13-5]

9. What force is applied to the suspension during acceleration? [13-5]
10. How does the suspension control braking torque? [13-5]
11. What different coil spring mounting arrangements are used on the front suspension? [13-5]
12. What different torsion bar mounting arrangements are used on the front suspension? [13-5]
13. Describe the operation of a stabilizer bar. [13-5]
14. How do suspension bushings allow the suspension to move? [13-5]
15. What is the function of the shock absorbers? [13-6]
16. Describe the operation of a typical shock absorber. [13-6]
17. Why do shock absorbers become warm as they are used? [13-6]
18. How is aeration of shock absorbers prevented? [13-6]
19. How is the shock absorber control of jounce and rebound specified? [13-6]
20. How does the action of a kingpin and a ball joint differ when used on the front suspension? [13-7]
21. How does the end play of a load-carrying and a nonload-carrying ball joint differ? [13-7]
22. What is the purpose of air-assist shock absorbers? [13-8]
23. How do air shock absorbers differ from standard shock absorbers? [13-8]
24. What happens to the shock absorber action if the air-assist part fails? [13-8]
25. How is the air pressure adjusted in manual-adjust level control systems? [13-8]
26. How do operator-controlled and automatic-controlled air leveling systems differ? [13-8]
27. Describe the operation of the vacuum-operated air compressor. [13-8]
28. How is the maximum air-lift assist controlled? [13-8]
29. Why does the level control system require a delay mechanism? [13-8]
30. What periodic maintenance is required on leveling systems? [13-8]

14
Suspension Control

The suspension system, like all other parts of the vehicle, is designed to meet specific requirements. These requirements consider the number of passengers to be carried, the useful load, the type of ride desired, and the required handling characteristics. As is generally true, when one suspension property is improved, the quality of the other properties is reduced.

The first priority of suspension design is to carry the vehicle weight. It must hold the wheels and their supports in the proper position with suitable restraints during acceleration, braking, and turning. At the same time it must flex sufficiently to absorb road bumps and dips so that the passengers and load, and even the vehicle itself, will not be subjected to harsh shaking. Shock absorbers must quickly stop the vehicle oscillations after jounce and rebound.

Controlling the vehicle in the desired track down the road is called *handling*. This assumes that the tires are in constant contact with the road surface. Handling becomes more critical as the vehicle speed increases or when the turns become more abrupt. The location and design of the suspension linkages, the pivot points, and the steering linkage greatly affect the way the vehicle handles. The handling characteristics of the vehicle are modified by the type of tires being used.

The vehicle suspension, combined with the steering linkage and tires, is designed to provide the

Suspension Control

driver with safe positive vehicle control that is free of irritating vibrations. The design should produce minimum wear on the tires and on other parts of the suspension system.

14-1 SUSPENSION CHARACTERISTICS

The objective of suspension geometry is to provide the best possible vehicle handling control while assuring maximum tire life and a comfortable ride. *Suspension geometry* is the arc followed by the suspension and the wheel movement as the suspension travels through jounce and rebound. An arc is always produced as the suspension member is rotated about a fixed axis. The sideways displacement of the wheel (the distance the wheel moves from the maximum outboard to the maximum inboard position as the suspension moves through the arc) is reduced as the length of the suspension arm is increased. This can be seen in Figure 14-1.

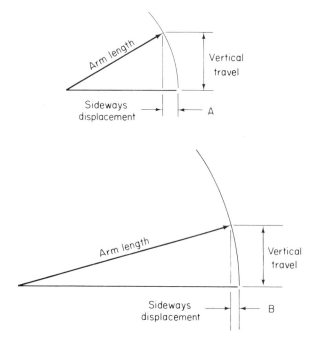

Figure 14-1 With equal vertical travel, a short arm length produces a greater sideways displacement (marked A) than the sideways displacement of a long arm length (marked B).

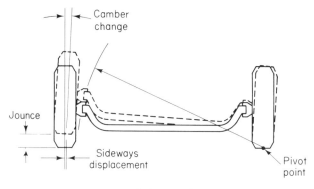

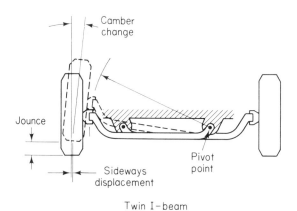

Figure 14-2 The different camber change that occurs on an I-beam axle and a twin I-beam axle.

I-beam Axle. A single I-beam axle will produce an arc when one wheel is moved through jounce and rebound with a small sideways wheel displacement. The small displacement is produced because the pivot point is around the wheel on the opposite side of the vehicle. The radius for the arc produced is equal to the tread width. The twin I-beam suspension produces an arc similar to the single I-beam axle. It has a slightly greater sideways displacement because the axle pivot point is inboard of the opposite wheel.

It can be seen in Figure 14-2 that as the wheel moves into jounce, the top of the wheel tilts inward and the bottom of the wheel tilts outward. This change in angle between the tire and the road surface is called *camber*. Camber angle is the angle formed between a line drawn through the tire centerline and a vertical line as viewed from the front of the vehicle. Camber is shown in Figure 15-1.

Long-Short Arm Suspension. The independent suspension using two control arms, or A-frames, is called a *long-short arm suspension*. It is so called

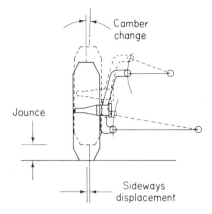

Figure 14-3 The effect of the same amount of jounce as shown in Figure 14-2 on the camber and sideways displacement of a long-short arm suspension.

because the lower control arm is always longer than the upper control arm. Due to the relatively short length of the lower control arm, approximately 24 inches (61 cm) compared to a single I-beam axle which may be in excess of 5 feet (152 cm), it can be seen in Figure 14-3 that the arc and the sideways displacement of the wheel is greater for the same jounce distance. When this displacement is transferred to the tire on the road surface, a greater amount of side travel may take place resulting in more tire wear.

Scuff Travel. The amount of side travel of the tire on the road surface is referred to as *scuff travel*. To reduce the amount of scuff travel in the long-short arm suspension, the angle of the arms and their lengths are critical factors. The upper control arm is shorter than the lower control arm. Therefore, its displacement is greater as it travels through the same jounce as the lower control arm. Through the proper placement of the arms and the selection of the length this produces the desired amount of scuff travel. This is illustrated in Figure 14-4. A small amount of scuff travel is useful because it produces some damping of the vehicle suspension action when the wheel encounters road surface irregularities. If no scuff travel were present, all of the damping would take place within the shock absorber, except for friction within the suspension components.

Effect of Camber Angle. The camber angle is important as the suspension travels through jounce and rebound. The desired operating characteristics are achieved through a compromise of scuff travel and camber change. Negative cam-

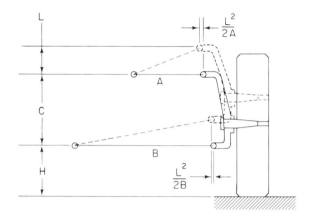

$\frac{L^2}{2A}$ — Sideways displacement of upper ball joint

$\frac{L^2}{2B}$ — Sideways displacement of lower ball joint

No scuff travel is present if the suspension is moved an amount L when $HB = (H + C)A$

Figure 14-4 Relative control arm lengths and pivot positions that produce minimum scuff travel.

ber during jounce and positive camber on rebound may be the design aim. As a vehicle enters a turn, centrifugal force and inertia will force the vehicle's center of gravity toward the outside of the turn. This will move the outside suspension into jounce while the suspension on the inside of the turn is moved into rebound. The accompanying twist or roll of the vehicle body moves the upper control arm pivot out further than the lower arm pivot. This action will produce a tendency for positive camber on the outside wheel which, in turn, will reduce the tire to road contact area. If the outside suspension is designed to produce negative camber, the tire will maintain maximum road contact at all times to maintain good handling as shown in Figure 14-5. The negative camber designed into the suspension tends to add support to the tire for the increased load in a turn when the suspension is in jounce. This same effect will occur on a crowned road that tends to place the outside suspension in jounce and the inside suspension in rebound. Camber will increase on the inside wheel to correct for road crown as shown in Figure 14-6. This suspension geometry produces excellent handling characteristics, but it may also reduce tire life. As the vehicle is loaded, the suspension goes

Suspension Control

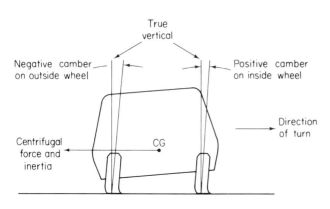

Figure 14-5 Long-short control arm suspension design to change camber as the body rolls to help the tires maintain maximum road contact.

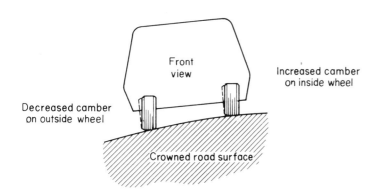

Figure 14-6 The effect of the long-short arm suspension design on a crowned road surface.

into jounce. This produces negative camber that causes tire wear.

Ride Quality. The suspension designer may choose to sacrifice some handling performance for improved ride quality. It is possible to produce a suspension with these two goals by designing a suspension having positive camber on jounce and negative camber on rebound. A vehicle with this design encountering a bump follows the road surface irregularity with little roughness felt by its passengers and small sideways deflection of the wheels. This can be observed in Figure 14-7. Other suspension geometries are also used. These depend upon the suspension designer's objectives. The MacPherson strut suspension shown in Figure 14-8 attaches the wheel spindle to the verticle spring-

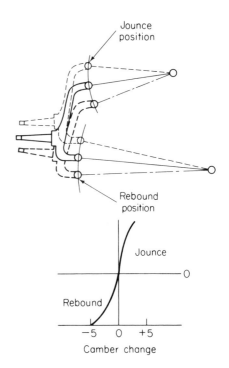

Figure 14-7 A long-short arm suspension that has been designed to produce positive camber on jounce and negative camber on rebound.

shock strut. Geometry in this type of suspension is predicted by the lower control arm pivot position, the length of the strut and arm, and the angle between the spring-shock strut and the lower control arm. A MacPherson strut suspension is usually designed to produce negative camber in jounce and positive camber during rebound.

Caster Angle. As the vehicle suspension travels through jounce and rebound, the upper pivot point of the steering knuckle (upper ball joint) on most suspensions moves rearward in relation to the lower pivot point (lower ball joint). The *caster angle* is an angle between an imaginary line drawn through the steering pivots and a line perpendicular to the road surface, as viewed from the side of the vehicle. If the top of the line tilts rearward, as shown in Figure 14-9, the vehicle is said to have *positive* caster. If the top of the line through the wheel pivots tilts forward, the vehicle is said to have *negative* caster.

The caster change during jounce-rebound on vehicles using I-beam axles and semielliptical springs can be predicted by the fore and aft position of the axle on the leaf springs, the amount of arc of the springs at ride height, and the horizontal relationship of the front-to-rear spring eye height. Vehicles employing I-beam axles and leading radius arms

Suspension Characteristics

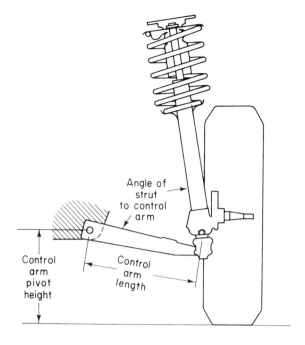

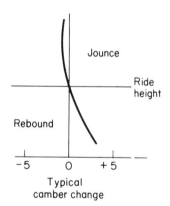

Figure 14-8 The effect of the MacPherson strut suspension on camber during jounce and rebound.

have increased caster in jounce and reduced caster during rebound. This is true because the arc in which the radius arms travel produces a rearward tilt of the top of the axle in jounce.

During braking the forward shift of the center of gravity tends to lower the front and raise the rear of the vehicle. The displacement removes some weight from the rear wheels and increases the effective weight on the front wheels. This forward weight transfer places the front suspension in jounce and the rear suspension in rebound. Because the suspension is attached to the vehicle frame, the change in the angle of the frame to the road surface temporarily reduces the amount of caster during braking.

The suspension is usually designed so that the effect of the vehicle frame angle to the road surface is offset by the tendency of the suspension to increase caster during jounce. The caster angle then remains unchanged during braking and this maintains the vehicle stability. The radius arm on twin I-beam suspension is lower in the front where it is attached to the axle and higher at the rear frame attachment point. The angle that the arm produces as the axle swings upward during jounce increases caster as well as reduces *dive* during braking. This can be seen in Figure 14-10.

On vehicles with long-short arm suspension, the upper control arm pivot shaft is angled upward at the front. This angle of the upper control arm tends to move the upper ball joint rearward in relation to the lower ball joint as the suspension goes into jounce. This movement results in an increase in caster during jounce and a decrease in caster during rebound. The angling upwards of the pivot axis produces an upward reaction on the axis to help reduce brake dive. This action is illustrated in Figure 14-11. The caster change on MacPherson strut suspension is largely dictated by the angle that the strut makes with the vertical, as viewed from

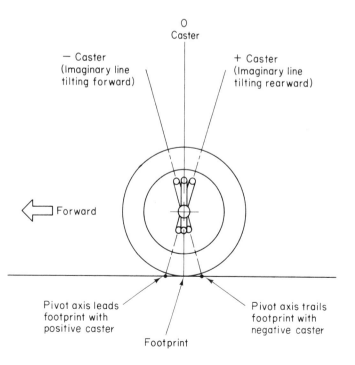

Figure 14-9 Caster angle.

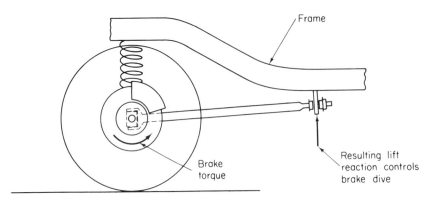

Figure 14-10 Control of brake dive on a twin I-beam suspension.

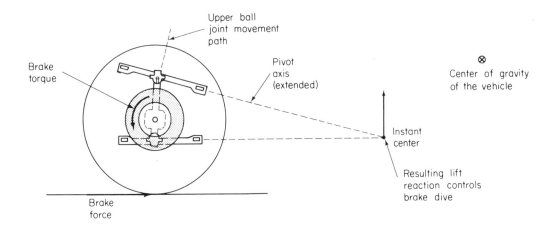

Figure 14-11 Control of brake dive on a suspension with control arms.

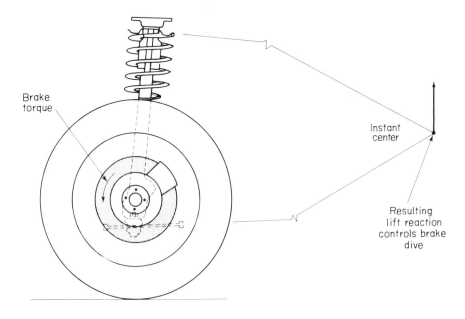

Figure 14-12 Control of brake dive on a suspension with a MacPherson strut.

the side of the vehicle, and the arc through which the stabilizer bar swings through while traveling through jounce and rebound. The caster change is illustrated in Figure 14-12.

14-2 STEERING LINKAGE CHARACTERISTICS

The camber change during jounce and rebound is caused by the different arcs through which the suspension arms travel. The camber change would turn the wheel either in or out if the steering linkage for the wheels was connected at a random location. The inner pivot for the steering linkage is usually placed directly in line with the inner lower control arm pivot. If the outer steering linkage connection were in the same horizontal plane as the ball joint and if it were the same length as the lower control arm, the arc followed by the steering linkage would be nearly identical to the arc followed by the lower control arm. When the arc followed by the steering linkage and control arm are nearly identical, no toe change will result when the wheels are in the straight ahead position. This can be seen in Figure 14-13. *Toe* is the difference in distance between the front of the two tires and the rear of the tires at the spindle height. *Toe-in* occurs when the tires are closer together in front than they are in the rear. If the two tires are further apart in front than they are in the rear, they have *toe-out*. This is shown in Figure 14-14.

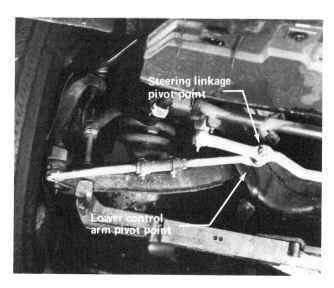

Figure 14-13 The steering linkage pivot point is in line with the lower control arm pivot point.

A typical steering linkage can be seen in Figure 14-15. The outer link of the steering linkage is called a *tie rod*. A ball and socket, similar to the ball joint, is attached to each end of the tie rod. As the vehicle enters a turn, the steering linkage is moved to provide the force required to turn the front wheels. A vehicle entering a turn causes the suspension to react. The outside suspension will go into jounce and the suspension on the inside of the turn will go into rebound. A suspension that is designed with toe-out in jounce and toe-in during rebound will generate a particular steering effect while making a turn. Such a design reduces the steering angle as the vehicle develops a large

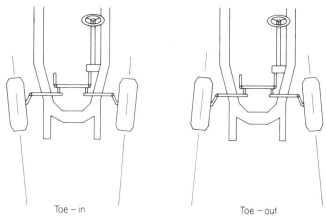

Figure 14-14 Steering toe.

Figure 14-15 A typical linkage ahead of the control arms.

Suspension Control

amount of body roll during a turn. This automatically provides the vehicle with a greater steering input effort from the operator with less steering angle as the severity of the turn and body roll increases. The toe change that occurs during a turn due to body roll is referred to as *roll steer*. Roll steer on domestic automobiles reduces the steering angle of the wheels. As a vehicle goes through rises and dips in the road surface, each suspension may go into jounce and rebound at the same time. This will turn the wheels in and out, producing toe changes. These changes are referred to as *wheel fight*. Wheel fight produces tire scuffing, which results in tire wear and increased rolling resistance of the tire. In racing circles, terms such as *bump steer* are mentioned. Bump steer is the steering effect that results as the wheel moves through its suspension travel. On racing vehicles, bump steer is reduced to a minimum, so that body roll does not change the steering angle.

14-3 RIDE HEIGHT AND HANDLING

It is now possible to see the problems caused by changing the ride height of a vehicle to something other than manufacturer's specified height. The geometry of the suspension provides completely different handling characteristics when ride height is changed. The most drastic effect is created when the vehicle is raised above ride height. The caster, camber, and toe curves begin to change rapidly when the chassis is above ride height as shown in Figure 14-16. Ride height is established with neither passengers (nor a driver aboard), so whenever the vehicle is driven it is slightly below the specified ride height. In a severe turn the wheel

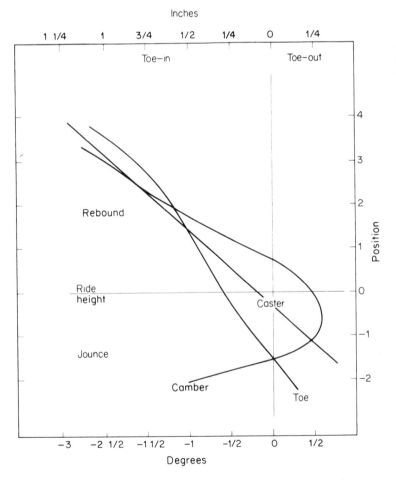

Figure 14-16 A graph of the change in suspension geometry of a typical automobile as it goes through jounce and rebound.

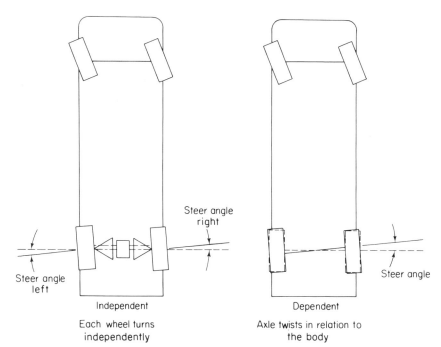

Figure 14-17 The angular steer effect of the rear suspension.

on the outside of the turn is carrying a greater load than the wheel on the inside of the turn. The increased load makes the outside wheel support more of the turning force. At the same time the inside wheel will provide less turning force. It is, therefore, more important to have the suspension geometry of the wheel on the outside of the turn more correct than that of the wheel on the inside of the turn.

While cornering, the rear axle centerline of a vehicle with a dependent rear suspension may twist in relation to the body centerline. When a vehicle with independent rear suspension is in a turn the centerline of each wheel turns independently, causing an angular change between the wheel centerline drawn through the tread of the tire and the body centerline. This can be seen in Figure 14-17. On front suspensions it is possible to design the suspension and steering system so the tires will turn either into or out of a turn as required. In this way it is possible to increase or decrease the effective steering input from the driver. The changes that take place in front and rear suspension systems are due to the geometry of the linkages and the amount of body roll. The tires introduce another input force into the steering of the vehicle. Tires have a *slip angle* which is the angular difference between the center line of the wheel and the line the vehicle is following tangent to the turn. Slip angle is illustrated in Figure 14-18. Slip angle at a given speed is dependent upon tire construction. All tires have some slip angle while making a turn. The greatest slip angle occurs at the maximum turning force. This happens just prior to the tire sliding sideways. It is governed by the rubber compound of the tire tread, tire construction, and the road surface, as discussed in Chapter 3. The rubber composition of the tire itself provides varying degrees of road adhesion (coefficient of friction) and varying degrees of maximum cornering power.

14-4 VEHICLE STEER

If the front suspension, rear suspension, and tire slip angles combine to increase the steering input required by the driver to make a turn as speed is increased, the vehicle is said to have *understeer*.

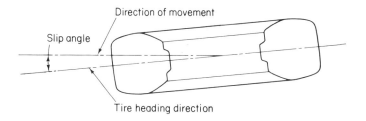

Figure 14-18 Slip angle.

Suspension Control

Understeer is the tendency for a vehicle to continue to increase the diameter of the turning circle as vehicle speed increases while holding a given steering wheel position.

If the front suspension, rear suspension, and tire slip angles combine to produce a vehicle that responds directly to the steering input effort by the driver, the vehicle has *neutral steer*. Neutral steer will produce a turning circle that remains constant as the vehicle speed increases, until the speed is reached at which the tires completely lose traction and the tires slide.

If the suspension geometry and tire slip angles cause a vehicle to continue to turn sharper as the vehicle speed increases while holding the steering wheel constant, the vehicle is said to have *oversteer*.

The understeer, neutral steer, or oversteer tendency of a vehicle making a turn is independent of the driver. It depends on suspension geometry, the amount of body roll in a turn, and the slip angle of the tires. Incorrect suspension ride height affects suspension geometry. Worn or incorrect shock absorbers and springs affect the amount of body roll in a turn. Incorrectly or underinflated tires, affect the tire slip angle. All of the above, singly or combined, will drastically affect the steering tendency of a vehicle making a turn.

Standard domestic passenger cars are designed with understeer. Understeer provides the average driver with a large safety margin when entering a turn because the vehicle has a slow response. Understeer prevents the unskilled driver from over-reacting in an emergency situation. Late-model, more precise handling vehicles are designed with less understeer tendency.

Tire slip angles increase with speed and any change in slip angle will change the handling characteristics of the vehicle. The slip angles of the tires will not affect the *steer angle* when the front and rear tires are matched and have equal loads.

Steer Angle. Steer angle is the vehicle heading angle plus the rear slip angle less the front slip angle. It can be seen in Figure 14-19 that when the front and rear slip angles are equal, the steer angle equals the heading angle. The problem becomes more complex when loads on the tires are not the same. The average front engine passenger vehicle has *front weight bias*. A greater percentage of the vehicle weight is on the front wheels than on the rear wheels. The weight bias may produce a greater slip angle on the front tires than on the rear tires. This will create a steering angle less than the heading angle which is understeer. The situation may be reversed when the vehicle is loaded with luggage and passengers. Loading may produce a rear tire slip angle that is greater than the front tire slip angle. The steer angle is now greater than the heading angle and the vehicle will oversteer. Oversteer is considered to be dangerous and the vehicle may be uncontrollable during evasive maneuvers.

It is now evident that there is a problem caused by mixing tires having different body construction (radial, bias belted, bias ply) or even mixing tires of a similar body construction but from different manufacturers. This is discussed in Section 3-3. Each construction type and tread compound has a

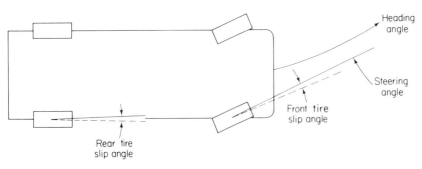

Figure 14-19 When the front and rear slip angles are equal, the steering angle equals the heading angle.

different slip angle. If radial tires with low slip angle were placed on the front of a vehicle and bias ply tires with large slip angle were placed on the rear, the resulting steer angle would be much greater than the heading angle and this would produce oversteer. If bias ply tires were placed on the front and radial tires on the rear, the heading angle would be much less than the steer angle and this would cause understeer. Tires of similar construction from different manufacturers may produce a slight variation in steer characteristics when placed on the same vehicle. Tire effects on steer are shown in Figure 14-20.

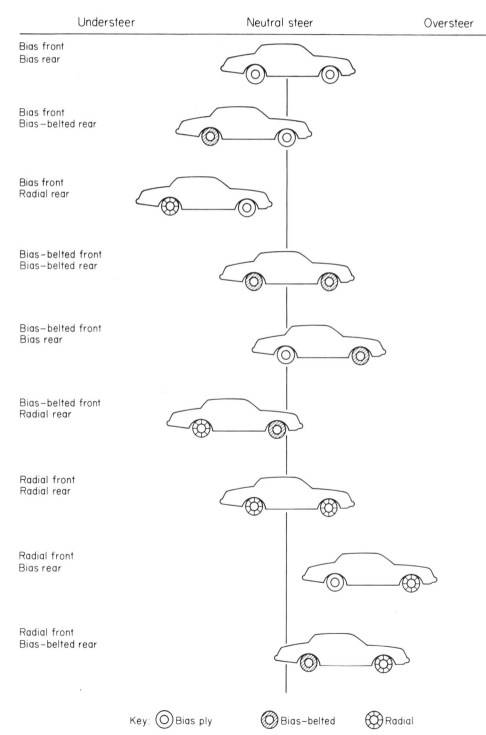

Figure 14-20 Steering effects caused by tire construction type combinations.

Suspension Control

Suspension Design Versus Tire Types. Vehicle suspension is designed by the manufacturer for use with specific tire types. The vehicle will produce the handling characteristics the manufacturer has so carefully engineered into the vehicle only when the specified tires are used. The manufacturer may design a vehicle suspension with a relatively large degree of understeer and then equip the vehicle with bias belted tires. This will produce a safe vehicle with the desired handling characteristics. If the owner exchanges the original tires for radial tires the vehicle will approach a neutral steer condition. Some drivers consider neutral steer as essential to "better handling." If a vehicle is manufactured for use with radial tires, the amount of steer built into the suspension is different than it would be for bias belted tires. If the radial tires are exchanged for bias belted tires, a great amount of understeer may be present and this will result in a "poor handling" vehicle. Some vehicles have the suspension designed to produce neutral steer with bias belted tires. These vehicles may oversteer if radial tires are installed. This could produce a vehicle with poor or possibly dangerous handling characteristics.

14-5 RIDE QUALITY

Ride quality is an important factor in suspension design even though it is the result of subjective evaluations. It was noted in the discussion of springs that a leaf spring has interleaf friction. This friction will not remain constant throughout the life of the vehicle as a result of varied environmental conditions and wear. Because of this, the characteristics of a leaf spring will continuously change throughout its service life and this will change the ride quality. The internal friction of a leaf spring was a desirable quality prior to the advent of the modern shock absorber. The internal friction of the leaf spring supplied the system with the damping required to provide a satisfactory ride. Coil and torsion bar springs have since become popular. They have a good ride quality and they require little space compared to leaf springs. Coil and torsion bar spring suspension systems have little static friction, so they readily flex to absorb smaller road surface irregularities. Modern shock absorbers can handle all of the suspension damping required by these springs. The mounting position of the shock absorbers is critical to reduce side sway, and thus provide better handling and ride quality on all road surfaces.

Road Clearance. Improved highways have allowed the required road clearance to be reduced. This reduces the ride height giving the vehicle a lower center of gravity. A low center of gravity will produce less body roll during a turn, for better passenger comfort. It gives the vehicle a lower overall height and it improves handling characteristics. The independent front suspension was an important step toward reducing the height of the center of gravity. The clearance needed for solid axle movement beneath the frame was eliminated and thus the frame could be lowered.

The reduction of the body-to-road clearance required an increased tunnel height on the floor inside the vehicle to provide clearance for the drive shaft under the body. This is the most undesirable result of a lower vehicle having a front engine location and rear drive wheels. The reduction in the size of the wheels to further lower ride height is limited by the size of the brakes that are required to stop the vehicle. As the wheel size is decreased, the brake diameter of the wheel mounted brakes must be reduced proportionately. This results in lower brake efficiency. The automotive engineer must compromise the ride height while considering passenger comfort, vehicle handling, and braking requirements. Rear-engine rear-wheel drive, and front-engine front-wheel drive vehicles do not have all of the compromises for handling and comfort that are present in front-engine rear-wheel drive vehicles. The tunnel required for drive shafts in front-engine rear-wheel drive vehicles is unnecessary. This results in more room for the passengers. This is important in compact vehicles. It has lead to the increased use of front-wheel drives.

Engine/Drive Characteristics. Cost, durability, and wear must be considered by all design engineers. A front-engine rear-wheel drive vehicle is an economical design where passenger space permits.

A front-engine front-wheel drive vehicle provides the most passenger space in the subcompact vehicle. The cost of the drive train is expensive, however, and it is only used where it is required.

Four universal joints, usually of the constant velocity type, are required on front-wheel drives because they must both drive and steer. The strength of the suspension components has to be increased to provide strength for drive torque and durability. Care must be used to assure that the forces from the driving wheels are not transferred through the steering linkage to the driver. As the complexity of the suspension system increases in front-wheel drive, the initial cost is increased as well as the wear rate of the suspension components.

A midship-mounted engine, forward of the rear axle, has ideal handling characteristics but it reduces the interior space for the passengers and load. It also increases the interior noise level of the vehicle. A rear-mounted engine behind the rear axle is less expensive than a front-wheel drive but it usually has poorer handling characteristics than midship engine vehicles. Engine cooling and accessibility of components are additional drawbacks of both rear and midship-engine vehicles. Figure 1-2 in the first chapter shows the different drive train combinations discussed here.

Space Availability. Space considerations, especially in small vehicles, will dictate the type of suspension used as well as the type of drive train. Front suspension of the MacPherson strut type usually requires less space than a long-short arm suspension. The space required by the rear suspension can be reduced by using a short torque tube or a torque reaction arm as shown in Figure 13-22 and 13-23. on page 192. These are used to eliminate the longer upper control arms usually required for absorbing brake and acceleration torque on vehicles with a rear-wheel drive.

Tire Considerations. Vehicles designed to use radial tires require special consideration by the suspension designer. Radial ply tires have a characteristic low-speed harshness that is not present in bias belted or bias ply tires. Radial ply tires often tend to generate a steering wobble which is an oscillation of the steering linkage. This is due to the higher cornering forces that are developed at the lower tire slip angles. Radial ply tires have a lower radial spring rate which changes the natural frequency of the unsprung weight. This is illustrated in Figure 14-21. This shift of frequency of the radial tires is the major cause of their low-speed harshness. Vehicles designed for radial tires have their suspension components designed to dampen vibration at this frequency range. The frequency of the vibrations will usually change when unsprung weight is added to the system. This may change the vibration frequency to a range that is not as noticeable. An objectionable vibration or a vibration resulting in noise may either be *isolated* at its source or the frequency of the vibration may be *shifted* to a frequency range where it is no longer noticeable. Suspension damping components include larger spring insulators, large suspension bushings with voids in them, isolation of the suspension components on a cross-member, and rubber mounting between the cross-member and the vehicle.

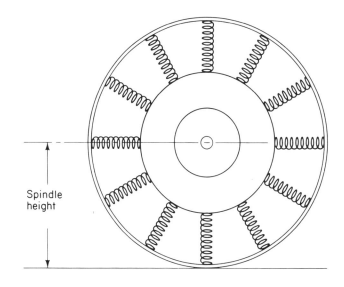

Figure 14-21 Tire flexing compared to springs to describe the tire spring rate.

REVIEW QUESTIONS

1. What is the suspension system's first design requirement? [INTRODUCTION]
2. How does camber change when one wheel on a I-beam axle goes into jounce? How does it change when going into rebound? [14-1]

3. What causes scuff travel? What are the results of scuff? [14-1]
4. Why is it an advantage to design a suspension that will cause the wheel to have negative camber in jounce? [14-1]
5. What happens to caster during braking? [14-1]
6. Why do the steering linkage pivots nearly match the pivots of the lower front control arms? [14-2]
7. What happens to toe when the front suspension goes into jounce? [14-2]
8. How does roll steer affect the steering angle of the wheels? [14-2]
9. What causes wheel fight? [14-2]
10. How is suspension geometry affected by ride height? [14-3]
11. During a turn, why is it more important to have the suspension geometry correct on the outside wheel than on the inside wheel? [14-3]
12. What causes tire slip angle? [14-3]
13. Define understeer, neutral steer, and oversteer. [14-4]
14. How does slip angle affect steer angle? [14-4]
15. What causes the front weight bias? [14-4]
16. How does tire construction affect steer angle? [14-4]
17. What determines vehicle ride quality? [14-5]
18. What are the advantages and disadvantages of front-wheel drives? [14-5]

15
Steering and Wheel Alignment

The primary purpose of the steering system is to give the driver directional control of the vehicle. This control condition can only occur if the tires maintain contact with the road and the steering efforts of the tires on each side of the vehicle compliment each other. The vehicle steering and wheel alignment angles work together to achieve the maximum possible directional control of the vehicle.

The vehicle is usually designed to have a neutral sense. A neutral sense is the tendency for the vehicle to go in a straight-ahead direction with no driver input effort and to return the wheels to a straight-ahead position after a turn is made. The steering system is designed to require minimum corrective input effort from the driver and to transmit as little road shock as possible to the steering wheel when encountering irregular road surfaces. It is most important that maximum safety be designed into each steering system component.

Wheel alignment, at the service level, is the reestablishment of all interrelated suspension angles to achieve maximum vehicle control while maintaining easy steering, good handling characteristics, free and true running wheels, and maximum tire mileage. Periodic wheel alignment is necessary because, in time, high mileage, poor driving habits, and rough roads cause changes in wheel alignment angles. These changes may be due to the wear of the suspension and steering components or to a relaxing of the vehicle frame structure and suspension components.

Steering and Wheel Alignment

Dynamic loads will cause slight deflections of the suspension and steering components as the vehicle is moving. Wheel alignment is checked in the service shop with alignment equipment while the vehicle is not moving. The alignment specifications given for the tire and wheel assembly while the vehicle is stationary are not the same as the angles that occur as a result of dynamic loading while the vehicle is in operation. The static specifications compensate for dynamic loads that occur as the vehicle moves. This results in the desired operating angles for a free-rolling, stable vehicle.

15-1 STEERING SYSTEM GEOMETRY

This section contains a description of steering systems and all interrelated suspension angles. It is essential that the automotive service technician know and understand the alignment angles (*camber, caster, toe-in, steering axis inclination, toe-out on turns*) and be able to relate these angles to the suspension components and handling problems that result from incorrect steering and alignment settings.

Camber. Camber is the inclination of the centerline of the wheel from the vertical when the wheel is viewed from the front of the vehicle. This is illustrated in Figure 15-1. Camber is said to be positive when the top of the centerline of the wheel inclines outward away from the vehicle. When the top of the centerline of the wheel inclines inward toward the vehicle, the camber is called negative. The camber angle is always measured in positive or negative degrees.

A wheel with camber will provide a vehicle with a directional sense. The vehicle will go toward the side which has the wheel with the most positive camber. It is the normal tendency of the wheel to roll around the center of a circle when the top of the wheel is inclined toward the center of that circle as shown in Figure 15-2. Wheels with positive

(a)

(b)

Figure 15-1 Camber illustrated: (a) positive; (b) negative.

Steering System Geometry

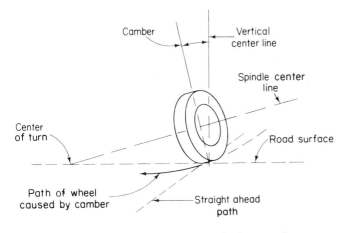

Figure 15-2 A wheel tends to turn in the same direction toward which the top of the wheel is leaning.

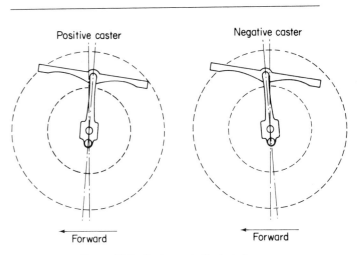

Figure 15-3 Caster angle illustrated.

camber tend to turn outward on the vehicle. If both wheels have equal camber, the turning tendency of the wheels oppose each other so the effective steering lead will be neutral or nondirectional. Radial tires tend to be less susceptible to camber settings.

Most vehicles utilize a static camber setting range from zero to slightly positive. The slight positive static setting will prevent camber from becoming negative when dynamic forces cause the suspension to deflect as the vehicle is moving or when it is carrying additional weight as discussed in Chapter 14. While under dynamic loading, either positive or negative camber will cause wear to occur on one side of the tire tread, thus shortening tire life. Camber settings are more critical for uniform tire tread wear when using wide tires. As the width of the tire increases, the importance of correct camber increases to provide equal weight distribution across the entire tire tread for even tire tread wear.

Caster. Caster is the angle formed between a line drawn through the steering axis and a vertical line through the axle spindle, as viewed from the side of the vehicle. The weight of the vehicle effectively acts through the vertical line. The steering and the forward vehicle movement act through the steering axis. The steering axis is an imaginary line drawn through the upper and lower ball joint centers or, in the case of MacPherson strut, the upper pivot and the lower ball joint. The steering axis and caster angle can be seen in Figure 15-3. Caster is positive if the top of the steering axis is inclined rearward and negative if the top of the axis is inclined forward. Caster is an angle measured in positive or negative degrees.

The effects of positive caster are seen by noting where the projected steering axis strikes the road surface within the tire contact patch, as shown in Figure 15-4. The vehicle weight is effectively concentrated in the center of the tire contact patch. If the steering axis strikes the road surface in

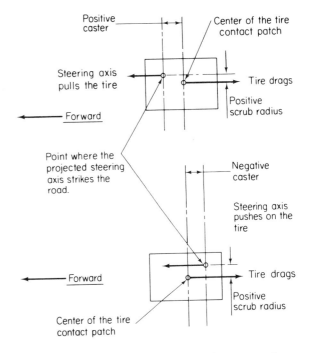

Figure 15-4 The directional effect of caster on the vehicle steering.

225

the front part of the tire contact patch, as it does in positive caster, it should be clear that the tire will be pulled by the steering axis. The tire drag on the vertical line at the center of the contact patch will always try to follow that axis contact point within the contact patch. This aids directional stability. At the same time, a greater steering effort is needed to turn the tire from a straight-ahead path than would be required if caster were zero. The forces directed within the tire contact patch area during a turn will try to return the wheels to the straight-ahead position. Weight transfer during braking will cause the front of the vehicle to lower and the rear of the vehicle rise. This change in the angle of the body-to-road surface will decrease the caster angle. Angling the upper control arm pivot points to prevent dive during braking, as discussed in Chapter 14, tends to control caster during braking. The effect of the upper control arm pivot angle is usually equal to, or greater than, the amount of change in body angle caused by braking. These counter forces tend to cause the caster to remain unchanged or to increase slightly during braking. Increased caster resulting from forward weight transfer will increase the directional stability of the vehicle while braking.

While making a turn, a suspension with normally positive caster will produce negative camber on the wheel that is on the outside of the turn and positive camber on the wheel that is on the inside of the turn. These camber changes provide increased tire support as the turn is made. During a turn, positive caster will cause the body to rise slightly on the inside and to lower slightly on the outside of the turn. This will produce body roll while making a turn. The amount of effective body roll from caster is slight because small caster angles are used.

The effects of negative caster can again be determined by noting where the projected steering axis strikes the road surface in the tire contact patch. With negative caster, the major portion of the tire contact patch is forward of the projected steering axis at the road surface. With the large portion of the contact patch forward of the projected steering axis, the normal forces developed on the tire will assist in turning the wheels in the direction being steered. This reduces the steering effort required from the driver. A vehicle with negative caster exhibits a self-correcting steering tendency during windy, gusty driving conditions. The wind acts on the vehicle body, forcing it sideways and this, in turn, forces the trailing steering axis sideways. The force will cause the wheel to steer into the direction of the wind. Negative caster counteracts the effect of the wind upset.

Negative caster will also affect the camber. Negative caster will force the camber to increase on the wheel that is on the outside of a turn. This results in reduced handling quality. At the same time the wheel on the outside of the turn is forced downward. The wheel is on the road so that corner of the body is forced to rise. This reaction reduces body roll. The body lifting tendency due to caster is usually negligible because of the small caster angles used on production vehicles.

It would appear that instability is unavoidable with negative caster because the tire contact patch is leading the steering axis. But instability is prevented by using a high steering axis inclination (abbreviated SAI). A large SAI tends to keep the wheels in a straight-ahead position. Therefore, it counteracts the instability that occurs from negative caster. The vehicle manufacturer carefully designs the SAI to correspond with the most ideal caster setting in order to produce the desired ride and handling characteristics in the vehicle.

Steering Axis Inclination (SAI). Steering axis inclination as illustrated in Figure 15-5 is the inward tilt of an imaginary line drawn through the upper and lower ball joints as compared to the vertical, when viewed from the front of the vehicle. Steering axis inclination is also referred to as the ball joint angle and as kingpin inclination when I-beam suspension is used. SAI is an angle and it is, therefore, always measured in degrees.

SAI is a stability angle. The vehicle body will be closest to the road surface when the wheels are pointed straight ahead as a result of SAI. A spindle with SAI will have the outer end of that spindle at the highest point when the wheels are pointed straight ahead. Therefore, as the weight of the vehicle pushes downward the spindle will always attempt to move up to return the wheels to the straight-ahead position. This effect is shown

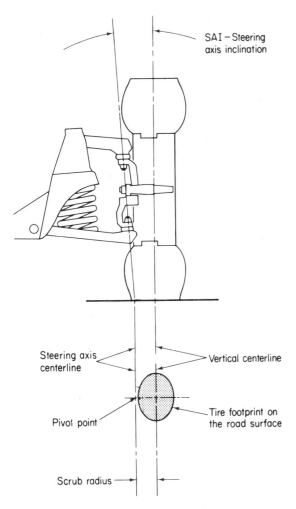

Figure 15-5 Steering axis inclination illustrated.

in Figure 15-6. After a turn, SAI aids in straightening the wheels. SAI also aids in directional stability by resisting road irregularities that attempt to turn the wheels away from the straight-ahead position. SAI produces many of the same benefits that improve steering stability as does positive caster. SAI therefore reduces the need for high positive caster. The effect of SAI on directional stability is usually greater than caster. Poor steering returnability and instability caused by negative caster is overcome, within limits, by using SAI to provide steering stability. One drawback with SAI is that it causes the wheel to have positive camber in a turn and this may reduce the maximum cornering power of the tire and wheel. Vehicles with power steering require a greater amount of steering wheel returning force than those with manual steering. SAI is often used with positive caster on vehicles equipped with power steering to provide the returnability needed. The power steering system supplies the additional power required to turn the wheels from the straight-ahead position for directional control. SAI with negative caster may be used with standard steering. This SAI angle is established between the wheel spindle centerline and the steering axle, as shown in Figure 15-7. This makes SAI a nonadjustable angle. SAI will only change if the spindle has been bent. A bent spindle is never straightened; it must be replaced.

SAI causes the steering axis to be inclined outward at the bottom. As a result the projected steering axis will strike the road surface at a point close to the center point of the tire footprint. The distance between these two points is called the *scrub radius.*

Scrub Radius. Scrub radius is the term used to describe the distance between the projected steering axis and the tread centerline at the road surface.

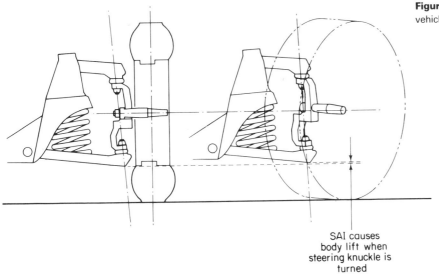

Figure 15-6 The directional effect of SAI on the vehicle steering.

Steering and Wheel Alignment

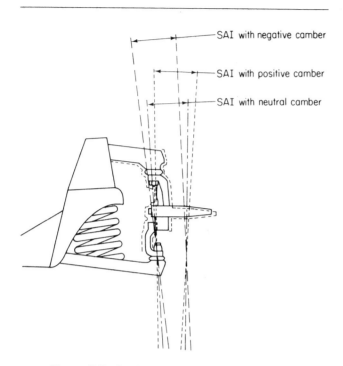

Figure 15-7 An illustration that shows the SAI remains constant as the camber changes.

Scrub radius is positive when the centerline of the tire lies outside the projected steering axis. This was shown in Figure 15-5. It is negative when the centerline of the tire is inside the projected steering axis. The scrub radius is a distance measurement and it is, therefore, measured in inches or millimeters.

The scrub radius is so named because the tire scrubs on the road surface as the wheel direction is changed. The wheel and tire will swing in an arc rather than pivot on a single point. The arc in which the tire swings is dependent upon the radius of the circle produced as the tire footprint center point swings around the projected steering axis as illustrated in Figure 15-8. This radius is called the scrub radius.

The size of the scrub radius depends on the steering axis inclination, the wheel offset, and the distance the spindle centerline is above the road surface. The vehicle manufacturer designs the required amount of scrub radius into the suspension by carefully choosing the correct SAI and the proper wheel offset for the designed spindle height.

Even though the spindle height has an effect on the scrub radius, little can be done to change this height because tire height is limited by the clearance space under the fender and body.

Aftermarket tires and wheels, those not supplied by the vehicle manufacturer, can drastically affect the scrub radius that the manufacturer has so carefully designed into the vehicle. Using wheels with increased offset in place of original equipment wheels can double the scrub radius. An example of this is shown in Figure 15-9. Tires with lower section height will also increase the scrub radius a small amount. Tires having camber wear will also position the effective centerline and footprint of the tires at a different place and this will affect the scrub radius.

Since all handling sensations pass between the tire and the road, the scrub radius provides the necessary feedback to give the driver road feel. All rolling resistance forces on the front wheels are working on the tire and attempting to force it rearward. This force is concentrated on the tire contact patch centerline. With the tire centerline outside the steering axis, the rearward force acts to turn both tires outward. A slight amount of scrub radius on each wheel is necessary to maintain constant tension or compression on the steering linkage to steady the linkage and prevent steering wobble or shimmy. As the amount of scrub radius increases, greater torque

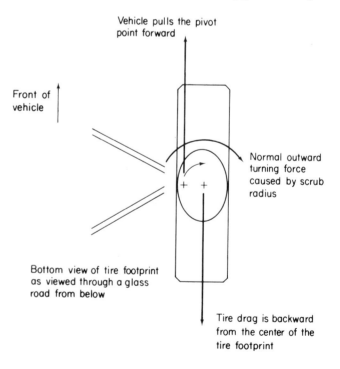

Figure 15-8 The directional effect of scrub radius on the vehicle steering.

Steering System Geometry

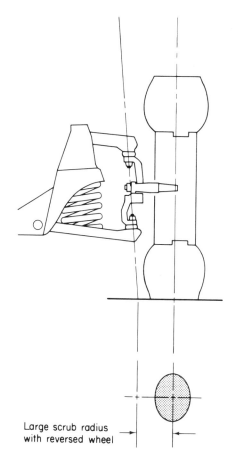

Figure 15-9 The effect of wheel offset on the scrub radius.

is applied to the tire to turn the wheel outward, and this will place higher loads on steering linkage and suspension components. The result will be increased wear and the probability of premature failure. The effect of unequal braking on each wheel is exaggerated with large scrub radius; this may make the vehicle dangerous to drive.

Some manufacturers are placing the tire centerline inside the projected steering axis (negative scrub radius). This is most effective when using dual diagonal braking (see Chapter 6) to minimize the wheel pull when braking if one-half of the brake system fails. It is also used on some front-wheel drive vehicles. Negative scrub radius is helpful when a tire fails at highway speeds. Tire forces normally cause the failed tire to immediately turn outward. Less steering pull is evident with less scrub radius.

One can see that the amount of scrub radius required is dependent upon many factors. Those factors include vehicle weight, desired steering wheel feedback, type of braking, the type of roads for which the vehicle is designed, and whether the vehicle is front- or rear-wheel drive. With front-wheel drive, traction effects are also fed back to the driver. This makes proper scrub radius even more critical.

Toe. Toe is the difference in distance between the extreme front and back of the tires as illustrated in Figure 15-10. Toe-in is the measurement in fractions of an inch or millimeters that the wheels are closer together in front than they are in back. Some manufacturers measure toe as the angular change from straight ahead in degrees. Slight toe-in is preferred over toe-out on most vehicles because steering is aligned while the vehicle is stationary. While under dynamic loads with the vehicle moving, linkage components flex causing a change in alignment angles. The static toe setting will differ from the desired running toe. For maximum tire life and least rolling resistance, the running toe must be zero. The usual tendency is for the tires to turn outward while under dynamic load so most vehicles are designed with a static toe-in setting. The toe will become zero as the linkage is dynamically loaded when the vehicle is in motion.

Incorrect toe setting is one of the alignment factors that causes tire wear. Although a running toe-in will produce no greater tire wear than running toe-

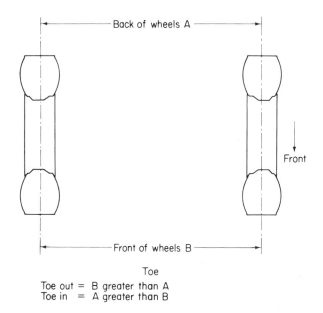

Figure 15-10 Toe illustrated.

Steering and Wheel Alignment

out, a slight running toe-in is preferred. A vehicle with a toe-out condition under dynamic loading will tend to cause the vehicle to wander. The reason for a static toe-in setting, then, is to maintain running toe between a slight toe-in and zero.

Toe-Out on Turns. When the vehicle is in a turn, the outside wheels are turning on a different, larger radius circle than the inside wheels. The tire centerline is tangent to the turn circle. The outside front wheel must, therefore, be turned at less angle than the inside wheel as illustrated in Figure 15-11. This will keep both wheels tangent to their respective turning circles and prevent tire scuffing in the turn. A steering linkage that is designed to provide the correct steering angle for each wheel in a turn is called an *Ackerman layout*. Ackerman layout is the principle that two wheels linked together, traveling on different diameter circles must be turned at different angles. If a tangent is drawn to the inner and outer turning circles at the tire contact patch, the ideal steering angle for each wheel is shown. Steering angle is the angle of the wheel in comparison to a line drawn through the front to back centerline of the body. The greater steering angle of the inside wheel will create a greater distance between the extreme front of the tires compared to the rear of the tires. This increased distance will change the static toe setting to toe-out, thus producing toe-out on turns.

Toe-out on turns, sometimes referred to as turning radius, is controlled by turning radius requirements that are determined in the vehicle design. Control of toe-out on turns is built into the steering arms by their angle on the knuckle. The turning radius specification for a vehicle is based on wheel base, wheel track, angle of the turn, and tire slip angle. As the angle of the turn is increased, the toe-out on turns must also increase. *Tire slip angle* has the effect of decreasing the effective steering angle, which in turn increases the diameter of the turn. The angle between the tire centerline and the tangent to the

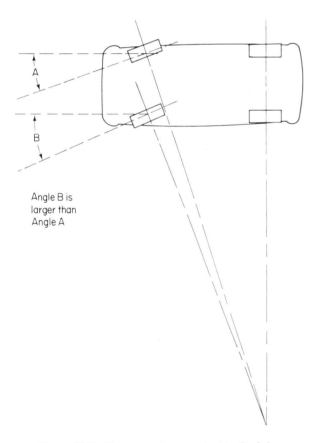

Figure 15-11 Toe-out on turns required by the Ackerman layout of the steering system.

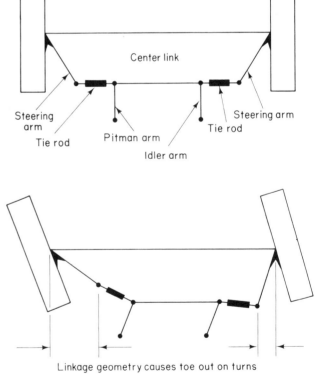

Figure 15-12 The steering arm angle produces the required toe-out on turns.

effective turning circle is the tire slip angle. This was shown in Figure 14-18. As the effective steering angle is reduced, the amount of toe-out on turns is reduced.

The correct amount of toe-out on turns is produced through the proper angling of the steering arms. This can be seen in Figure 15-12. Due to the position of the steering arms in their respective turning arcs during a turn, the toe-out will be proportional to the steering angle. The greater the steering angle, the greater the toe-out on turns.

The turning radius or toe-out on turns of the vehicle is a nonadjustable angle. It is always checked during a wheel alignment. If the turning radius is out of specification, a steering arm may be damaged. Damaged steering arms are replaced to correct an out-of-specification turning radius. An incorrect idler arm or Pitman arm would also create an out-of-specification turning radius.

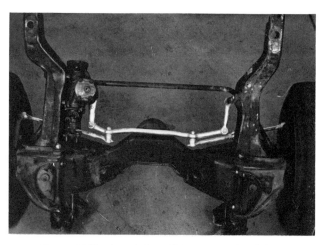

Figure 15-13 A typical parallelogram steering linkage located ahead of the suspension.

15-2 STEERING LINKAGES

Steering linkage provides the mechanical link between the steering gear and the steering arm and knuckle. The steering linkage gives the driver the ability to control the direction of the vehicle.

Steering linkage, depending upon design, will consist of a number of different components. Tie rods, center links, bell cranks, drag links, and idler arms are some of the common components used in various steering linkage systems. A typical parallelogram steering system, such as illustrated in Figure 15-13, will be used to describe the functions of the linkage components.

The steering linkage is designed so that there is very little toe change as the vehicle suspension travels through jounce and rebound. This is accomplished by making the axis of the inner pivot point of the steering linkage very close to the axis of the lower control arm inner pivot. This can be seen when examining the steering linkage of any vehicle.

Parallelogram Steering Linkage. Parallelogram linkage is the type most commonly used on domestic passenger vehicles. It uses two tie rods, a center link or drag link, and an idler arm. Tie rods connect the steering arm to the center link. They consist of an inner and an outer tie rod end with a tie rod sleeve for connecting the ends and adjusting the tie rod length.

The tie rod end ball studs allow both vertical wheel movement and steering. The ball stud is fastened to the steering arm with the socket being attached to the tie rod as shown in Figure 15-14. The steering gear Pitman arm and the idler arm connect on opposite sides of the vehicle. This provides a connection to, and facilitates movement of, each end of the center link. The center link is connected indirectly to the frame so it is not affected by vertical suspension motions. Its sideways movement is controlled solely by the Pitman and idler arms. The idler arm provides symmetry between the left and right side of the vehicle steering

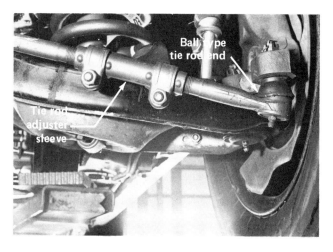

Figure 15-14 Ball tie rod end.

Steering and Wheel Alignment

linkage so that similar length tie rods can be used on both left and right sides.

Haltenberger Steering Linkage. Haltenberger linkage is one of the simplest linkage systems. It is used in conjunction with twin I-beam independent front suspension. This system consists of a tie rod and a drag link, illustrated in Figure 15-15, minimizing the number of linkage components. The drag link connects the Pitman arm of the steering gear to the right side steering arm. The tie rod has an inner and outer tie rod end and these are connected together with an adjusting sleeve. The tie rod connects the left steering arm to the drag link. Only one adjusting sleeve is used to make the toe-in adjustment.

The Haltenberger linkage provides the length necessary to minimize toe changes during jounce and rebound. The inner pivot of the steering linkage for each side falls relatively close to the inner pivot for each I-beam. Proper selection of the pivot points by the design engineer is of the greatest importance on this type of linkage so as to provide similar turning response for each wheel.

Bell Crank Steering Linkage. Bell crank linkage is a more complex linkage that is slowly being phased out of passenger vehicle and light-duty truck use. This linkage system employs two tie rods, a bell crank, and a drag link as shown in Figure 15-16. Although complex, the advantage of this type of steering linkage is that it permits greater freedom in the location of the steering gear assembly. On the Haltenberger and parallelogram linkages, the Pitman arm must be in alignment with the linkage. On the bell crank steering linkage the *drag link* connects the Pitman arm to the bell crank to provide rotation of the bell crank around its pivot. When properly positioned, the bell crank transforms the drag link movement into a side-to-side motion that is required for steering the wheels. Tie rods connect the bell crank to the steering arms. The bell crank forms the inner pivot point of the tie rods. The tie rod adjustment method is similar to that of the parallelogram linkage.

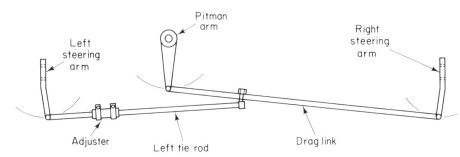

Figure 15-15 A Haltenberger steering linkage.

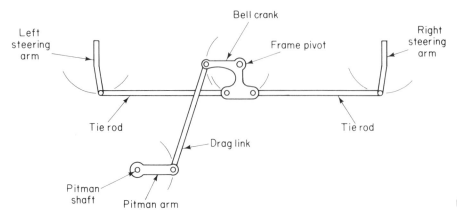

Figure 15-16 A bell crank steering linkage.

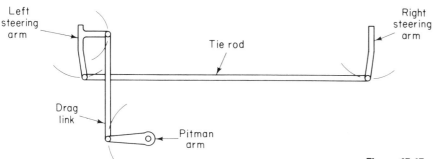

Figure 15-17 A drag link steering linkage.

Drag Link Steering Linkage. On solid axle dependent front suspensions, a single tie rod connects the left and right steering arms. A drag link is used to connect the Pitman arm of the steering gear to one of the steering arms. This is illustrated in Figure 15-17. With this type of linkage and axle, no toe change occurs when the suspension travels through jounce and rebound.

Rack and Pinion Steering Linkage. When rack and pinion steering is used, a linkage system similar to the parallelogram linkage is used. The steering rack and two tie rods comprise the complete assembly. This can be seen in Figure 15-18. The tie rod inner ends connect directly to the end of the steering rack. The outer ends connect to the steering arms at the knuckles. The steering rack takes the place of the center link in a parallelogram linkage. The tie rod inner end is usually a ball that is fitted into a socket on the end of the rack. Tie rod length adjustment is provided by threading the tie rod into or out of the tie rod end sleeve. Proper positioning of the steering rack is of primary importance when adjusting the linkage. It must always be centered in its travel with each tie rod equal in length.

Figure 15-18 A rack and pinion steering linkage, shown without the tie rod ends.

15-3 WHEEL ALIGNMENT

Wheel alignment is checked whenever abnormal handling or tire wear exists, as a preventive maintainance measure, or after parts that affect suspension and steering geometry have been replaced. Wheel alignment is the measurement and adjustment of suspension and steering angles. The measured angles are compared to the manufacturer's specifications for that vehicle. The alignment angles are adjusted when they do not fall within the specifications. Prior to measuring the alignment angles, a thorough inspection of the condition of the vehicle suspension and linkages must be done. This preliminary inspection is necessary to ensure that everything affecting alignment angles is in normal condition.

Preliminary Inspection. Preliminary inspection must include a check to assure that the vehicle is at *curb weight*. Curb weight is established when a vehicle has all liquid levels filled, including a full fuel tank, one spare tire, tires properly inflated, no occupants, and no luggage or cargo. If the fuel tank is not full, 7½ pounds (3.4 kg) properly positioned over the fuel tank may be substituted for each gallon of fuel required to fill the tank.

After establishing curb weight, the tires are inspected. Tire wear indications are shown in Figure 15-19. The tire size must correspond to tire manufacturer's specifications. Tread wear should be equal on both front and rear tires. The wheels should be of the size and offset specified. Most importantly, they must be matched side-to-side in axle sets. Tire pressures are adjusted to recommended values and the tires are inspected for any abnormalities or conditions that might result in tire

CONDITION	RAPID WEAR AT SHOULDERS	RAPID WEAR AT CENTER	CRACKED TREADS	WEAR ON ONE SIDE	FEATHERED EDGE	BALD SPOTS
CAUSE	UNDER INFLATION	OVER INFLATION	UNDER-INFLATION OR EXCESSIVE SPEED	EXCESSIVE CAMBER	INCORRECT TOE	WHEEL UNBALANCED
CORRECTION	ADJUST PRESSURE TO SPECIFICATIONS WHEN TIRES ARE COOL			ADJUST CAMBER TO SPECIFICATIONS	ADJUST FOR TOE-IN 1/8 INCH	DYNAMIC OR STATIC BALANCE WHEELS

Figure 15-19 Tire wear indications caused by abnormal conditions (Courtesy of Chrysler Corporation).

failure. The tires and wheels are inspected for excessive radial and axial runout as discussed in Section 4-1 while the vehicle is raised to inspect the suspension and linkages.

Since alignment angles change with suspension height (see Chapter 14), proper ride height is essential for correct wheel alignment and satisfactory handling characteristics. The ride height is measured and the distance is compared with specifications. A typical measurement location is shown in Figure 15-20. Most vehicles using torsion bar suspension are provided with height adjusters. When required, the height is adjusted at this point in the inspection. When the suspension height is not correct on vehicles with coil and leaf springs, the suspension must be serviced to set ride height prior to wheel alignment.

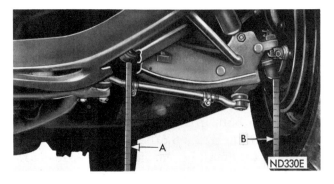

Figure 15-20 Typical front suspension height measurement location (Courtesy of Chrysler Corporation).

Steering linkage inspection should include an inspection of the tie rod ends, idler arm, and drag link. They should be checked for wear and binding. If the vehicle has power steering, the fluid level and belt tension should be checked and corrected as required. Although these items do not affect alignment, they must be correct to insure proper steering when the vehicle alignment is completed.

Ball joints and shock absorbers are inspected and repaired as needed. Wheel bearings must be adjusted to specifications. Wheel bearing adjustment is critical, because wheel alignment equipment is either fastened to the wheel or the wheel hub. The readings of the alignment equipment will be affected by improper bearing adjustment.

Alignment Procedure. After all the preliminary checks are made, the alignment angles are measured. Various types of wheel alignment equipment are in use. Each manufacturer of wheel alignment equipment has their own method of measuring the wheel alignment angles. Common methods for measuring the camber, caster, and SAI include the use of a bubble gauge (similar to a carpenter's level) with an adjustment and a scale for readout, light beams, and digital readouts that automatically display numerical readings. Toe readings employ a mechanical scale that compares the distance between the front and back of the tire, a light beam, or a digital readout. Turning angle is indicated by a scale on the radius plates on which the front wheels

Figure 15-21 Typical portable wheel alignment equipment.

Figure 15-22 Rack-type wheel aligner.

are placed while checking the alignment. The radius plates allow the wheels to be turned freely inward or outward and to determine the amount of the turn in degrees for each wheel.

Alignment equipment may be either portable or permanently mounted. Portable equipment of the type shown in Figure 15-21 allows a work area to be used for other service as well as wheel alignment. Facilities with limited space and a limited volume of alignment work will generally use this type of equipment. Portable equipment consists of radius plates on stands placed under each front tire and stands that are placed under each rear tire to level the vehicle. Portable alignment heads are attached to the hub or wheel where the readings are taken. Permanently installed equipment measures the alignment while the vehicle is on a level platform. The vehicle may be driven up ramps to the platform, the ramps may be raised to provide a level platform (Figure 15-22), or the vehicle may be driven onto alignment equipment at floor level. A "pit" or recessed area in the floor (Figure 15-23) will provide space for the technician to work under and around the vehicle. The alignment platform, often called an *alignment rack*, may be used in conjunction with either portable alignment heads or they may use alignment heads that project light beams forward of the vehicle onto a screen that is marked with a scale for the reading. The equipment either raises the vehicle from the floor surface or provides room beneath the vehicle where the technician can make under-the-vehicle adjustments.

The vehicle is placed on the center of the radius plates, prior to measuring any of the alignment angles. The locking pins on the plates are then removed to allow the plates to move (Figure 15-24). The parking brake and a brake pedal depressor, shown in Figure 15-25, are firmly set to

Figure 15-23 Pit-type wheel aligner.

Figure 15-24 Lock pins being removed from the radius plate after the vehicle is positioned on the wheel aligner.

Figure 15-25 Typical brake depressor in place.

prevent any vehicle movement or rotation of the front wheels.

Camber. The first angle to be measured is camber. The camber is measured with a bubble-level gauge. The gauge compares the centerline of the wheel with true vertical. This is easily accomplished

Figure 15-26 Bubble on a wheel alignment spindle adapter.

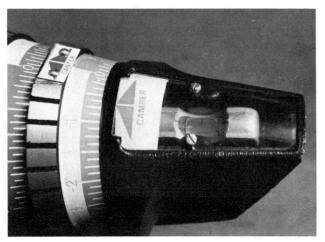

because the bubble-level gauge will always rest at zero if the gauge is attached to a true vertical plane. A bubble gauge is shown in Figure 15-26. Camber is always measured with the wheels in the straight-ahead position. The degree of camber is indicated by the gauge showing that the wheel is inclined from the vertical. Camber is positive if the tire is inclined outward at the top or negative if the tire is inclined inward at the top. This is illustrated in Figure 15-27. When a light beam alignment is used

Figure 15-27 Camber shown by using a rod attached to the spindle. The rod angle can be compared to the horizontal line in the background. (a) The rod slopes downward, indicating positive camber; (b) the rod is level indicating no camber; (c) the rod slopes upward indicating negative camber.

Wheel Alignment

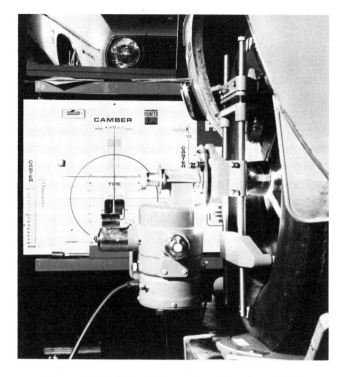

Figure 15-28 Wheel alignment machine using a light beam projecting on a viewing screen scale.

the angle of the light beam is compared to a graduated scale on the viewing screen to indicate the degree of camber. This is shown in Figure 15-28.

Caster. Caster is measured as the wheel is turned through a specified arc angle. This arc may vary from 40° on a typical bubble gauge alignment head to as low as 20° on a light beam or digital alignment head. Turning the wheel through an arc produces a camber change and a spindle height change as a result of caster. This is illustrated in Figure 15-29. The alignment head senses the camber change produced as the wheels are turned through the specified degree arc. The change in camber is converted through scales to indicate caster. The procedure may vary between manufacturers, although the common method of measuring caster is through camber change. Caster is measured by turning the wheel inward, zeroing the scale and then turning the wheel outward a specified number of degrees. The caster can then be read on the caster indicator scale.

Steering Axis Inclination. SAI is measured in a manner very similar to caster. The wheel is turned inward 20° (10° on some machines) and the caster and camber scales referenced. The wheel is then turned outward 40° (20° on some models) and the SAI is read directly from the SAI indicator scale.

The camber scale is not rezeroed as was done when measuring caster. SAI produces an up and down rotation of the wheel spindle as illustrated in Figure 15-30. The change is sensed by the alignment head to indicate the SAI. Recall that SAI is the angle

Figure 15-29 The rod attached to the spindle indicates camber change produced by positive caster as the wheel is turned through an arc: (a) inward 20°; (b) outward 20°.

(a)

(b)

Steering and Wheel Alignment

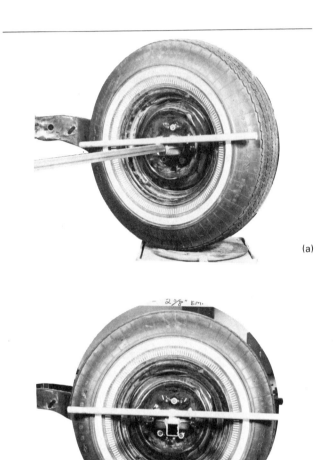

Figure 15-30 The spindle movement caused by SAI. (a) Spindle is tipped downward as the wheel is turned inward; (b) spindle is level when the wheel is straight ahead; (c) spindle is tipped downward as the wheel is turned outward.

between the steering axis and true vertical, while camber is the angle formed between the wheel centerline and true vertical as viewed from the front. These angles will change as the upper or lower ball joints are moved in or out to adjust camber during a wheel alignment. The sum of the SAI and the camber angles or the angle formed between the centerline of the wheel and the steering axis will always be constant. This combined angle, called the *included angle,* does not depend on the position of the ball joints. During wheel alignment the SAI reading is measured along with the camber reading for each wheel. These readings are added to determine the included angle. The measured value is compared to that specified, when the included angle is specified. If SAI is specified, camber is subtracted from the measured included angle to determine the vehicle SAI so this value can be compared to the alignment specifications. Because the alignment head is usually attached to the hub or wheel rather than the spindle, any rotation of the hub will alter the SAI reading. Having the brake pedal depressor firmly set will prevent wheel rotation so that accurate readings can be measured on each side of the vehicle. This will remove any slack that may exist in the brake components when measuring SAI as the wheel is turned 20° inward and then turned 20° outward. SAI, as measured during an alignment check, is used to detect a bent spindle or steering knuckle.

Toe-Out on Turns. Toe-out on turns or turning radius is checked with the vehicle front wheels on the radius plates. The wheels are set in the straight-ahead position and a check is made to assure that both radius plates are set on zero. One wheel is turned outward 20° and the turning angle is read on the radius plate degree scale under the other wheel. The turning radius on the inner wheel is normally between 21° and 22½° when the outer wheel is turned 20°. Toe-out on turns is the difference between the degree-reading on the inner wheel radius plate and the outer wheel radius plate. Toe-out on turns is not adjustable. It is used in service to detect bent or incorrect steering arms.

Toe. Toe is measured in a variety of ways, depending on the manufacturer of the alignment equipment. Although varying in technique, all methods eventually end up with a measurement corresponding to the difference in distance between the centerlines at the front and the rear of the tires at their spindle height.

Mechanical measurement techniques include jacking each wheel and chalking a band around the tread surface of the tire. A scribe is then used to mark a fine continuous line in the chalk as the wheel is rotated as shown in Figure 15-31. This will prevent any runout in the tire-wheel combination from affecting the toe measurement. The tires are then lowered to the radius plates and the vehicle is jounced to settle the suspension. The distance is then measured between the scribe lines at the rear and at the front of the tires at the spindle height. If the distance at the front of the tires is less than the distance at the rear, the vehicle has toe-in. The difference in the measured distances is the amount of toe-in.

Another mechanical method used to measure toe is to mark a point on the inside front of each wheel rim at an equal distance above the floor at the spindle height. The distance between the two points is measured. The tires are rotated so the same mark is at the rear at the spindle height. The distance between the two marks is again measured after jouncing the vehicle. The distance between the front and back measurements is the toe.

Electronic measurement of toe includes a light beam projected across the front of the vehicle from an instrument attached to one wheel spindle to another instrument on the opposite wheel. The amount of toe is read on a scale illuminated by the light beam line in inches or degrees.

Another method used to measure toe is to project a light beam forward from each wheel alignment head. The beam is then reflected rearward to a scale on the projector head. The total toe of the vehicle is the sum of the toe measurements obtained at each wheel.

Adjusting Alignment. Camber and caster may use the same or separate means of adjustment. The adjustment methods used for camber include shims, cams, slots, and eccentric bushings. Caster adjustments include shims, cams, slots, and adjustable lower control arm struts. Both camber and caster are affected by the position of the ball joints in relation to each other and to the control arms or A-frames. The adjustment of the ball joint position to make a change in either angle will affect both angles. This is extremely important when vehicles using shims, cams, or slots for adjusting both camber and caster are aligned. The interdependent relationship of camber and caster adjustment is minimized when the caster adjustment is made by adjusting the brake strut. Camber is adjusted first on vehicles that use cams, shims, and eccentric bushings for making camber adjustment and a strut adjustment for caster. Both camber and caster are adjusted at the same time on vehicles using shims, cams, or slotted bolt holes on the inner end of the upper A-frames. Before making any change the service technician should decide which *one* adjustment can be used to correct both camber and caster. Shims and cam adjustments may be provided on either upper A-frames or on the lower control arms. Slot adjustment is found only on the upper A-frame. Eccentric bushings on the upper ball joint, as shown in Figure 15-32, have been used to make the camber adjustment on one manufacturer's vehicles.

Special tools are used to provide control while making the caster and camber adjustment on suspensions with a slot type adjustment. This can

Figure 15-31 Measuring the toe between lines scribed on a chalk marked tire.

Figure 15-32 An eccentric ball joint bushing for alignment.

Figure 15-33 A tool used for slot adjust alignment.

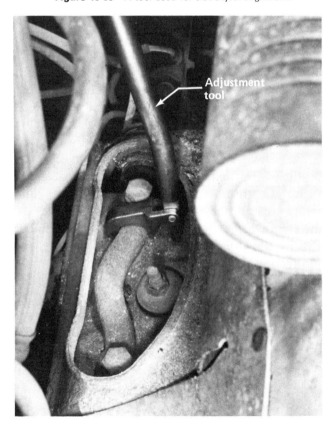

be seen in Figure 15-33. The special tool is attached at the end of the upper control arm cross shaft that must be moved to correct both camber and caster. Only that bolt is loosened to make the adjustment with the special tool supporting the cross shaft. It may be necessary to retighten the bolt and repeat the adjustment procedure on the other bolt, while holding the cross shaft in place with the special tool to obtain the correct camber and caster.

Shim adjustment of camber and caster is accomplished by loosening the bolts that fasten the control arm to the frame and adding or removing shims, as required, to bring the suspension to specifications. This adjustment location is shown in Figure 15-34. Before making the alignment adjustment the technician should decide where a shim must be added or removed and if the shim change will make both camber and caster adjustments at the same time.

Cam adjusters, one on the front and one on the back inner arms, provide a means to adjust both camber and caster. A typical cam adjuster is shown in Figure 15-35. After loosening the locking devices, the cams are rotated to provide the required suspension adjustment. Some vehicles use a cam adjustment on the lower control arm for camber adjustment and an adjustable strut bar for caster adjustment. The camber on these vehicles is adjusted before the caster is set.

Caster is changed by increasing or decreasing the distance on the brake strut from the fixed point on the frame to the control arm. The brake strut is adjustable when a separate means is provided for

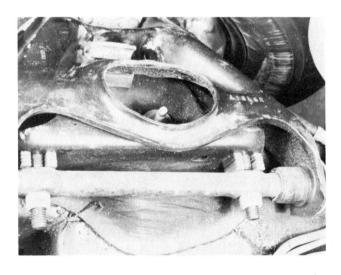

Figure 15-34 Alignment adjusted with shims.

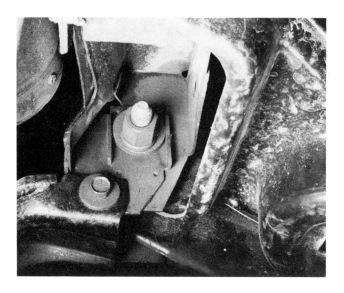

Figure 15-35 A cam for adjusting alignment.

camber and for caster adjustments (Figure 15-36). Struts may be positioned either in front of or to the rear of the control arm. The technician should decide prior to making an adjustment whether the strut must be lengthened or shortened to provide the correct caster.

Toe adjustment can be made by changing the effective distance between the inner and outer tie rod ball pivots. On most domestic automobiles the adjustment involves loosening two clamp bolts on the adjuster sleeve and rotating the adjuster sleeve, as shown in Figure 15-37, to obtain the correct toe setting. Vehicles with rack and pinion steering will require a locknut to be loosened on the steering linkage next to the tie rod end. The tie rod is then rotated to adjust the toe setting. Adjustment of one tie rod will change the toe setting but it will also change the straight-ahead center position of the steering wheel. Rotating both tie rod sleeves an equal amount in the *same* direction will change the toe setting and will not change the steering wheel position. Adjustment of both tie rods, equal amounts but in the *opposite* directions, will produce no toe change but it will rotate the steering wheel to a new position. It is important that the steering wheel and the steering gear be in its center position when the wheels are aimed straight ahead. The conventional steering gear is designed to have the least amount of free play when it is centered. All steering linkage is designed to be symmetrical from side to side. If the overall length of the tie rods differs on each side of the vehicle, as would be possible with an uncentered steering gear, erratic vehicle handling will result. Proper tie rod adjustment gives the operator more positive, predictable steering control.

Steering wheel position may be checked visually or it may be measured to assure proper positioning. Using either method the front wheels are first set in the straight-ahead position, then the

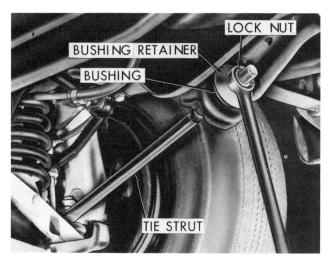

Figure 15-36 Strut adjustment to set caster (Courtesy of Cadillac Motor Car Division, General Motors Corporation).

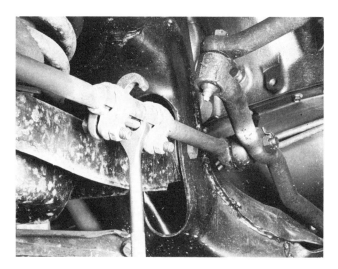

Figure 15-37 Toe is set by rotating adjusting sleeves on a tie rod.

Steering and Wheel Alignment

steering wheel position is checked. If the wheel is not centered and the toe adjustment is incorrect, the technician may be able to correct both malfunctions by adjusting only one of the tie rods.

Toe is the last adjustment made during wheel alignment. The tie rod sleeves and clamps must be properly positioned to provide operating clearance and to prevent any binding in the tie rod pivots during movement of the vehicle body and steering linkage. The tie rod adjuster sleeve clamp bolts are then tightened. The proper position is shown in Figure 16-34.

Prior to moving the vehicle from the alignment area, the vehicle is rechecked to assure that all suspension and steering attachments have been properly tightened and cotter pins put in place. The technician should test drive the vehicle to make sure that there is no directional drift, vibrations, or other alignment and handling conditions the operator might associate with alignment or suspension problems.

15-4 PROBLEM DIAGNOSIS

After alignment, it is possible to have problems arise that may appear to the operator to be caused by improper wheel alignment. It is possible that alignment settings were made incorrectly and these are causing the problem. If so, these should be corrected.

Directional Drift. When a directional drift problem exists, the camber and caster should be rechecked. If camber and caster are within specifications, the tires and wheels should be reinspected to make sure they match. Tire wear that is due to incorrect camber will cause the tread to wear unequally across the tread width. After camber adjustment, this unequal wear will move the *effective* tread centerline away from the center of the wheel and this will change the *effective* scrub radius. With unequal scrub radius on each tire, the wheel will pull the vehicle in one direction while operating on a level road surface. This pull will be exaggerated during braking. Drift from wheel pull can be caused by problems within a belted tire. If a belt shifted during tire construction, the stiffness across the tire tread width is not symmetric. This becomes most critical with radial tires. The unsymmetric loads in the tire footprint will create the same effect as incorrect camber setting. If this is suspected, it can be checked by replacing the suspected tire with the spare tire and road-testing the vehicle. It can also be checked by changing the tires, left to right, on the front of the vehicle and road-testing. If the pull changes to the opposite direction, replace the faulty tire.

Directional drift may be created within the steering gear or by a problem in the linkage. A preloaded steering linkage bushing or a malfunctioning power steering gear will act to steer the vehicle toward one side.

Wheel Pull In Braking. Pull during braking may be the fault of the front suspension or of the vehicle brakes. The problem must first be isolated, either suspension or brakes. The brakes can be inspected after all visual checks of the tires, wheels, and suspension have been made. Pull during braking can be caused by a different scrub radius on each side. The scrub radius may not be the same because different wheels are used on each side of the vehicle. Caster may cause wheel pull during braking, even when the caster is within specifications if the caster on one wheel is positive and the other is negative. Caster usually has maximum and minimum limits; therefore it is possible, on some vehicles, for these limits to allow either a positive or a negative setting. Manufacturers will recommend a maximum cross-caster specification, usually ½°. This is the maximum allowable side-to-side variation. While setting alignment, it is always a good policy to avoid going across 0° on caster (+ ¼° on one wheel and − ¼° on the other wheel) even though all adjustments are within specifications.

Wander. Wander is another problem that may be encountered. When this occurs, the first thing to check is the tires for proper inflation. Next the wheel alignment angles are rechecked, paying particular attention to the toe and caster setting. Wander can occur when excessive caster exists or if the wheels toe-out while under dynamic loads. Center link nonparallelism may also cause wander especially when the vehicle is driven over rolling roads. Nonparallelism is present when the inner tie rod socket side-to-side heights are unequal. This problem will exist if either the steering gear or idler arm are not properly positioned. When out of position, the steering linkage will pivot away from the lower control arm pivot point. If the distance between the linkage pivot points differ from side to side more than ½ inch (13 mm), an adjustment is required. Unequal length tie rods can cause problems similar to nonparallelism. This will cause a different toe change, from wheel to wheel, while making a turn. These toe changes will produce steering pull. The steering wheel must be centered or wander will occur due to looseness in the off-center section of the steering gear.

Front-Wheel Shimmy. Front-wheel shimmy is often blamed on front-wheel misalignment. Common factors contributing to shimmy include tires, loose steering linkage, incorrectly adjusted wheel bearings, shock absorbers, and possibly wheel alignment. Most shimmy problems are related to tires and wheels, either out-of-balance or out-of-round. Shock absorbers, if excessively worn, may not control wheel motions and this can create wheel tramp. Excessive caster under certain conditions will allow the wheels to shimmy. If caster is not excessive, then it should be ruled out as a possible cause of shimmy.

REVIEW QUESTIONS

1. When is wheel alignment required? [INTRODUCTION]
2. Which way will a wheel with positive camber lead? [15-1]
3. How do dynamic loads affect wheel alignment geometry? [15-1]
4. What are the visible signs of improper camber? [15-1]
5. What is the effect of positive caster on steering? [15-1]
6. How does caster affect body roll? [15-1]
7. How does caster affect camber during a turn? [15-1]
8. What two points produce scrub radius? [15-1]
9. Make a sketch to show how scrub radius can be negative with a positive SAI. [15-1]
10. What causes the tension on the steering linkage while the vehicle is moving? [15-1]
11. What running toe setting produces minimum tire wear? [15-1]
12. Why are wheels set with a slight toe-in? [15-1]
13. Why do vehicles use toe-out on turns? [15-1]
14. What type of steering linkage is most commonly used on domestic passenger vehicles? [5-2]
15. How does rack and pinion type steering compare to the parallelogram steering linkage? [15-2]
16. List the items checked in the wheel alignment preliminary inspection. [15-3]
17. Describe the method used to measure camber, caster, toe, and SAI on a specific alignment machine. [15-3]
18. When using shims, how can caster be changed without changing camber? [15-3]
19. When using cams, how can camber be changed without changing caster? [15-3]
20. How is toe adjusted? [15-3]
21. What is checked during an alignment inspection to help diagnose a problem? [15-4]
22. List the abnormal operating conditions that can be caused by improper wheel alignment. [15-4]

16
Front Suspension Service

The chief purpose of the steering and suspension system is to provide maximum vehicle control under all operating conditions. Suspension system wear occurs gradually and will cause slow deterioration of the vehicle ride quality, handling, and steering responsiveness. Because of this slow deterioration process, the operator of the vehicle will usually become accustomed to the gradual change to abnormal handling and inferior ride quality and accept it as normal.

The automotive service technician must be able to recognize a vehicle with abnormal handling characteristics. The general area of the suspension responsible for the vehicle operator's complaints can usually be located during a road test. This will enable the service technician to experience the complaint described. Often the problem will be reflected as excessive or abnormal tire wear. The service technician must be able to relate tire wear characteristics to suspension and steering angles or to components that will cause this wear.

Components require service or replacement when wear that makes the vehicle either unsafe or difficult to handle has occurred. Suspension parts that maintain alignment will become worn as a result of road conditions, neglected maintainence, and high mileage. This will cause the springs to sag, control arm bushings to deteriorate, and ball joints to loosen. Wear of any of these components will contribute to changes in the wheel alignment.

A thorough inspection done in an orderly sequence, so as not to overlook any part, is important to a complete and satisfactory suspension repair job. A safe vehicle will result when careful inspection and correct repair procedures are properly completed.

16-1 SYSTEM INSPECTION

A visual inspection to identify worn parts should always precede any alignment measurement. Raising the vehicle by placing it on a lift or supporting it, both front and rear, with jack stands will allow the technician to examine the suspension and steering system. It is important to check the condition of the rear suspension components as well as front suspension components since the rear suspension condition also affects vehicle handling. Details of rear suspension service are covered in Chapter 17.

Inspection Procedure. During the visual inspection, the technician checks for broken, bent, or worn components. Faulty main suspension components that are usually visible from beneath the vehicle include broken *springs*, worn spring eye *bushings* (leaf springs), broken or leaking *shock absorbers*, *stabilizer bar* link or bushing failure, *control arm* or bushing failure, and *brake strut* bending or bushing failure. The steering system inspection should include a check for loss of *fluid* from the power steering gear and hoses. It also includes checking for *Pitman arm* damage or wear, *idler arm* damage or bushing wear, and *tie rod end* or *linkage* damage or wear. While visually inspecting the steering components, any excess play will generally be visible when shaking the wheels vigorously from side to side. It may also show up when grasping each component and attempting to move it in both vertical and horizontal directions. Tie rod ends and other steering linkage pivot points are zero clearance components; thus, no play should be present. While shaking the wheel, any wheel bearing looseness will be evident.

A thorough inspection of the tires and wheels should follow a visual inspection of the suspension components with the vehicle raised from the floor. It is possible to visually check the tire and wheel assembly for tire tread run out. While making these inspections the service technician should listen for any abnormal noises originating from the bearings.

The vehicle should now be lowered to the floor surface for further checks. These include the front suspension upper control arms and bushings, ball joints, vehicle ride height, and a check for any suspension noises or binding. While the vehicle is cycled through jounce and rebound, excessively worn or internally damaged shock absorbers can be located. A correctly functioning shock absorber should halt the up-and-down motion of the vehicle body in approximately two cycles after release. It should be noted that this is not a positive test of the shock absorber condition. Any shock absorber that is of questionable condition should be removed from one end or from both ends and rapidly extended through its full travel by hand to determine its effectiveness.

On vehicles having the upper control arm pivot shaft fastened to the frame member inside the fender well, a visual inspection of the shaft bushings can be made by raising the hood and observing their condition through the engine compartment. The ride height of the vehicle can now be measured, both front and rear. Typical measurement points (Figure 16-1) are referenced between the inner and outer control arm pivots and the floor for the front suspension, and between the

Figure 16-1 Typical ride height measuring points.

Front Suspension Service

axle housing and frame for the rear suspension. An applicable shop manual should be consulted for specific measurement points and dimensions for the vehicle being measured. At the correct ride height, the lower control arm is usually slightly higher on the inside end than on the outside end. The ride height measurement, when compared with specifications, will allow the technician to determine if the springs need to be replaced, or if the torsion bars need to be adjusted.

Ball Joint Inspection. The condition of the ball joints is crucial in the inspection of the front suspension. Each ball joint is inspected for lateral (sideways) and axial (up-and-down) movement. The vehicle must be raised, one side at a time, in a manner that will relieve all vehicle weight and spring force from the load-carrying ball joint. The technician can determine which method to employ by noting whether the upper or lower ball joint is the vehicle-loaded joint. The loaded ball joint is the joint that is on the control arm to which the spring is attached.

If the lower ball joint is the loaded ball joint, the jack should be placed under the lower control arm as shown in Figure 16-2. When the upper ball joint is the loaded ball joint, a spacer block is placed between the upper arm and frame; then the jack is placed under the vehicle frame. This can be seen in

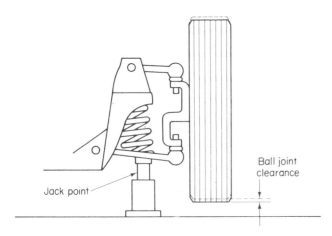

Figure 16-2 Jacking for checking the clearance of a lower loaded ball joint.

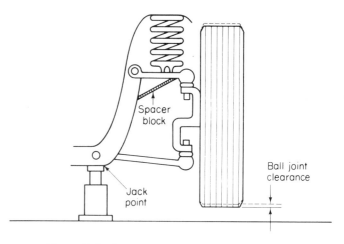

Figure 16-3 Jacking and blocking points for checking the clearance of an upper loaded ball joint.

Figure 16-3. This block will hold the upper control arm in its normal position to assure a true ball joint clearance measurement. With a MacPherson strut type front suspension, the jack is placed under the frame; this allows the strut to fully extend to take the load off the joint. The MacPherson strut acts in a manner similar to a spring on an upper control arm; therefore the strut receives the spring force. If there is any doubt about which method of inspection to use, consult the applicable shop manual for the specific vehicle being inspected. With the vehicle raised, a dial indicator or ball joint wear indicator is used to determine if the ball joints are satisfactory for continued service.

A ball joint wear indicator is a tool with a pin that is inserted into the grease fitting socket after the grease fitting is removed. Any movement of the ball in the socket is detected and measured with the tool. When a dial indicator is used it is fastened to the control arm with the indicator anvil positioned against the steering knuckle, as shown in Figure 16-4, to allow measurement of the movement between the two parts. A bar is placed beneath the tire of the suspension being checked. The bar is lifted and moved sideways as readings are being taken to check maximum lateral and axial movement. The friction-loaded ball joint should have zero lash while the vehicle-loaded ball joint should have a specified clearance. A specification chart must be consulted to determine the normal ball joint clearance. A ball joint with clearance that is out of specification should be replaced in order to properly set alignment angles and to promote safe operation.

System Inspection

Figure 16-4 Using a dial gauge to measure ball joint clearance.

One manufacturer's procedure for checking the friction-loaded ball joint includes separating the joint from the steering knuckle. The ball joint is then rotated in its socket with a torque wrench and the amount of torque noted. If the torque falls below a minimum specified value, the ball joint should be replaced. If a faulty ball joint is kept in service, it could end up like the ball joint pictured in Figure 16-5.

Figure 16-5 Failed ball joints: (a) cracked; (b) separated.

The majority of manufacturers on their late model vehicles are using a ball joint wear indicator incorporated into the ball joint to provide a quick and simple inspection while the vehicle is raised on a lift. Special jacking or measurement is not required. A visual inspection of the protrusion of the nipple into which the grease fitting is threaded will indicate excessive wear. As wear occurs, the nipple recedes into the ball joint housing. Replacement is necessary when the nipple shoulder is recessed below the lower portion of the housing. Figure 16-6 shows an external view of the protrusion. A section view of the ball joint was illustrated in Figure 13-43.

Control arm bushings are inspected for looseness while the vehicle is jacked for the ball joint inspection. The load-carrying control arm will have to be supported to relieve all loads, enabling the technician to check the bushings for excessive wear and deterioration.

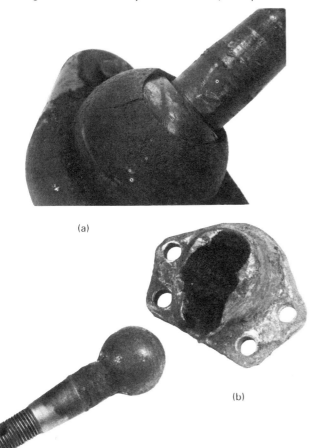

(a)

(b)

Figure 16-6 The ball joint is satisfactory when the grease fitting nipple shoulder extends below the ball joint housing.

Front Suspension Service

16-2 FRONT SUSPENSION SERVICE

When inspection of the suspension and steering system indicates that a repair is required, an orderly repair procedure should be followed. A systematic approach to repair will save the service technician time and the vehicle owner money.

Shock Absorbers. The shock absorbers are removed before any suspension work requiring removal of the coil springs or control arms can be done. This is necessary because the shock absorber is connected between the vehicle frame or body and one of the control arms. Front shock absorbers are usually located inside the coil spring.

Shock absorbers are one of the more common suspension replacement items and they are often replaced even though they are still working properly. Two of the more common, although often unnecessary, reasons for shock absorber replacement are slight fluid seepage around the ram and seal, or noise produced during jounce and rebound. Slight fluid seepage is a normal condition and necessary in shock absorber operation to lubricate the seal and ram. Noise may be produced by aeration of the fluid within the shock absorber. This air can be purged from the fluid. When replacement is required, shock absorbers should always be replaced in pairs to provide uniform suspension control.

Removal of the shock absorbers is accomplished in much the same manner for all current-model passenger vehicles. One exception is the MacPherson strut suspension which combines the strut, shock absorber, and spring into one unit. The following discussion will cover the procedure used to service the conventional shock absorber.

Servicing Conventional Shock Absorbers. The front shock absorber is connected between one of the control arms and the frame or body. The uppermost connection point is usually accessible from within the engine compartment while the suspension arm connection is most accessible from beneath the vehicle. The shock absorber may, on some vehicles, provide the rebound limit for the suspension. In this situation, the suspension is jacked to remove the spring force before disconnecting the shock absorber.

The vehicle is raised to a convenient height and supported with safety stands or a hoist to permit access to the suspension components. If suspension fasteners appear corroded, it is wise to spray penetrating oil on the threads of all fasteners requiring removal. Due to the potential danger of releasing a coil spring from its seat and harming the technician or a fellow employee, always use a safe procedure such as placing a jack under the lower control arm to limit the suspension travel while the shock absorber is being removed. (Note: *The coil spring is highly preloaded and all precautions to prevent its uncontrolled release should be followed.*) The fasteners holding the upper and lower shock ends are disconnected and the shock absorber is lifted from the vehicle as shown in Figure 16-7. On some vehicles it is necessary to remove a plate or bracket along with the shock absorber as it is lifted from the vehicle and then remove the shock absorber from it.

If the shock absorber has been removed for a bench test it should be carefully mounted in a vise pointing in the same direction it had when it was mounted on the vehicle. It can then be extended and compressed and its action noted. If the action is erratic, it is possible that aeration of the fluid has occurred. Purging may correct the problem so the

Figure 16-7 The front shock absorber being removed through the lower control arm.

Front Suspension Service

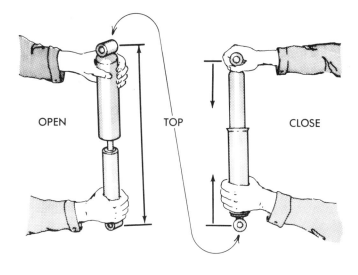

Figure 16-8 Purging air from a shock absorber (Courtesy of Chrysler Corporation).

shock absorber can be placed back in service. This is accomplished by fully extending the shock absorber, then turning it upside down and fully compressing it. Purging is shown in Figure 16-8. This cycle is repeated at least three times. Replace the shock absorber in its normal position on the bench and extend and compress it as fast as possible. If the operation of the shock absorber is still erratic, the shock absorber should be replaced. Purging and bushing replacement are the only service repairs that can be made on current conventional design shock absorbers. Replacement is necessary for any other type of abnormal operation.

Figure 16-9 Typical front shock absorber mounting ends.

Shock absorbers may have a spike end, in which one bushing is used on either side of the mounting hole. Shock absorbers may fit over a stud which will require a bushing fitted on either side of the shock eye. It may have a flanged end with a rubber mount that is made as a part of the shock absorber. These mounting styles, shown in Figure 16-9, are made to prevent the road vibration and noise from being transmitted into the vehicle frame. For this reason it is important that the shock absorber be mounted on insulators in such a way that there is no metal-to-metal contact between the shock absorber and its mount.

Servicing MacPherson Strut Shock Absorbers. The replacement of the MacPherson strut shock absorber differs from conventional shock absorbers. The strut assembly connects to the steering knuckle or lower control arm on the bottom end and to the fender housing on the upper end. Most MacPherson struts have load supporting coil springs surrounding the shock absorber as shown in Figure 13-30. It can be removed when the vehicle is raised and supported so that the front suspension is fully extended. A modified MacPherson strut has the load supporting coil spring between the lower control arm and the frame. This is illustrated in Figure 16-10. Removing this type of strut is done with the vehicle raised and supported under the lower control arm to hold the spring under compression. If camber adjustment is provided by the lower attaching bolts in either type of strut, mark the adjusting cam position as shown in Figure 16-11 before disassembly to aid in alignment after reassembly. The lower clamp bolts or ball joint can then be freed from the knuckle or lower control arm.

On the upper end of the strut remove only the attaching plate bolts that connect the plate to the upper fender housing. Do *not* remove the upper shock absorber rod nut. If the rod nut is removed the coil spring will be released suddenly from the shock absorber. This could cause personal injury and damage the vehicle.

The spring can be easily removed from the strut after the strut has been removed from the vehicle. The coil spring is compressed, as shown in Figure 16-12, using a spring compressor to release the spring load from the strut. After the spring is removed the shock absorber portion of the strut assembly may either be repaired, rebuilt, or replaced, depending on the problem and the strut design.

Repairing and rebuilding requires that the strut be completely disassembled. Strut repair kits are available for some types of struts. These kits include new seals, cups, valving, piston rod, and the required fluid. The replacement parts are installed following the instructions that apply to that strut. Both parts and the labor required to recondition the strut are charged to the customer's repair cost. This has lead some manufacturers to market replacement cartridges for the strut shock absorber assembly. The strut is disassembled and the cartridge assembly is inserted in the strut in place of the many separate shock absorber parts removed from the strut. The cartridge assembly costs somewhat more than the repair kit but the labor cost is less. Some of the companies that market replacement shock absorbers produce only complete replacement shock strut assemblies. New replacement shock struts are more costly than the

Figure 16-10 A modified MacPherson strut.

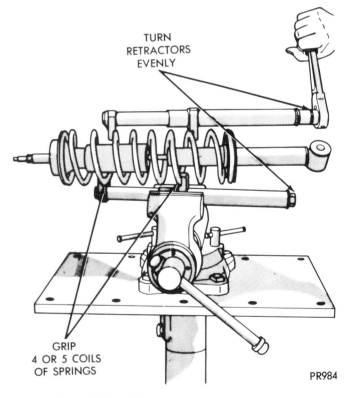

Figure 16-12 Using a coil spring retractor to compress the spring on a MacPherson strut for disassembly (Courtesy of Chrysler Corporation).

Figure 16-11 A camber adjustment cam on a MacPherson strut front suspension.

cartridge repair but the labor cost is the lowest of the three reconditioning methods described.

With the strut on the bench the coil spring is mounted over the reconditioned strut assembly and the bearing plate is bolted on to retain the spring. Installation on the vehicle is performed by reversing the removal procedure. The wheel alignment must be checked to ensure proper installation and trouble-free operation.

Stabilizer Bar Link and Bushing. The most common repairs associated with the stabilizer bar are link and insulator replacement. The outer link and insulators, used on many vehicles to connect the stabilizer bar to the control arm, will experience wear problems when the vehicle is frequently driven over rough road surfaces. The link insulators wear and the link itself may break and require service. The stabilizer bar may break or bend in service due to a vehicle accident and will have to be replaced.

The outer link is replaced as an assembly including the insulators, retainers, through bolt, and sleeve. To replace the outer link, the technician first removes the through bolt, then the sleeve and the four associated insulators (Figure 16-13). The link assemblies are normally replaced in pairs to prevent preloading of the bar due to unequal sleeve lengths.

Stabilizer Bar Replacement. Replacing the stabilizer bar requires that the outer connections at the lower control arm be removed. The connection may either be a through bolt or a clamp fastened on the strut bar near the lower control arm. The clamp-type connector is disconnected by taking out the bolt holding the clamp together. The stabilizer bar-to-frame mountings are then disconnected. These mountings may be a bracket attached to the frame with a bushing around the bar or it may be a link with insulators that attach the bar to the frame. In either case, the frame mounting is disconnected and the bar removed. It is necessary on some vehicles to remove one front tire and wheel assembly to provide clearance for the stabilizer bar to be removed from the vehicle.

Stabilizer Bar Frame Bushings. Stabilizer bar frame bushings can be replaced without removing the complete bar. The frame mounts are discon-

Figure 16-13 Removing the stabilizer bar link.

nected and the brackets removed. The bushings are then removed from the bar. The new bushings can then be installed over the stabilizer bar, the brackets positioned, and bolted into place.

Reassembly. Reassembly is accomplished by first installing the bushings on the stabilizer bar. The bar is fitted beneath the vehicle and the mounting brackets correctly positioned and loosely attached with bolts. The outer links are then installed and their assembly bolts are torqued to specifications. The tire and wheel assembly is installed and properly torqued. The frame bracket mounting bolts should be torqued to specifications *after* the wheels are at their normal ride height position. The vehicle must be supported by the suspension before the mounting brackets are torqued. This prevents preloading the bushings in an unnatural position. Preloading would shorten the bushing service life. Wheel alignment will not be affected by repair of the stabilizer bar.

Strut Bar, Strut Bushing, and Radius Arm. Strut bar, strut bushings, and radius arm replacement requires removing the bar or arm from the vehicle. Bushing replacement is the most common strut repair, although a vehicle collision may cause a

Front Suspension Service

bent bar or arm, requiring its replacement. These suspension components are highly stressed in use; therefore heat should *not* be used to straighten a strut.

The strut bar or radius arm is disconnected from the vehicle frame by removing a nut and washer from the end of the bar protruding through the frame or through the bracket which is attached to the frame as shown in Figure 16-14. On most vehicles, the two bolts that fasten the strut to the lower control arm are removed to disconnect the strut. On some vehicles the lower control arm and strut are removed as a unit in order to have the clearance necessary to remove the frame end of the bar from the vehicle. The bar is separated from the control arm for service. A radius arm on twin I-beam suspension is removed by disconnecting the frame end of the bar and removing the bolt which fastens the radius arm to the axle. The arm can then be lifted from the vehicle.

Bushing designs vary according to manufacturer, although replacement of all bushings is quite similar. Strut bushings are designed with elastomer to prevent direct contact between the bar and frame during fore-and-aft loading. The bushing has an internal sleeve to provide the correct amount of compression when installed. A typical worn suspension bushing is pictured in Figure 16-15. The bushing is positioned either on the bar

Figure 16-15 A worn suspension arm bushing.

or into the support before the bar is put in place. The fasteners are installed and torqued to specifications. All required cotter pins are installed. After assembly the vehicle suspension must be aligned to correct any change in camber or caster setting that occurred during disassembly and reassembly.

Ball Joint Removal. Ball joint replacement is necessary whenever excessive wear or damage renders the joint unserviceable. Wheel alignment will always be necessary after replacing the ball joints because the actual position of these suspension components determines the alignment angles.

Ball joint removal begins by jacking the front of the vehicle and supporting it with jack stands. The stands should be positioned rearward of the front wheels and suspension members so the service technician will have full access to the suspension components that require removal.

One wheel stud and a corresponding lug hole are marked and the tire and wheel assembly is removed and set aside. The brake drum or disc brake caliper and rotor must be removed to gain access to the bolts holding the ball joint to the steering knuckle. The removal procedures discussed in Section 2-5 should be followed.

Separating the Ball Joint From the Steering Knuckle. The bolts or rivets attaching the lower ball joint to the steering knuckle are accessible with the brakes

Figure 16-14 Typical location of strut suspension bar bushing attachment to the frame.

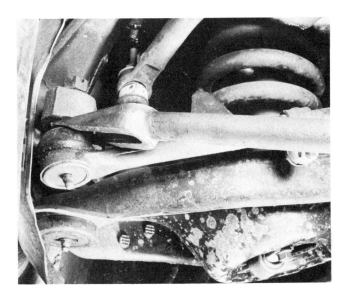

Figure 16-16 Removing a tie rod end with a tie rod spreader.

Figure 16-17 Typical arrangement of ball joints.

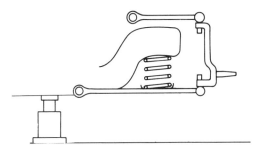

Figure 16-18 Suspension design with both ball joints under tension when jacked on the frame.

removed. The steering arm is connected to the lower ball joint on some vehicles. On these vehicles the tie rod end must be separated from the steering arm. To do this, first the cotter pin is removed and the castellated nut is loosened. The tie rod end is then separated from its socket using a tie rod spreader (Figure 16-16) or a sharp blow to the tie rod socket with a hammer in the same manner as separating ball joints. Once the tie rod stud has been separated from its socket, the castellated nut is removed and the tie rod lowered.

The nut that holds the ball joint stud to the steering knuckle is loosened. If both upper and lower ball joints are being removed, both nuts may be loosened at this time. *Caution: Do not remove the nut completely at this time. If the nut were removed completely, the ball joint and steering knuckle might separate and the unrestrained motion of the spring and the control arm may harm the service technician.* The ball joint stud has a tapered shoulder that will fit into the steering knuckle, as shown in Figure 16-17. An alternative design is to have the ball joint fastened to the steering knuckle and the ball joint stud fit into a bracket in the lower control arm.

Ball joints that are under tension from spring forces of the suspension and the weight of the components, as shown in Figure 16-18, can be separated from the steering knuckle. The tension will help them separate. *Caution: Make certain the ball joint nut is installed on the stud with clearance to allow separation of the ball joint stud and the steering knuckle.*

Ball joints that are arranged with one ball joint in tension and one in compression (Figure 16-19), or both in compression (Figure 16-20), will require the spring force to be removed from the compression joint before the joint is separated. Systems with the coil springs mounted on top of the upper control arm will not have any forces on the compression joint with the spring held as

Front Suspension Service

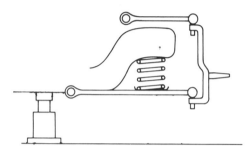

Figure 16-19 Suspension design with the upper ball joint under tension and the lower ball joint under compression when jacked on the frame.

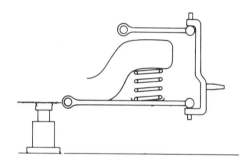

Figure 16-20 Suspension design with both ball joints under compression when jacked on the frame.

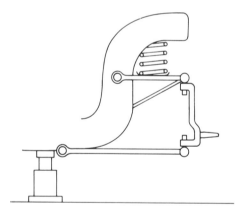

Figure 16-21 Suspension design with both ball joints unloaded when jacked on the frame.

illustrated in Figure 16-21 and so the joint can be separated.

To separate the ball joint from the steering knuckle on vehicles with the spring force working to hold the joint together, the spring force must be released. A coil spring compressor, or, more commonly in service, a jack placed under the lower

Figure 16-22 A spring compressing tool in place to hold the spring as the loaded ball joint is separated.

control arm would be used. The extent of the repairs would determine whether to use the spring compressor or the jack. If control arms are to be removed, it is advantageous to use the spring tool and remove the spring. Figure 16-22 shows one type of spring tool. If repairs are limited to ball joint replacement and the jack is used, it should be positioned toward the outside of the vehicle under the lower control arm as shown in Figure 16-23. The

Figure 16-23 Jack position to lower the suspension as the ball joint is separated. The frame is supported on stands.

Front Suspension Service

Figure 16-24 A tool used to push the ball joints out of the knuckle (Courtesy of Chrysler Corporation).

Figure 16-25 Separating the ball joint from the knuckle with a ball joint spreader.

jack is raised until the rebound snubber clears the frame.

Various methods are employed to separate the ball joint stud from its socket. A ball joint tool that applies a force between the upper and lower ball joint studs may be used to separate them. This is pictured in Figure 16-24. If the ball joint is not to be used again, a ball joint spreader, often called a "pickle fork" by mechanics, may be used. It is shown in use in Figure 16-25. The spreader will usually damage the seal and for this reason this tool is not recommended if the ball joint is to be reused. Another method, one that must be used with care and discretion, is to tap the ball joint stud seat with a hammer as illustrated in Figure 16-26. With the ball joint loosened, the tapered socket and the spring forces work together to cause the ball joint to separate. *Never* use heat to free ball joints. Heat will weaken the components.

On vehicles with the spring force acting on the lower control arm, the spring force must be relieved from the ball joint to remove the steering knuckle assembly from the suspension. The jack is positioned to support the spring force so the ball joint is unloaded. Raise the jack just enough to move the rebound snubber away from its contact point. Do not jack it so high that it will raise the vehicle off the jack stands.

The ball joint nuts are completely removed and the steering knuckle assembly is moved aside. The knuckle must be handled carefully and supported to prevent damage to the flexible brake line, steering linkage, or brake assembly. Some vehicles may require the separation of the brake line from the wheel cylinder to allow the suspension to move far enough to provide space for the service technician to work.

Figure 16-26 Loosening a ball joint by hitting the knuckle fastening with a hammer while it is backed up with a second hammer.

Front Suspension Service

With the steering knuckle assembly out of the way, the ball joints may be removed. Ball joint removal is most often done by removing rivets that attach the joint to the control arm or A-frame. Some lower ball joints are bolted to the steering knuckle and the ball stud fits into a socket in the lower control arm. Other alternatives are ball joints that do not separate from the control arm or that are a press fit into the control arm. The two parts of the nonseparable type are replaced as a unit. Check the applicable shop manual to verify the type with which you are working.

When the riveted type is used, the rivets must be chiseled or drilled out. Care must be taken when drilling or chiseling to avoid damage to the control arm. Replacement ball joints use bolts to attach the joint to the control arm or A-frame. The bolts provide a solid and secure mounting for the service replacement.

When reassembling suspension components, care should be used to prevent damage to the boot seals on the new ball joints. They are needed to retain the grease. After assembly, suspension bolts and nuts are torqued to specifications with a torque wrench. Proper torque is important to properly seat the tapered stud used on many of the suspension components. All cotter pins that have been removed must be replaced with new ones.

Control Arms, A-Frames, and Bushings. Control arm and A-frame removal is accomplished in much the same manner as removing ball joints. The disassembly procedure for ball joint removal is followed, up to the point of removing the ball joint from the arm. This will expose the control arm and/or A-frames. The ball joint does not have to be removed.

Upper A-Frame Removal. If the unloaded upper A-frame is the only suspension member to be removed from the vehicle, the spring-loaded lower control arm should be jacked to support the spring force. It is not necessary to remove the coil spring or the torsion bar to service the unloaded A-frame. The required disassembly would look like Figure

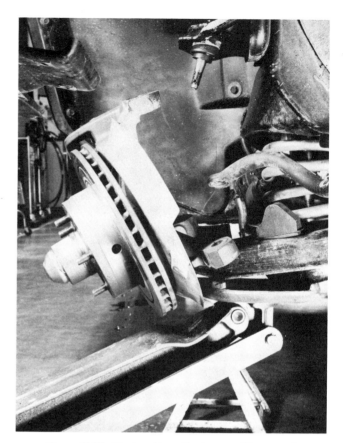

Figure 16-27 The knuckle and brake assembly removed from the ball joints with the spring remaining in place.

16-27. Conversely, on vehicles using a spring force on the upper A-frame, the coil spring does not have to be removed if the lower control arm or A-frame is the only suspension member that requires service.

Spring Removal. The coil spring or torsion bar must be removed to service the spring-loaded lower control arm or A-frame. First, the shock absorber is removed. The stabilizer bar is disconnected from the A-frame or control arm. If the arm is to be removed, the coil spring is removed using a spring tool which compresses the spring. The spring tool fits up through the coil spring and attaches to the highest and lowest accessible coils of the spring. The center bolt of the compressor is tightened. This will shorten the effective distance between the attaching points to compress the spring. Figure 16-28 shows a spring held with a compressor ready to be removed.

Parallel torsion bars are removed by loosening the height adjuster bolt on the lower control arm. The retaining clip inside the rear torsion bar bracket is removed. The bar is driven rearward using a torsion bar removing tool designed for this

Front Suspension Service

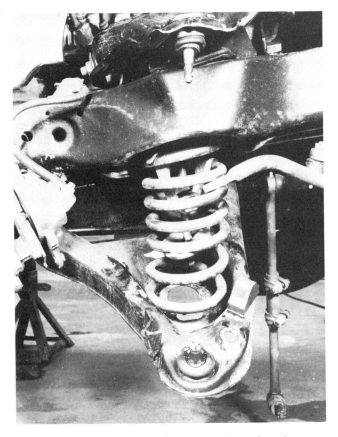

Figure 16-28 The control arm lowered away from the spring. The spring is partly compressed with a spring compressor.

purpose as shown in Figure 16-29. This method prevents damage to the torsion bar during removal.

Vehicles using spring force on the upper A-frame must have the spring removed prior to servicing the spring-loaded upper control arm. The spring is removed in the same manner as previously described using the spring compressing tool.

Lower Control Arm Removal. The lower control arm is removed by first removing the brake strut. It may be attached to the frame with a pivot held by two or more bolts. If the bolts attaching either upper or lower control arms to the frame are equipped with eccentric adjustments for camber, the adjuster position should be *marked* to speed alignment after reassembly. If shims are used, they are taped together and marked to prevent loss or mixing with shims from the opposite side. With a slot adjustment, marking will also aid reassembly.

A-frames are connected to the frame using a cross-shaft inserted through the bushing and then bolted to the frame (Figure 16-30). They often will have bolts passing through brackets and through the control arm bushings. When all of the load is off the control arm, the bolts may be removed.

Bushing Replacement. Once the arms are removed, the bushings in the control arms or A-frames are easily replaced. Worn bushings are pressed from the control arm or A-frame (Figure 16-31). Removal of the complete bushing is necessary. Care should be taken to prevent damage to the A-frame or control arm during removal. A-frames employing a cross-shaft are serviced with either new bushings or an entire bushing and cross-shaft assembly.

Figure 16-29 Driving a torsion bar rearward using a torsion bar removing tool clamped around the torsion bar.

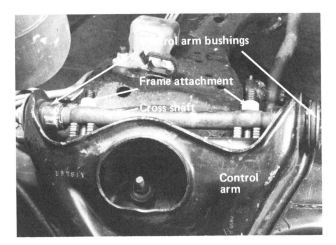

Figure 16-30 Typical control arm attachment to the frame.

257

Front Suspension Service

Figure 16-31 A clamp-type tool used with adapters to remove and install control arm bushings.

It is important that bushings be *neutralized* during assembly. The purpose of neutralizing the bushing is to allow the bushing rubber to be in a relaxed condition when the vehicle is standing and to allow it to twist within its natural limits during operation. If the bushing is installed and tightened with no reference to the A-frame position, the bushing will have to be twisted during assembly to properly position the suspension. This is called *preloading a bushing*. Once this occurs, the bushing is destroyed. To prevent shearing the bushings, the A-frames are positioned on the vehicle in the center of their normal travel. The end bolts that press on the inner sleeves are then tightened. For bench assembly, service manuals will give reference angles that must be used to set them for optimum service life.

Assembly. The upper and lower control arms or A-frames are properly positioned on the vehicle. The bolts and any shims that were present on disassembly are reinstalled. The control arms are placed at their normal ride height position and then the bolts are tightened. Eccentric cams or bolts in slots are positioned as they were prior to disassembly.

On suspension systems having the coil spring above the upper A-frame, the steering knuckle is re-fastened to both the upper and the lower ball joints. The brake strut is then fastened to the lower control arm. The spring is inserted in place and the spring tool removed. *Note: The end of the coil fits into a specific recessed area in the spring seat. The coil spring must be positioned properly in its seat.*

Vehicles that have the coil spring seat in the lower A-frame are assembled in a slightly different manner. After the A-frames are mounted on the vehicle, the coil spring is positioned in the spring tower. The lower A-frame or control arm is raised into position with a jack. The steering knuckle is attached to the upper and lower ball joints making sure the nuts are finger tight. The jack and spring compressor can now be removed, making certain the coil spring is properly located in its seat. When a spring compressor is not used, the lower A-frame is carefully jacked into position as the technician guides the spring into the spring seat. The steering knuckle can then be connected to the upper and lower ball joints and the nuts made finger tight.

Vehicles using torsion bar suspension are assembled in much the same manner. After connecting the ball joints to the steering knuckle, the torsion bar is installed making sure the left and right torsion bars are placed on the correct side.

The brake struts, when used, are connected to the lower control arms. With the suspension in its normal position, all bolts and nuts used for assembly are torqued to correct specifications. New cotter pins are installed where required. If the tie rod end was disconnected during disassembly, it is reconnected, the nut torqued, and a cotter pin installed. Shock absorbers and stabilizer bars are installed and their fasteners secured. *Note: Care must be used to align the stabilizer bushings in their proper holes to prevent metal to metal contact.* If front wheel bearings require repacking with grease, follow the procedure in Section 2-5. The brake drum or rotor is then installed on its spindle and the bearings properly adjusted. With the spindle nut safetied the dust cover is replaced. The brake caliper is installed on the rotor then the assembly bolts are put in place and torqued to specifications. The tire and wheel assemblies are installed and the wheel nuts are properly torqued. The suspension components are lubricated as required, the jack stands removed, and the vehicle is lowered to the floor to complete the assembly procedure.

Wheel alignment should follow assembly to complete the repair. This is most important due to the close relationship of the suspension components to alignment angles.

Ride Height Adjustment. The position of the control arm in relation to the vehicle frame has a bearing on front suspension angles. Because of this close relationship, the ride height of the vehicle must be measured; it may require adjustment before alignment is performed.

Before measuring the ride height, remove any excess cargo from the vehicle. Tire pressures and tire sizes are inspected and changed if incorrect. The fuel tank must be filled for an accurate ride height measurement. The ride height is measured, following manufacturer's recommendations, and compared to specifications.

On vehicles using coil spring front suspension, shims can be placed beneath the coil spring seat to obtain a maximum of one-half inch (13 mm) change in the ride height. If the suspension height is more than one-half inch out of specification, both coil springs will have to be replaced.

Torsion bars, equipped with height adjusters, are reset for correct ride height. To adjust height, the vehicle is positioned on a level floor surface. Each adjusting bolt is turned until the specified measurement height is achieved. Figure 16-32 shows a typical torsion bar adjuster.

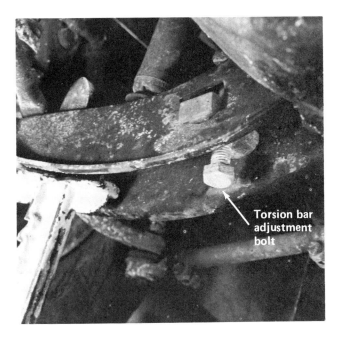

Figure 16-32 Typical location of the vehicle suspension height adjustment bolt on a torsion bar-equipped vehicle.

16-3 STEERING LINKAGE SERVICE

The steering linkage is repaired with the vehicle raised. It should be inspected to verify the cause of the problem. Any time that work is done on the steering linkage the toe should be adjusted and for most satisfactory operation, a complete wheel alignment should be done.

Tie Rod Service. Tie rod ends and tie rod assemblies are often replaced as a result of wear or bending. For replacement, the tie rod adjusting sleeve clamp is loosened. Penetrating oil sprayed on all fasteners to be removed will aid in disassembly. Remove the cotter pin and loosen the castellated nut at the tie rod ball stud. Using either a tie rod spreader or the two hammer method, the tie rod stud is freed from its tapered socket. With the two hammer method, as illustrated in Figure 16-33, slight force should be applied to help separate the stud. When the entire tie rod assembly is to be replaced, both ends of the tie rod are removed. When only one end is being replaced, the castellated nut is fully removed and the tie rod end is unthreaded either right or left handed from the adjuster sleeve. To aid reassembly, measure the length from pivot to pivot or count the number of turns required to remove the tie rod end. The replacement tie rod end is threaded into the sleeve with approximately an equal number of threads on both inner and outer tie rod ends. The tie rod ball stud is placed into its tapered socket and the nut installed. The steering wheel is positioned so that the wheels point straight ahead. The tie rod stud nut is torqued to the correct specifications and a new cotter pin is installed. The tie rod ends must be positioned with each ball stud in the center of its free movement to prevent binding. The adjuster sleeve clamp is rotated on the adjuster sleeve so that it is clear of any suspension components and in the proper position to best hold the sleeve, as shown in Figure 16-34, and then the bolt is tightened. The part is lubricated and checked to make sure that

Front Suspension Service

Figure 16-33 Loosening a tie rod end from the steering arm with two hammers, one used to support the arm and the other used to hit the arm.

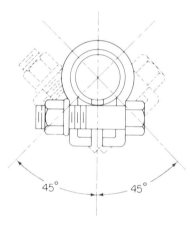

Figure 16-34 Typical tie rod sleeve clamp position range.

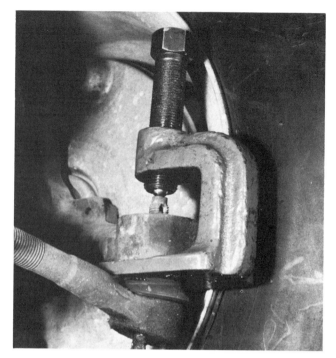

Figure 16-35 Use of a tie rod end puller.

Figure 16-36 Typical idler arm.

there is no interference with the steering linkage as it moves through its full travel.

Center Link Replacement. To replace the center link, disconnect both inner tie rod ends from the link. The link is also removed from the Pitman arm and the idler arm. Most often this is done by taking out a cotter pin and loosening the castellated nut from each end and using a spreader to separate the joint. If any of the tie rod ends or other connections are to be reused, a puller similar to the one in Figure 16-35 is used to prevent damage to the seal at the joint. A tie rod spreader will damage the tie rod end seal, so it should not be used if the joint is to be placed back in service. An alternative design may be found on the idler arm end, which would use a rubber bushing, as shown in Figure 16-36, eliminating the need for using the spreader to separate the connection. All loosened nuts and the center link can be removed from the vehicle.

Installation of the replacement link is accomplished by installing all connections, centering the steering wheel, and torquing the nuts to specifications. The cotter pins are replaced and all grease fittings lubricated. The linkage is inspected for interference with the suspension components anywhere through their full travel.

Idler Arm Service. Idler arm service may consist of only bushing replacement or it may require replacement of the complete assembly, depending upon the idler arm design. The serviceable idler arm incorporates a replaceable press-fit bushing. The nonserviceable idler arm is usually of the ball and socket design or it incorporates bearings. Both types require complete assembly replacement. Some idler arms may be serviceable on one end but they require replacement if wear occurs on the other end.

On the serviceable idler arm, the arm is disconnected at the center link and at the frame mounting bracket. The fastener is removed to disconnect the arm from the frame bracket. The bushings may be removed using a suitable puller or by carefully driving the bushing out with a suitable driver. The new bushing is pressed in using properly fitting adapters. The idler arm is now replaced and all connections are torqued to specifications. New cotter pins are installed in all studs equipped for their use. The wheels must be in the straight-ahead position while the nuts are being tightened to prevent preloading the new bushing.

The nonserviceable type arm is replaced when worn. It is removed and the replacement part is set in place and the mounting bolts torqued to specifications. The center link is then positioned on the stud. With the wheels in the straight-ahead position, the nut is torqued to specifications. Cotter pins are installed in all locations requiring them. Any lubrication fittings are greased to assure proper service life.

Pitman Arm. The Pitman arm does not normally require replacement. Causes for replacement would be accidental damage or, if equipped with a bearing stud joint, wear of the joint. The Pitman arm is removed by first disconnecting the steering linkage. To do this, take out the cotter pin, remove the castellated nut and then, use a suitable puller to separate the joint stud from its socket. The arm is disconnected from the Pitman shaft by first removing the nut and lock washer. The Pitman arm is then pulled from the shaft with a puller similar to the one shown in Figure 16-37. Hammering on the arm or shaft must *not* be done because the steering gear may be damaged. When replacing the Pitman arm, install it on the Pitman shaft with indexing splines correctly aligned. These can be seen in

Figure 16-38. The lock washer and nut are installed and torqued to specifications. With the wheels in the straight-ahead position, connect the steering linkage and then install and torque the castellated nut. The cotter pin is replaced and the linkage checked for binding.

Figure 16-37 Pitman arm puller in use.

Figure 16-38 Pitman arm and Pitman shaft indexing splines.

Front Suspension Service

Steering Arm. The steering arm should be replaced if, during initial suspension and steering linkage inspection, the arm showed visible damage. Excessive tire squeal on turns, while other components appear normal, would prompt the technician to check toe-out on turns. An out-of-specification reading for toe-out on turns indicates a bent steering arm.

Depending on the manufacturer, steering arms may be bolted to the steering knuckle, may be a part of the steering knuckle, or be a part of the lower ball joint. These are shown in Figure 16-39. Due to the high forces placed on steering arms in service, heat must not be used to straighten them. Heat will alter their heat-treated strength. The vehicle is raised and the wheels and brakes are removed as previously described to remove the steering arms. The tie rod is then removed.

Steering arms that are not an integral part of the knuckle or the ball joint are now readily replaced. After disconnecting two bolts and nuts passing through the steering arm and knuckle, the arm is removed. The new steering arm is positioned in place and the bolts inserted. The nuts are installed and torqued to specifications.

If the steering arm is an integral part of the knuckle or ball joint, the procedure for ball joint and steering knuckle replacement described earlier in the chapter should be followed.

After the steering arm has been properly attached, the tie rod end is reconnected to the steering arm and the nut torqued. The wheel bearings are inspected for wear and repacked if necessary. The hub and rotor, or hub and drum, are installed on the spindle and the wheel bearing properly adjusted. Install the cotter pin and dust cover. The tire and wheel assembly is installed using the lug and wheel alignment marks to locate them in the same position they were in before disassembly. The wheel nuts are torqued to specifications and the wheel cover installed. The vehicle is lowered to the floor. A wheel alignment must follow the repair to assure correct alignment angles, steering, and vehicle handling.

(a)

(b)

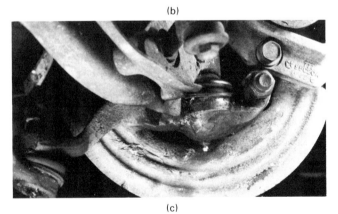

(c)

Figure 16-39 Steering arms: (a) bolted to the knuckle; (b) part of the knuckle; (c) part of the ball joint.

REVIEW QUESTIONS

1. How can the automotive service technician recognize abnormal vehicle handling characteristics? [INTRODUCTION]
2. What causes suspension parts to become unsafe? [INTRODUCTION]
3. How are shock absorbers checked? [16-1]
4. How are load-carrying ball joints checked when the spring is on the lower control arm? [16-1]
5. How are the load-carrying ball joints checked when the spring is on the upper control arm? [16-1]
6. How is a friction-loaded ball joint checked? [16-1]
7. What precautions should be observed while removing shock absorbers? [16-2]
8. Describe the procedure used to remove a MacPherson strut to check the shock absorber. [16-2]
9. Describe the procedure used to replace a bushing on the front stabilizer bar. [16-2]
10. Describe the procedure used to replace a loaded ball joint. [16-2]
11. Describe the procedure used to replace inner support arm bushings. [16-2]
12. What precautions should be observed when installing front coil springs and torsion bars? [16-2]
13. How is the ride height adjusted with coil spring suspensions? [16-2]
14. What service can be performed on the steering linkage? [16-3]

Rear Suspension Service

The rear suspension plays an important role in vehicle handling. Vehicle ride is as dependent on the rear suspension as on the front suspension. Most of the useful load the vehicle carries is supported by the rear springs. The amount of load being carried will change the ride height. Ride height change at the rear will affect the front end alignment angles as discussed in Chapter 14. Wear in the rear suspension components will produce a tendency for steering pull during braking and acceleration if the rear axle moves beyond normal limits. Worn suspension components on independent rear suspensions will affect the alignment of the rear wheels as well as the front wheels. Vibration that is caused by misaligned drive line components may be a direct result of a rear suspension that is out of alignment. Because of these and other rear suspension problems, service of the rear suspension is periodically required.

17-1 SHOCK ABSORBERS

Rear shock absorbers are usually replaced with the rear axle positioned at or near its normal ride height. The vehicle is raised and supported in a position so that the suspension will hold the vehicle weight and still allow the technician adequate room to remove the shock absorbers.

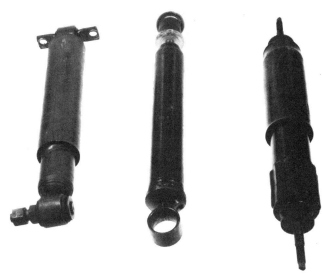

Figure 17-1 Typical rear shock absorbers.

Figure 17-3 Rubber insulators and retainers on a shock absorber stud.

Removing Rear Shock Absorbers. The upper end of the shock absorber mounting may be accessible from beneath the vehicle or through access holes placed either in the trunk or behind the rear seat back. The lower end of the shock absorber mounting is accessible from the underside of the vehicle. Some shock absorber mounting ends are shown in Figure 17-1. The lower end is disconnected first, enabling the technician to cycle the shock absorber when there is a question about its effectiveness. This procedure is the same as testing and purging front shocks as described in Section 16-2.

A "nut cracker" becomes an effective tool for stubborn shock absorber nuts in a situation where the shock absorber may be replaced in service. With this tool, as pictured in Figure 17-2, the technician can remove the end of the shock absorber without breaking the mounting stud or spike. The upper end can then be disconnected and the shock absorber removed from the vehicle.

Installing Rear Shock Absorbers. To prepare shock absorbers for installation the required retainers and insulator bushings (Figure 17-3) are placed over the spike end of the shock, or into the eyes of the shock absorber, depending on design. The upper end is mounted and the insulators, retainers, bolts, and nuts are installed loosely. The lower end is then connected using the proper retainers and insulators. The upper and lower connections are tightened and when used, pal nuts are installed to complete the job.

Figure 17-2 A upper shock absorber mount nut being split with a "nut cracker."

17-2 STABILIZER BAR

Rear stabilizer bar service includes the replacement of links, bushings, and the entire stabilizer bar. The technician should not attempt to straighten a damaged stabilizer bar.

Rear stabilizer bars may be bolted to the trailing arms, mounted to the frame with links connected to the suspension, or mounted to the suspension with links connected to the frame or underbody. A typical rear stabilizer bar installation was illustrated in Figure 13-34.

Rear Suspension Service

Rear Stabilizer Bar Service. For service, the vehicle is raised and supported on a hoist or jack stands to provide room for work on the stabilizer bar. The position of the suspension is not critical while servicing the stabilizer bar.

Stabilizer bars bolted to the trailing arms do not employ bushings or linkage. Removal is accomplished by removing fasteners connecting the bar to the suspension components.

Stabilizer bars mounted to the body or axle housing use links with insulators and bushings. Broken linkage, damaged bushings, and worn insulators will result in poor vehicle handling. These faulty parts will cause rattles or noise as the vehicle maneuvers. Connecting linkage insulators and retainers can be replaced while the stabilizer bar remains mounted on the vehicle. A through bolt holds the assembly of insulators, retainer, and spacer. They are disassembled to allow replacement of the worn components. If the pivot bushings are to be replaced, the bar is disconnected by removing the bolts holding the bushing bracket in place. Bushings may either be split to allow easy removal or it may be necessary to slide the bushing in from the end of the bar.

The bushings are placed on the bar before the stabilizer bar is installed. The bar is placed in position and secured loosely until the weight of the vehicle is supported by the springs. The fasteners are torqued to specifications with the suspension at its normal ride height. Finally the linkage is assembled and the link bolt tightened. Jouncing the vehicle will assure the service technician that no binding or noise problems exist in the installation.

17-3 TRACK BAR

A track bar, sometimes called a Panhard rod, is used with some coil spring rear suspensions to prevent excess lateral body movement during turns (Figure 17-4). If the bar breaks or the bushings wear, dangerous handling, uncomfortable ride,

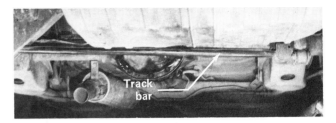

Figure 17-4 A typical example of a rear track bar.

and noise will result from the sideways body movement. In extreme cases this movement will allow the tires to rub the fender wells.

If the track bar is to be serviced when it is found to be faulty, the fasteners are removed from both ends of the track bar with the vehicle raised and supported on the rear axle. The bar can then be lifted from the vehicle.

The bushings in the bar ends are replaceable. The old bushings are pressed out and the new bushings pressed in. If the bar is bent or broken, it is replaced with a new one having new bushings already in place.

The bar is placed over the frame lug first, then, by pushing on the body slightly, the lug on the rear axle is aligned. The retainer washers and nuts are installed and torqued to specifications. The vehicle is then lowered to the floor.

Recall that the shock absorbers, stabilizer bars, and track bar have no effect on ride height. Springs establish the ride height. If the rear ride height is out of specification, all components of the suspension should be inspected even though damage to only one of them may be apparent.

17-4 SPRING REPLACEMENT

Spring service is required whenever the ride height is not within specifications, or when there is visual damage to the spring. Coil spring removal may be necessary to correct damage or wear that has occurred to the trailing arms.

Coil Spring Service. To remove the spring the vehicle must be raised on a frame contact hoist or have jack stands placed under the frame *forward* of the axle assembly. A floor jack is positioned beneath the center of the rear axle housing and raised to support the rear axle weight, but not enough to lift the vehicle from the jack stands. The lower end of the shock absorbers are disconnected. The jack is then lowered, allowing the rear axle to drop below its normal travel limit to relieve all of the coil spring force as illustrated in Figure 17-5. The coil spring positioner, if used, is unbolted from the center of the coil spring. On many vehicles, the coil spring may then be removed from the vehicle by lifting it from its spring seat. If the coil spring remains under load after the rear axle assembly is lowered, a spring compressor will have to be used to shorten the spring. The spring compressor is inserted in the spring and the coils compressed so the spring can be removed in a manner similar to front spring removal. If the springs are to be placed in service again, mark each spring so it can be returned to its original location.

Figure 17-6 A typical spring locator on a rear axle housing.

Figure 17-5 A coil spring being removed from the rear suspension.

Coil springs may be shimmed to adjust the side-to-side ride height when there is less than one-half inch (13 mm) difference. Coil springs that vary more than this will have to be replaced. The insulators used on the top half of the springs are replaced if they show signs of damage. When replacement is required, coil springs are always replaced in pairs to assure a uniform level ride height.

When necessary, the coil spring is compressed for installation. The insulator is then placed on the top of the coil spring and the spring is positioned on the spring seat. The location of the end of the top coil is critical. It must be positioned to coincide with the recess in the spring seat. The spring compressor can then be removed. The opposite side coil spring is installed in the same manner. The rear axle housing is raised into position and the shock absorbers reconnected. The spring locator is shown in Figure 17-6.

The service technician may find it easier to disassemble and reassemble one spring at a time. If he chooses to do so, the jack is positioned on the side being serviced. This procedure will eliminate any chance of placing the parts on the wrong side

Rear Suspension Service

of the vehicle and will maintain major component alignment.

Leaf Spring Service. Removal of the leaf spring is accomplished with the vehicle raised and supported on the frame. Jack stands are placed on the frame *ahead* of the springs. A floor jack is positioned under the rear axle housing close to the spring being removed. It is recommended, when working with leaf springs, that one side be serviced at a time. The opposite side will give support to the suspension and aid in locating the axle housing on the spring being serviced.

The axle housing is raised slightly and the lower end of the shock absorber disconnected. The jack is now lowered so the leaf spring does not place any force on the rear axle housing. The U-bolts or bolts holding the iso-clamps together are removed as shown in Figure 17-7. The lower spring seat is removed and the axle housing jacked up from the leaf spring.

The rear spring shackle nuts are the next items to be removed. The spring must be supported as pictured in Figure 17-8 to prevent it from dropping and harming the technician before removing the rear shackle or the front spring eye bolt or bracket. The spring front attachment may have a mounting bracket bolted to the underbody. It is removed from the frame to free the front of the spring. An

Figure 17-8 Removing the spring shackle while the spring is supported with a jack.

alternative design is to have the spring front eye bolt pass through the vehicle frame. This mounting requires the technician to remove the spring eye bolt to free the front of the spring. When both ends of the spring are free, the spring can be lowered to the floor.

Leaf spring service includes replacing the spring eye bushings, the interleaf separators, and the center bolt. The complete spring assembly is often replaced when leaves are broken. With the spring on the workbench, eye bushings are readily pressed out of the main leaf and new bushings pressed in. The spring may be disassembled by carefully clamping C-clamps on either side of the spring center bolt as shown in Figure 17-9. The center bolt is removed and the C-clamps loosened to allow the leafs to separate.

New interleaf separators, center bolt, and required spring leaves are positioned for reassembly. It is common practice to purchase center bolts approximately one inch (25 mm) longer than the original equipment bolts to aid in assembling the spring. The excess bolt length can be cut off with a hacksaw after the spring has been completely reassembled.

The assembled leaf spring is positioned under the vehicle and the front of the spring connected.

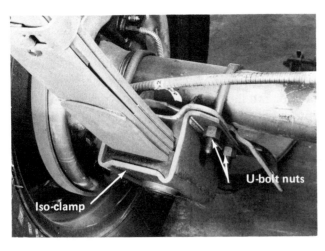

Figure 17-7 U-bolts loosened to free a leaf spring.

Spring Replacement

Figure 17-9 Holding the spring together as the center bolt is serviced.

On some vehicle spring designs a spring spreader is required to aid in connecting the rear spring shackle to the spring. The spreader, as shown in Figure 17-10, increases the span between the spring eyes to allow the spring shackle to be connected. In some installations the front bracket is connected and the rear shackle is pulled rearward to align with the frame mounting location as shown in Figure 17-11. If a spreader is not available the front end of the spring is connected. Then the rear axle housing brackets are positioned so the housing can be fastened to the spring. The center of the spring is jacked upwards and the rear end of the spring is pulled down, until the shackle bolt will pass through the spring eye. This forces the arch in the spring to decrease, thus increasing the distance between the spring eyes so the shackle can be attached.

The axle housing must be properly centered over the spring center bolt. The U-bolts and nuts used for clamping the spring to the axle housing are installed with the iso-clamps in place and torqued to specifications. Overtorquing the U-bolts may result in excessive noise transfer, or damage to the axle housing or spring. The shock absorbers are reconnected. The wheels and tires, if removed, are now replaced and the wheel covers installed to complete the repair.

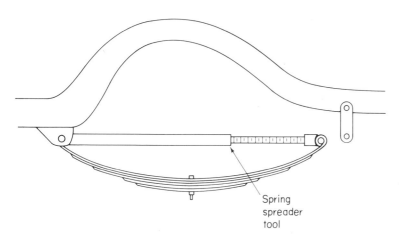

Figure 17-10 Use of a typical spring spreader tool to align a leaf spring eye with the shackle holes.

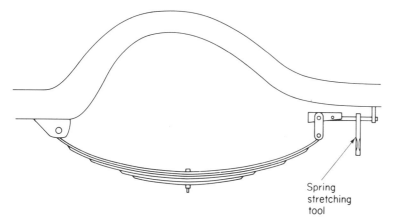

Figure 17-11 Using a typical spring stretching tool to align a leaf spring shackel bracket with the frame holes.

Rear Suspension Service

17-5 CONTROL ARM SERVICE

The control arms or trailing arms used with coil spring (Figure 17-12) and transverse leaf rear suspension may occasionally require service. Service involves the replacement of the bushings or the entire control arm, depending on design.

The vehicle is raised and supported on the frame so the suspension is fully extended. When only the upper control arms are to be serviced, jack stands must also be placed beneath the axle housing to provide support and hold the rear suspension in alignment while the control arms are being serviced.

Upper Control Arm Service. Upper control arms are removed from the vehicle by removing the bolts passing through the control arms at the frame and at the axle ends. Servicing one side of the vehicle at a time will make it easier for the technician to realign the components during assembly. It is dangerous to attempt to service both sides of the vehicle at the same time because the axle housing may rotate. This could result in damage to the vehicle and injury to the technician. Control arm bushing replacement is accomplished on serviceable control arms by removing the defective bushing with a suitable puller. The new bushing is properly positioned and pressed into place. Some service bushings and bolts are provided with an eccentric adjustment to allow for drive line angle adjustment after service, as shown in Figure 17-13. The repaired control arm is positioned in the vehicle and the fasteners are loosely installed. The service is repeated on the control arm located on the other side of the vehicle, if required. The bolts and nuts are properly torqued when the entire vehicle weight is on the springs.

Lower Control Arm Service. Lower control arm service is usually performed with the coil spring removed. One side of the vehicle is serviced at a time to prevent rotation of the axle housing assembly. With the vehicle properly supported, the bolts and nuts that pass through the control arm are removed. The control arm is removed from the vehicle and serviced in the same manner as the upper control arm. After service, the control arm is positioned under the vehicle and the bolts and nuts loosely installed.

If a drive line working angle adjustment is not specified in the applicable service manual, the technician can torque the control arm bolts to specification while the full vehicle weight is on the rear axle. This will provide neutral bushing tension at normal ride height.

Adjusting Drive Line Working Angle. If the vehicle has an adjustment for drive line working angle, the angle is adjusted prior to tightening the control

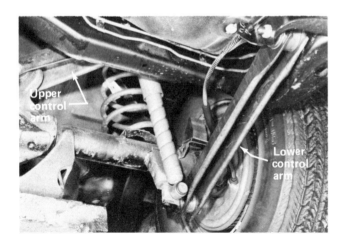

Figure 17-12 Typical control arms used with rear coil spring suspensions.

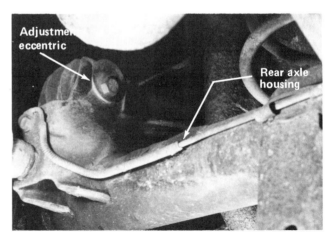

Figure 17-13 Typical control arm eccentric adjustments on a rear suspension.

arm bolts. The working angle of the universal joints on the drive shaft (Figure 17-14) is always checked after rear suspension service has been completed to minimize the possibility of drive line vibration. The rear working angle of the drive line is the angle between a center line drawn through the drive shaft and the center line drawn through the rear axle drive pinion. The front working angle is the drive shaft center line compared to the angle of the transmission main shaft centerline.

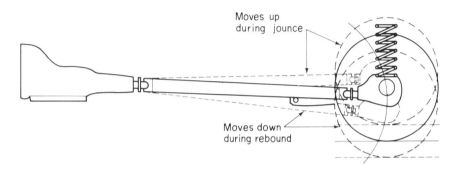

Figure 17-14 The working angle of the propeller shaft changes as the rear suspension moves from jounce to rebound.

Working angle is important because of the characteristics of a nonconstant velocity universal joint. The universal joint working through an angle will slightly increase and decrease the speed between the input and output members twice during each revolution. The greater the working angle the greater this speed change will be. The speed or velocity change results in a vibration when the working angle is excessive or if the front and rear universal joint working angles are different. The working angle should be greater than zero to move the bearings in the universal joints. This increases their service life.

A constant velocity joint, as shown in Figure 17-15 and used on some vehicles, does not produce a speed change. The working angle is not as critical on the rear constant velocity joint and consequently there may be no provision for working angle adjustment when these joints are used.

After rear suspension work is completed, the nose of the differential may have a different angle than it had before disassembly. This angle must be corrected so that it is within specification limits to assure minimum drive line vibration. To check the universal joint working angle, the vehicle is raised and supported by the rear axle housing and the front suspension so the entire vehicle is level. This may be done easily on an alignment rack or on a twin post hoist. The rear axle housing must be in its normal ride height position in relation to the vehicle frame. An inclinometer that measures the angular degree from level is held against the drive shaft to measure the drive shaft angle. One type of inclinometer is shown in Figure 17-16. The inclinometer is then positioned against the rear universal joint cup with the cup positioned at the bottom to measure the differential angle. The drive

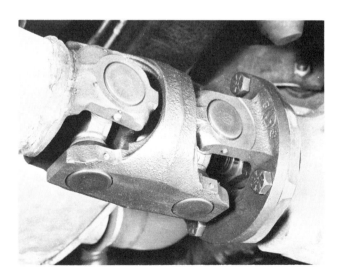

Figure 17-15 A typical constant velocity universal joint on a propeller shaft.

Rear Suspension Service

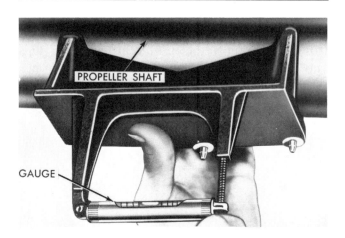

Figure 17-16 An inclinometer designed to measure the working angle of the universal joints (Courtesy of Chrysler Corporation).

shaft angle and the differential angle are compared. Their difference is the joint working angle. The working angle is then compared with the specifications that apply to the vehicle being serviced. If the working angle, usually 1° to 5°, is out of specifications, it will have to be adjusted so that it falls within the specifications.

Adjustment of the working angle, when leaf springs are used, is done using tapered shims placed between the spring and the axle housing pad. Shims are readily available for 1°, 2°, and 4° changes. A typical taper shim is shown in Figure 17-17. The amount of the angle change required is found by comparing the measured angles to the specifications. To install shims, the vehicle must be supported at the frame on stands. The rear axle is then jacked just enough to hold the axle so the U-bolts can be loosened. The taper shims or wedges are positioned between the axle housing and spring to

Figure 17-17 A typical taper shim used to adjust the universal joint working angle on leaf spring rear suspensions.

give the desired angle change as pictured in Figure 17-18. The U-bolts are retorqued and the vehicle is lowered, allowing the axle to again support the vehicle weight in a level position. The body of the vehicle is jounced to assure that the suspension is settled so the measurements of the working angle can again be checked.

The universal joint working angle on the coil spring rear suspension is adjusted by lengthening or shortening the effective length of the upper control arms. This is accomplished with slotted adjustments, shims, or eccentric bolts. The shims space the upper control arm bracket from the frame to change the working angle. The slots or eccentric adjustment provide a change in distance

Figure 17-18 Location of a taper shim on a leaf spring (Courtesy of Chrysler Corporation).

from the axle housing to the frame along the upper control arm to change the differential angle. After the required adjustments are made, all bolts on the control arms are retorqued to specifications. The vehicle is test driven to assure vibration-free operation.

17-6 REAR SUSPENSION ALIGNMENT

Rear suspension alignment problems can be divided into two types: problems that result in excessive tire wear and problems that cause a vehicle to track improperly. A vehicle that does not track properly is most visible when driving on a wet road as the tracks of the rear wheels follow at one side of the front-wheel tracks. Suspension alignment problem detection and correction analysis depends on whether the rear suspension is dependent or independent.

Alignment checks on rear-wheel dependent suspension vehicles include toe, camber, and track. Tire tread scuffing and rapid wear result from toe problems. The rear tires may show uneven wear across the surface of the tread. This usually identifies a camber problem. Tire wear conditions are shown on page 28. Tracking problems occur when the axle centerline is not perpendicular to the centerline of the vehicle as shown on Figure 17-19.

On dependent suspension vehicles with coil springs, tracking problems are most often the result of loose or broken control arms. The control arms maintain the axle position on these suspensions. The vehicle will not track properly if a bushing or arm fails.

Broken rear leaf springs will cause tracking problems on dependent rear suspensions. Track on a leaf spring rear suspension is maintained by the leaf springs themselves. The leaf spring is held together by a center bolt (see page 269). The axle is held in position by the center bolt fitting into a hole in the spring pad on the axle housing. Failure of either the spring or center bolt will allow the axle to move on the spring. This will cause the track to change. Replacement of rear suspension parts is discussed in Sections 17-4 and 17-5.

Rear Wheel Alignment. The axle housing will usually have to be replaced when toe and camber readings are out of specification. Toe measurements that are out of specification result from a bent axle housing. The housing can be bent by a sharp rearward blow to the tire and wheel as a result of a pot hole in the road surface or from an accident. Camber, that is the inward or outward tilt of the top of the wheel, may be incorrect. This is usually caused by severe overloads that will slightly bend the axle housing. In either case, axle housing straightening is usually not recommended. A bent housing must be replaced. Heavy-duty and off-the-road vehicles designed for rough use may be equipped with struts connecting the left and right sides beneath the differential carrier to add strength.

Vehicles equipped with an independent rear suspension often have provisions for making toe and camber adjustments. On this type of suspension, toe can affect vehicle tracking when one wheel is adjusted to point inward and the other wheel is adjusted to point outward. This will cause the vehicle rear end to steer in one direction or the other. Periodic adjustment of the rear suspension angles may be required as a result of sagging springs and wear of the independent suspension components.

Rear suspension adjustment methods differ. They depend upon the type of vehicle. Most

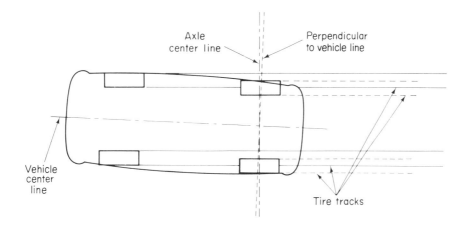

Figure 17-19 The effect of an axle that is not perpendicular to the centerline of the vehicle.

Rear Suspension Service

independent rear suspensions use either shim- or cam-type adjustments. Race cars frequently have adjustable length control arms. The adjustments are used to set toe, camber, and caster. An applicable shop manual should be used to determine the adjustment procedure for the vehicle being serviced.

Measurement of rear suspension angles as well as track can be done using alignment equipment or by using a tape measure and some masking tape or chalk. Some electronic front-end alignment machines are designed to measure rear-wheel camber and toe with the vehicle on the alignment rack. With wheel attachments in position, toe and camber readings can be read directly. Other types of equipment will require backing the vehicle on to the alignment machine and measuring the angles as described in Section 15-3. When this method is used, it is important to realize that toe-in is measured as toe-out. This is caused by the reversed position of the vehicle on the alignment machine. When alignment measurements are being made, it is important that the rear tires are on the radius plates to permit free movement of the tires.

Camber readings will be measured in degrees on the alignment machine, however, many manufacturers specify rear axle camber readings in inches. In this case the camber specifications will have to be converted to degrees. This can be done by using a trigonometric value called an *arcsine* ($\sin^{-1}$). The fractional specification is divided by the distance from the floor to the center of the wheel. The arcsine of this number is the angle in degrees equivalent to the specified distance. This is shown in Figure 17-20.

On dependent rear suspension the vehicle is supported on the frame so the suspension is extended for toe and camber checks. Short pieces of tape are placed on each rear wheel on one tread bar near the center of the tread. The wheels are rotated so the tape is positioned at the extreme front of the tire. The distance between the two pieces of tape is measured and the reading recorded. The wheels are then rotated until the two pieces of tape are at the extreme rear. The distance between the tapes is again measured and the readings recorded. The

Standard conversion of camber specifications

Inches　　Degrees

$\frac{1''}{16} = \frac{1°}{4}$

$\frac{1''}{8} = \frac{1°}{2}$

$\frac{3''}{16} = \frac{3°}{4}$

$\frac{1''}{4} = 1°$

$\frac{5''}{16} = 1\frac{1°}{4}$

$\frac{3''}{8} = 1\frac{1°}{2}$

Example: inches to degrees
Tire: 28" diameter = 14" radius
Specification: $\frac{1''}{8}$ positive

$$\text{Camber (degrees)} = \sin^{-1} \frac{\text{Specification (inches)}}{\text{Tire radius}}$$

$$= \sin^{-1} \frac{0.125 \text{ in.}}{14 \text{ in.}}$$

$$= \frac{1°}{2}$$

Example: degrees to inches
Tire: 28" diameter = 14" radius
Specification: $\frac{1°}{2}$ positive

$$\text{Camber (inches)} = \sin \frac{\text{Specification (degrees)}}{\text{Tire radius}}$$

$$= \sin\left(\frac{1°}{2}\right) \times 14 \text{ in.}$$

$$= \frac{1}{8} \text{ in.}$$

Figure 17-20 Converting degrees to inches for rear camber.

wheels are finally rotated so the tape is at the bottom and that measurement is taken and recorded.

Toe is the difference between the front of the tire readings (FTR) and the rear of the tire readings (RTR). Toe-in occurs when the front of the tire readings is smaller. If the front of the tire readings is larger the axle has toe-out. Thus,

$$\text{RTR} - \text{FTR} = \text{Toe}$$

For determining camber, the front of the tire readings and the rear tire readings are averaged. The bottom of the tire reading (BTR) is subtracted from the average. Thus,

$$\frac{\text{RTR} + \text{FTR}}{2} - \text{BTR} = \text{Camber in inches}$$

If the bottom of the tire reading is larger than the average, the vehicle has negative camber. If the bottom of the tire is less than the average, the camber is positive.

Vehicles with dependent and independent suspension can have the alignment checked another way. The vehicle must be raised and supported at the frame to mark the tire with chalk. This is done by rotating the tire and marking a straight chalk mark around the tire on one tread bar. After the chalk mark is made it is scribed with a sharp tool as the wheel is rotated. This will produce a fine smooth sharp line in the chalk. Measurements can be taken from this line without rotating the tires. Toe and camber on independent suspension vehicles can be measured easily using this method because the vehicle must have its full weight on the radius plates of the alignment machine. The same calculations are done to determine toe and camber readings as described above.

Track Alignment. Track measurements are easily checked with electronic alignment equipment that uses light beams. Measurement is started on each side of the vehicle at a point an equal distance back from the front wheels. The light projects forward from these points on each side of the vehicle as shown in Figure 17-21. The vehicle will track properly if the measurements are equal.

Other methods can be used to find out if the vehicle is tracking properly. One of these is to sight forward across the sidewall of the tire so the front and back of one tire sidewall appear as a single plane. A helper holds a yardstick in a horizontal position, touching the frame at a specific distance back from the front wheels. The assistant moves a marker across the yardstick until the plane of the tire and the marker line up. The position of the marker is noted and the procedure is repeated on the opposite side of the vehicle using the corresponding frame point. The measurements should be identical when measured from identical points on each side of the frame. If the measurements differ, track is incorrect. Track must be set while adjusting toe on independent rear suspensions. Another means of checking track can be made by placing a long board or similar object against the side of the tire. The distance is measured to a frame or body point at an equal distance forward on each side of the vehicle. Again the measured readings are compared.

Tire wear is usually least if the rear wheels have a little positive camber and there is slight toe-out. Power driving the wheels produces toe-in. The normal tendency of the vehicle weight is to cause camber to reduce. These conditions working together make the running toe and camber near zero. During braking, toe-out is produced and this tends to make the vehicle less stable. To compensate for this, some vehicles specify toe-in for the static setting.

If all rear-wheel angles are within specifications, tire wear should be normal. It is relatively uncommon to find excessive tire wear produced by rear-wheel angles in vehicles with dependent suspensions.

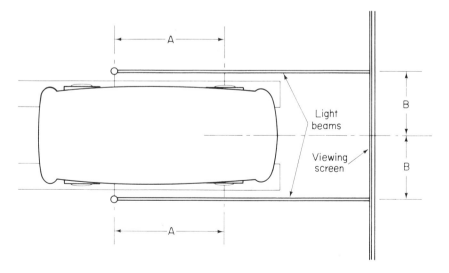

Figure 17-21 Checking vehicle track. The distances marked A are equal and the distances marked B are equal.

REVIEW QUESTIONS

1. What precautions should be observed when replacing rear shock absorbers? [17-1]
2. How do the rear stabilizer bar attachments differ from the front stabilizer attachments bar? [17-2]
3. How is a rear coil spring removed? [17-4]
4. How is a rear leaf spring replaced? [17-4]
5. What is the purpose of a leaf spring center bolt? [17-4]
6. How is the drive line working angle adjusted with coil spring suspension and with a leaf spring suspension? [17-5]
7. When it is necessary to check the alignment of the rear suspension? [17-6]

18-1 STANDARD STEERING GEAR

A standard steering gear unit consists of two gears, a *worm gear* and a *sector* or section of a spur gear. Another type of steering gear, *rack and pinion,* is finding popularity in small automobiles. The steering gear is normally designed to swing the front wheels through a 60° arc, from lock to lock. Turning effort on the steering wheel is multiplied through the steering gear to turn the front wheels, even when the vehicle is not moving. If a steering wheel requires five full turns to swing the wheels 60°, the steering ratio is 30:1. This is because each full turn of the steering wheel is 360°. In five turns, the steering wheel turns 1800°. Eighteen hundred degrees of steering wheel turn divided by 60° of wheel turn equals the 30:1 steering ratio.

Standard steering gears are based on the action between a worm gear and a spur gear section or a rack and pinion to give the required mechanical steering force. Therefore, the driver can easily control the front wheels and the front-wheel road shock will not twist the steering wheel from his hands.

Worm and Sector Interaction. Friction between the *worm* and *sector* in standard gears is usually quite large. This friction was reduced as steering gear designs improved, by replacing the sector with a *roller* (Figure 18-1), and then further reduced by placing a nut with exterior teeth over the worm as pictured in Figure 18-2. The threads between the *nut* and *worm* are modified into grooves that will take ball bearings. Selective fit balls, very close to the same size, are placed between the nut and worm so the turning action is as free rolling as a ball bearing. The ball groove may be cut for either right- or left-hand thread directions.

The balls are fed into what is the equivalent of two separate ball races. These races are on an angle so the balls move endwise as the worm turns. The balls are redirected by a *guide* from the end of each slot back to the entrance across the outside of the

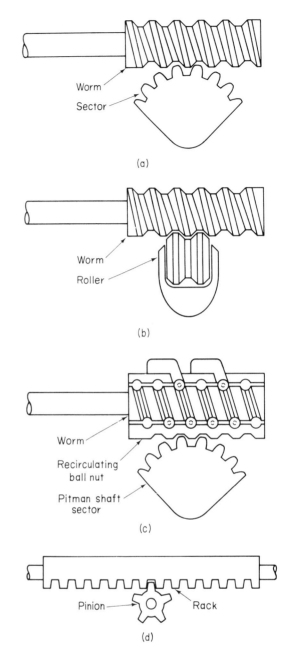

Figure 18-1 Worm-sector, worm-roller, recirculating ball nut and rack-and-pinion type steering gears.

nut. The guide can be identified in Figure 18-3. This type of gear is called *recirculating ball-type*.

The ball-nut gear teeth engage with the sector that, in turn, is part of the *cross-shaft, sector shaft,* or *Pitman shaft*. This whole assembly is enclosed in a case that can be attached to the chassis frame. A *Pitman arm* connects the Pitman shaft to the steering linkage.

Each end of the worm is supported in ball bearings. A worm lash adjustment provides a bearing

18
Steering Gears

The driver controls the direction of the front wheels of the automobile through the steering gear. Modern steering gears are made up of two major units, a *gear unit* and a *steering column*. The gear unit multiplies the driver's steering effort to provide adequate force for steering control. The steering column is primarily a supported shaft that connects the driver's steering wheel to the gear unit.

The power steering gear is similar to the standard steering gear but is has *pressure surfaces* upon which fluid pressure is applied to aid the driver's control of the front wheels. Power for the steering gear is provided by an engine-driven *pump*. The pump forces power steering fluid through a system controlled by a valve. This *control valve,* which is sensitive to the driver's steering effort, puts fluid (hydraulic) pressure against the various *pressure surfaces* in the steering system. This hydraulic effort assists the driver.

The steering column in the modern automobile is a complex mechanism. It is designed to collapse in a collision to protect the driver. In some installations it may be tilted and telescoped to place the steering wheel at a convenient angle for the driver. To reduce the chance of theft, it contains steering gear and transmission locks. Because it is easily accessible to the driver, the steering column may carry the transmission shift control, turn signal switch, flasher switch, head light dimmer, and windshield wiper control.

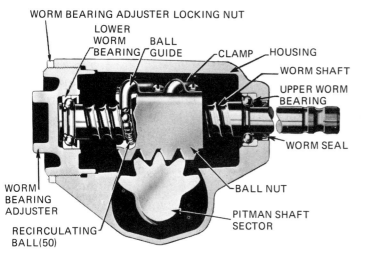

Figure 18-2 Typical worm and recirculating ball-nut type steering gear (Courtesy of Buick Motor Division, General Motors Corporation).

preload that holds the worm securely in position. On some steering gears, the worm shaft bearing preload adjustment is on the lower end of the gear case, while others have the adjustment on the upper end of the case.

Needle bearings or bushings support the Pitman shaft. Both may be located on the Pitman arm

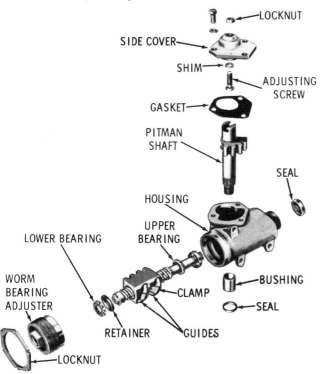

Figure 18-3 Exploded view of a typical worm ball-nut steering gear (Courtesy of Oldsmobile Division, General Motors Corporation).

side of the sector or one may be on each side of the sector. The ball nut gear teeth and the sector teeth are cut at an angle. Moving the sector gear into close mesh with the ball nut teeth reduces the looseness or backlash. This adjustment is made in the cover on the upper end of the cross-shaft. It can be identified in Figure 18-4.

The steering gear teeth are cut to provide a closer fit at the center teeth than at the end teeth as illustrated in Figure 18-5. This reduces gear backlash at the center position where most driving occurs. At the same time, it allows sufficient cost-reducing gear tolerance at the off-center positions where the close fits are not needed. Sector backlash adjustments are made at the tight center position.

The standard steering gear case is partly filled with the specified lubricant. A seal is used at the lower end of the Pitman shaft to keep the lubricant from leaking out the bottom of the gear case. In most standard steering gears, there is no seal on the upper end of the worm shaft because it is usually above the lubricant level, where only a dust seal is required.

Rack and Pinion Steering. *Rack and pinion* steering (Figure 18-6) is used on a number of small

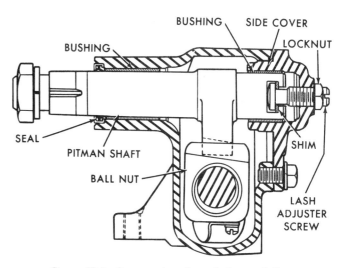

Figure 18-4 Cross-section of a typical worm ball-nut type steering gear showing angled teeth that allow lash adjustment (Courtesy of Chevrolet Motor Division, General Motors Corporation).

Figure 18-5 Typical steering gear sector-to-ball-nut contact (Courtesy of Buick Motor Division, General Motors Corporation).

automobiles. Its design is simple, light, and responsive. It uses few linkage parts. Road shock feedback and steering effort are greater than if a recirculating-ball-type steering gear were used in the same automobile.

The rack and pinion steering gear takes the place of the center steering linkage. The steering gear housing is fastened to the front frame member through rubber-type insulators. The rack ends, under a rubber boot, fasten to the tie rods with ball type ends. A pinion, controlled by the steering wheel, has teeth that engage with the teeth of the rack. These can be seen in Figure 18-7. A yoke bearing puts a preload on the rack directly across from the pinion.

18-2 POWER STEERING GEAR

Power steering is basically power-assisted standard steering. The driver supplies part of the steering effort. The power-assist portion of the unit supplies the remaining effort required. Reduction in driver steering effort allows the power steering gear ratio to be about ⅔ of the standard steering ratio. Power steering, therefore, provides much faster steering response than standard steering.

The power steering system consists of an engine-driven oil pump with reservoir, a control valve, and pressure surfaces that are used to assist steering effort. A schematic drawing of a typical power steering system is illustrated in Figure 18-8. The pump is usually belt-driven and located on the front of the engine. In some applications it is mounted at the front of the crankshaft. Pressure surfaces in most power steering systems are located in the steering gear case. This type of power steering gear is called an *integral type*. Some power steering pressure surfaces are located in an exterior power cylinder connected between the steering linkages and vehicle frame. These are called *link-type* power steering gears. A power rack and pinion is similar to a link-type power steering gear. This is shown in Figure 18-9.

Control Valve. Whenever the engine is running, power steering fluid keeps flowing through the system from the pump to the control valve, then back to the reservoir. Both pressure surfaces are exposed to the same system pressure when the wheels are straight ahead; the fluid flows freely through this open-center circuit. This can be seen in Figure 18-10. When steering effort is applied to the steering wheel, the control valve shifts. This directs fluid to one pressure surface and increases the size of the return passage opening from the opposite pressure surface to the reservoir. The amount of valve shift is proportional to the effort applied to the steering wheel. This, in turn, increases the pressure proportionally to provide the required power assist.

Figure 18-6 A rack and pinion steering gear.

Figure 18-7 The pinion removed to show the rack.

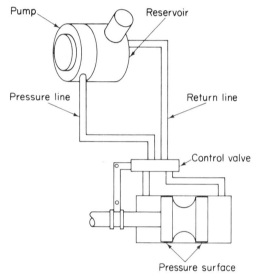

Figure 18-8 Schematic view of a typical power steering gear system.

The power steering control valve is either located inside or is attached to the exterior of the integral-type power steering gear. The link-type power steering may have the control valve built in-to the end of the power cylinder or it may be a separate unit, depending upon its design. The control valve on a rack and pinion power steering gear is built into the pinion housing.

The control valve is balanced between the mechanical input force applied by the steering wheel, and the mechanical-hydraulic resistive force of the steering linkage and tires on the road while the steering wheel is centered. As the steering wheel is turned, its mechanical input force moves the control valve toward the linkage resistive force. This is shown in the upper view in Figure 18-10. In this position, the control valve restricts the pressure area outlet flow so pressure will buildup on the pressure surface. This assists the steering wheel input force to move against the steering linkage resistance. During steering wheel movement, the mechanical input force always leads by first moving the control valve. The hydraulic assist force is always trying to catch up with the control valve position by helping to move the steering linkage. When the steering wheel arrives at the desired position, it is held steady and the linkage finally catches up. This will center the control valve to end the power assist.

The control valve is provided with some type of natural centering device, usually springs. When no effort is being applied to the system, the centering springs center the spool position to balance pressures on both pressure surfaces. This will hold the steering linkage in place.

If the steering wheel is held in position when the front tires hit an object that tries to deflect

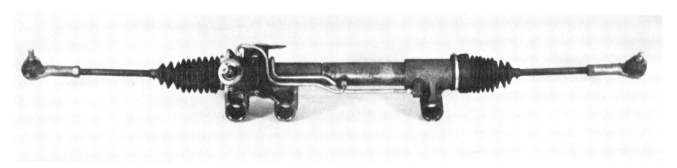

Figure 18-9 A typical power rack and pinion steering gear.

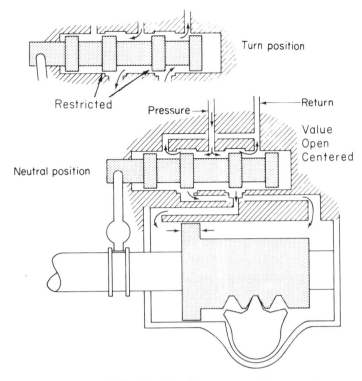

Figure 18-10 Principle of a power steering control valve in neutral (open-centered) and during a turn.

Two types of control valves are used, sliding spools and rotating spools. The *sliding spool* valve may be located between the Pitman arm and steering linkage in the link-type units, mounted concentric with the worm shaft, or placed parallel in a housing outside the steering gear case in integral types. When a *rotary spool valve* is used, it is mounted concentric with the worm shaft or around the pinion shaft. In all power steering gears the action of the spool control valve is similar.

Sliding Spool Valve. The sliding spool valve in the integral steering gear is controlled by either of two actuating methods. In one valve, movement is caused by a slight endways or axial movement of the worm shaft or steering linkage. The front wheels and steering linkages hold the sector so it resists movement. The worm shaft is moved axially against one of its bearings as the ball nut tries to move the sector. This slight movement is transmitted to the parallel spool valve with a pivot lever. This type is shown in Figure 18-11.

The axial movement of the concentric spool valve in Figure 18-12 is actuated by a torsion bar. The worm shaft is turned through a torsion bar, so the upper end turns slightly more than the lower

them, the control valve directs assist pressure to the pressure surface that opposes the deflecting force. This helps control steering feedback to maintain vehicle control.

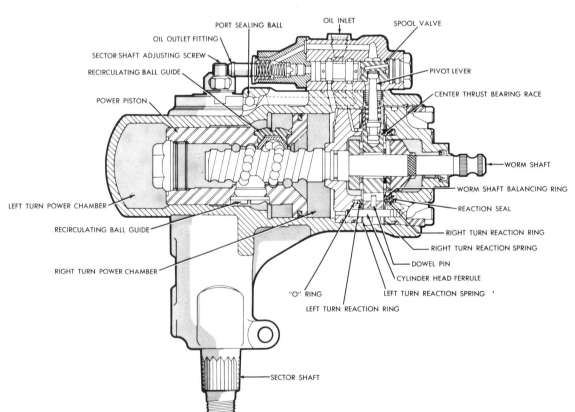

Figure 18-11 Section view of a sliding spool control valve in a power steering gear (Courtesy of Chrysler Corporation).

Power Steering Gear

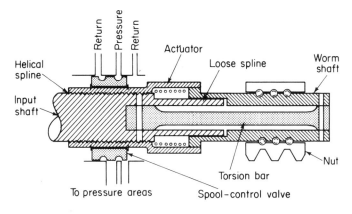

Figure 18-12 Section view of a torsion bar actuated sliding control valve.

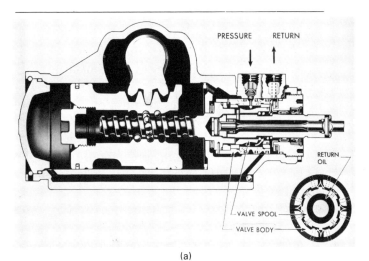

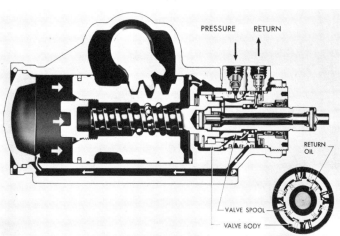

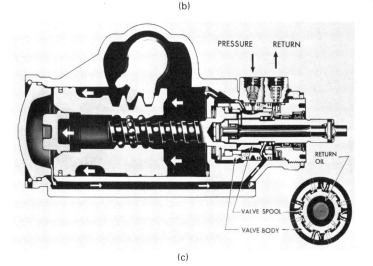

Figure 18-13 Section views of a torsion bar actuated rotating control valve: (a) neutral; (b) right turn; (c) left turn. (Courtesy of Buick Motor Division, General Motors Corporation).

end. The maximum amount of torsion bar flex is limited by a loose fitting spline between the input shaft and wormshaft. A short-length helical spline threaded around the input shaft engages an actuator. The other end of the actuator is splined to the worm shaft.

The steering wheel input effort twists the torsion bar. This twist changes the relative position of the worm shaft actuator assembly in relation to the input shaft position. This action will move the actuator along on the helical splines.

The spool valve is held in position on the actuator with retaining rings so any actuator axial movement also moves the spool valve endways.

Axial movement of the concentric spool valve controls the fluid flow just like the sliding spool type. It opens the fluid return from one pressure surface. At the same time, pressure is built up on the other pressure surface by restricting the return passage. This action provides steering effort assist.

Rotary Spool Valve. The rotary spool control valve, shown in Figure 18-13, is also operated with a torsion bar. The spool valve body surrounds the control valve. The valve spool is attached to the input shaft. Steering effort on the input shaft twists the torsion bar. This repositions the spool valve in the valve body which is attached to the worm shaft. Repositioning the spool valve within the valve body opens a return passage to one of the pressure areas and directs fluid under pressure to the opposite pressure area.

The power rack and pinion steering gear also uses a rotary spool valve around the pinion shaft. It operates in the same way as the valve just

Steering Gears

described. This action is shown in Figure 18-14. The effort required to turn the steering wheel is carried through a torsion bar to move against the resistance of the steering linkage and wheels. The twist of the torsion bar changes the position of the spool valve in the sleeve ports. This turning movement restricts one of the fluid return passages. The restricted passage causes the pressure to build up in one of the pressure chambers. The pressure in the chamber is applied to the pressure surface of the steering gear in the rack housing to assist the driver to make the turn. The same spool valve turning movement increases the opening to the reservoir from the other pressure surface chamber.

Reaction Control. Power steering gears are designed to give the driver some *feel* of the amount of effort he is putting into the steering system. This driver feel is called *reaction control*. The torsion bars provide driver feel when they are used. The pivot-lever-actuated type of control valve gives driver feel with centering springs and with fluid pressure developed in the steering system to provide driver feel.

Power Chambers. The pressure surfaces or areas of the link-type power steering are the surfaces on either side of the piston within the power cylinder. As pressure rises, the chamber fluid is pressurized and fluid is forced into the chamber. At the same time, fluid is returned from the opposite side of the piston.

The pressure surface of a link-type power steering gear is located on a double-faced piston within a cylinder. The cylinder is connected to the steering linkage and the piston rod is connected to the frame. Both connections are designed to allow the movement required. Metal lines or hoses are used to connect the sliding spool control valve to the power cylinder. Sometimes the valve-to-cylinder passages for the fluid are internal. The power piston has two seals on the piston, one facing each end of the cylinder. These seals form the pressure surfaces. Another seal is located around the piston rod as it comes out of the cylinder.

The pressure surfaces in the power rack and pinion steering gear are similar to the link-type. Two outward facing seals are placed on the rack to form the pressure surfaces. The rack housing becomes the power cylinder. In this gear, seals are required on each end of the power rack to prevent external fluid leakage. Metal lines connect the rotating spool control valve to the power chambers. A sectional view of a power rack and pinion steering gear is shown in Figure 18-15.

Ball-Nut Power Piston. In integral power steering gear designs, the ball-nut exterior acts as a piston and so it is called a *ball-nut power piston*. Each side of this piston is, therefore, a power chamber. The

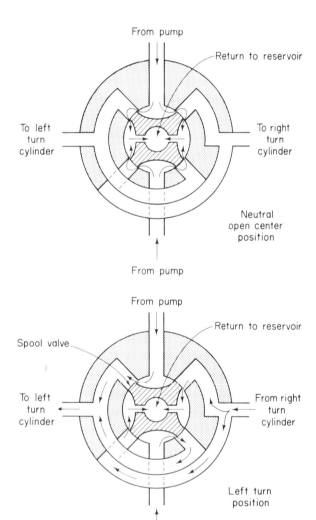

Figure 18-14 A torsion bar activated rotated control valve used in a power rack and pinion steering gear.

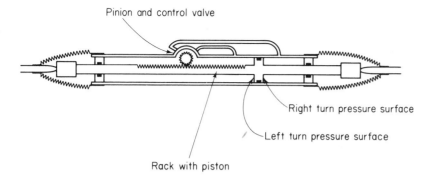

Figure 18-15 Pressure surfaces in a power rack and pinion steering gear.

control valve directs fluid pressure to the side of the piston which will assist the ball-nut in moving the sector in the desired direction.

The ball-nut piston seal ring may be above or below the rack teeth. In either case, the oil pressure can go between the worm shaft and ball nut to pressurize the interior when the ball nut is assisting in the downward direction. Pressure is applied to the ball nut end and piston flange when assisting in the upward direction. Reaction areas are used to balance the worm shaft and control valve in some power steering applications. Details of these pressure areas can be seen in Figures 18-11 and 18-13.

The ball-nut power piston has a rack of gear teeth that rotates a sector in the same way the standard steering does. Steering ratios for power steering gears are usually numerically lower than for standard steering gears, so they have a quicker response to steering wheel movement. A very fast response is not desirable when driving in the straight-ahead position since slight movements would cause too much correction. It is desirable when making a turn. Power steering gears are a compromise when using a constant ratio. Variable-ratio steering gears are used in some automobiles to give a slow ratio when steering in the straight-ahead position and a fast ratio on turns. This is done primarily by the way the sector teeth are cut, as illustrated in Figure 18-16.

18-3 POWER STEERING PUMPS

Three types of power steering pumps are in common usage: the vane type, the slipper type, and the roller type. These are shown in Figure 18-17. Their principles of operation and design are very similar. A power steering pump consists of a belt-driven rotor that is turned within an elliptically shaped cam insert ring. Vanes, slippers, or rollers are installed in the rotor slots, grooves, or cavities. Pressure thrust plates on each side of the rotor and cam will seal the pump. This assembly is placed in a housing that contains rotor bearings and oil passages. The pump housing is usually surrounded by an oil reservoir. The pump and reservoir are sealed with O-rings for easily assembled oil-tight joints (Figure 18-18).

Power Steering Pump Operation. In operation, the rotor spins, causing centrifugal force to throw the vanes, slippers, or rollers outward so their outer surfaces maintain contact with the cam. Slippers usually have a backup spring to aid in maintaining cam contact.

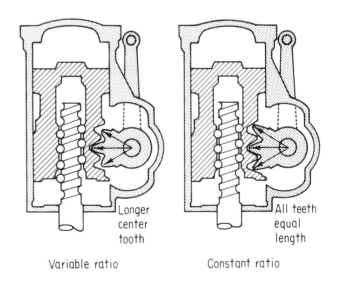

Figure 18-16 Constant steering ratio compared to a variable-ratio steering gear (Courtesy of Buick Motor Division, General Motors Corporation).

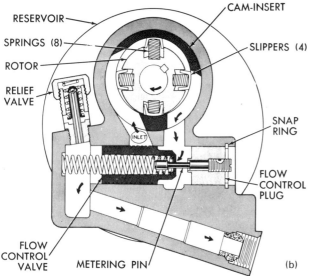

Figure 18-17 Power steering pump types: (a) vane (American Motors Corporation); (b) slipper (Chrysler Motors Corporation); (c) roller (American Motors Corporation).

The cam fits the rotor closely at two opposing locations. This is called a balanced pump. The spaces between the vanes, slippers, or rollers gradually move outward as the rotor turns them past the close fitting point. This portion of the pump is connected to the inlet passage from the reservoir, so fluid will flow from the reservoir into these expanding cavities. As the vanes, slippers, or rollers reach the widest part of the cam insert, they pass and close the inlet passage from the reservoir. They then come in contact with the pressure passage. Continued turning now decreases the volume of the cavity between the vane, slippers, or rollers, forcing the fluid into the pump pressure outlet passage.

Power steering pumps are positive displacement pumps. Each revolution delivers the same amount of fluid, no matter at what speed it is turning. The pump capacity is large enough to supply the required fluid volume and pressure required for steering assist during parking while the engine is idling. On some vehicles it is also supplies fluid pressure for power-brake assist.

Power steering assist requirements are very low when driving at highway speeds. At these speeds, the pump will deliver high fluid volume and pressure, unless the pump output is modified. Modification of the pump output is done by providing the pump with a flow control valve and a pressure relief valve.

Flow Control Valve. The greatest amount of steering effort is required as the steering wheel is turned with the vehicle not moving or while parking. This, therefore, is the operating condition that requires most power assist pressure. Unfortunately, the engine is idling during this time, so the engine-driven power steering pump is also running slowly. Therefore, it must be designed to produce high-pressure at low pump speeds. Pump speeds are fast when the vehicle is operating at highway speeds, when little or no power assist is required. A flow control valve is used to compensate for high pump volumes produced at highway speeds.

The flow control valve, as illustrated in Figure 18-19, is operated by small differences in pressure and a calibrated spring. The passage from the pump outlet contains a restricted opening. Pressure is greater on the pump end of the opening than it is on the steering control valve end. The pressure

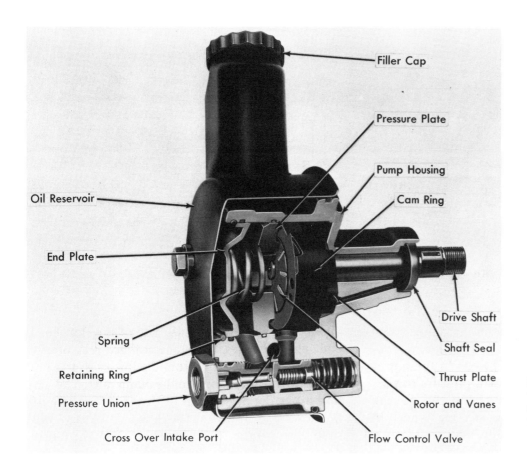

Figure 18-18 Sectional view of a typical vane power steering pump and valves (Courtesy of Cadillac Motor Car Division, General Motors Corporation).

difference across the restricting opening increases when the rate of oil flow increases.

High oil pressure from the pump side of the opening is directed at one end of the flow control valve. Low oil pressure from the steering control valve side of the opening is directed at the other end of the flow control valve. A calibrated spring also is located on the low-pressure side of the flow control valve.

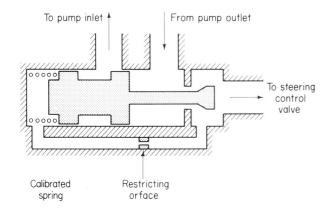

Figure 18-19 Flow control valve in the open position.

Flow Control Valve Operation. As the engine speed increases from idle, both the pump flow and pressure increase. This will cause a pressure drop across the restricted opening. The flow control valve moves toward its low-pressure end (to the left in Figure 18-20) when flow increases to the maximum required. This movement opens a passage between the pump outlet and pump inlet, to bypass a portion of the fluid back to the pump inlet. The opening gets larger as the pump speed increases, thus keeping the flow at the required rate. Recirculating the oil back through the pump reduces the pump power requirement and keeps the oil temperature low. Some control valves have a tapered metering pin that moves in the opening to change its size to match the flow to the steering gear demands. This reduces the effective opening size at high pump speeds. Another type of flow control valve does this by restricting a passage that parallels the opening. Both types provide precise flow control operation.

Pressure Relief Valve. The flow control valve just described operates when there is little restriction

Steering Gears

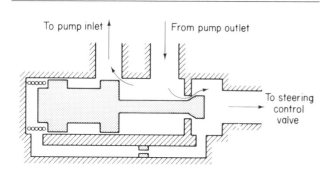

Figure 18-20 Flow control valve in the restricting position.

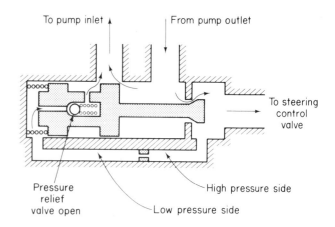

Figure 18-22 Pressure regulator valve in the regulating position.

and low pressure requirements in the power steering system. Flow is restricted by the steering control valve when it sends fluid to one of the power chambers. This causes the system pressure to rise. If the driver turns the wheels against the steering linkage stops and holds the steering wheel in this full turn position, pressure will build to a maximum. Pressure must be limited to a safe value to avoid damage to the power steering unit seals and hoses. The pressure relief valve limits the maximum pressure by opening a passage between the pump outlet passage and either the pump inlet passage or the pump reservoir. In some power steering pumps, the pressure regulator is a separate valve, while in others, it is built into the flow control valve and acts as a pilot valve as illustrated in Figure 18-21. The pressure relief valve opens when pressure on the low-pressure side of the flow control valve increases to a predetermined pressure. It allows the fluid to flow from the low-pressure side of the flow control valve to the pump inlet. This action drops the pressure in the low-pressure side of the flow control valve so the flow control valve will snap open (to the left in Figure 18-22). The pump output fluid can then flow freely into the pump inlet. This action lowers the pump output pressure. Pressure and spring force will balance the valve at the pressure for which it is set.

The power steering pump is connected to the power steering gear control valve with one high-pressure hose and one low-pressure return hose. The high-pressure hose is frequently made in two sizes that will dampen oil pulsations to minimize noise. The ends have screw-type tubing fittings to hold the high pressure. The low-pressure hose, on the other hand, is attached to the steering gear and reservoir with less expensive hose clamps.

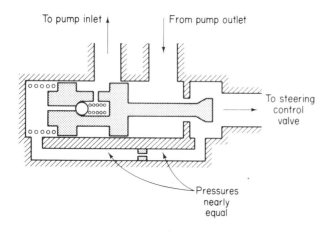

Figure 18-21 Pressure regulator valve during low-pressure operation.

18-4 POWER STEERING SERVICE

Power steering problems are either external fluid leaks or operating problems. Typical external leak points are shown in Figure 18-23. External fluid leaks are the result of damaged oil seals, damaged O-rings, cracked hoses, or cracks in the metal parts. These can be readily seen and corrected by replacing the damaged items.

Power Steering Service

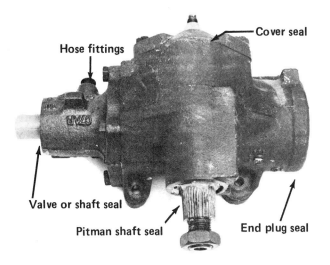

Figure 18-23 Location of external power steering gear leaks.

Problem Diagnosis. The first thing to do in any power steering test is to see that there is adequate fluid in the reservoir and that the drive belt has correct tension. Operational problems involving hard steering or lack of assist will involve the pump assembly, the steering gear assembly, or the steering linkage. The steering linkage should be checked for binding before deciding that the pump or steering gear is at fault. A pump pressure test is used to isolate the problem location. The pressure test unit consists of a valve with a gauge on the high pressure side. This unit is installed between the pump and the steering gear high-pressure hose, as pictured in Figure 18-24. The pressure should rise when the steering gear is turned to the extreme position. If it does not, the valve is closed. The pump is functioning satisfactorily if closing the valve causes the pressure to increase, so the problem is in the steering gear. If, on the other hand, the pressure still does not rise when the valve is closed, the pump is causing the problem.

The most likely cause of pump failure is either the flow control or the pressure valve sticking

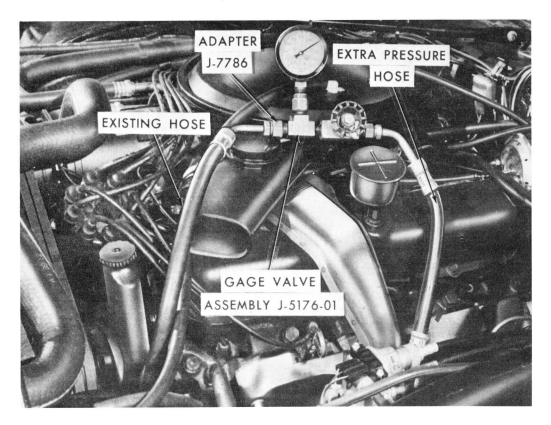

Figure 18-24 Power steering test valve and gauge (Courtesy of Cadillac Motor Car Division, General Motors Corporation).

Steering Gears

open. These can be removed, cleaned, and polished with crocus cloth. Care must be taken to avoid rounding the edges of these valves. If this does not correct the problem the pump is usually replaced.

Self-steering or unequal assist are other problems encountered. This usually results from an improperly centered control valve. Self-steering of the parallel sliding type control valve housing can usually be corrected from the outside by changing the position of the valve housing on the steering gear. Gears using the concentric-type control valve will have to be disassembled and inspected to correct the problem. Unequal assist may also result from internal fluid leaks in the steering gear. The steering gear will have to be disassembled to make either of these repairs. It is the general practice to replace all seals any time the gear is disassembled.

Service Procedure. Before disassembly, be sure to have the service manual that applies to the gear being serviced. Usually, special tools are required for alignment and installing seals without damage. These are *required* for satisfactory servicing. To avoid damage, the disassembly instructions must be carefully followed and all precautions observed.

The gear should be thoroughly cleaned externally before disassembly to avoid contaminating the interior. Fluid is drained and the gear is carefully disassembled, with the service technician noting the condition of parts, and making the required checks as the parts are removed. The parts should be thoroughly cleaned after disassembly, then inspected for looseness, binding, scoring, wear, etc., as described in the applicable service manual.

Any required repairs are made; replacement parts, including all new seals, are obtained; and the parts are lubricated with power steering fluid or other specified lubricant as they are assembled. All domestic passenger car manufacturers specify a special power steering fluid for use in their automobiles. This fluid is essentially a clear fluid with properties similar to automatic transmission fluid. Some manufacturers package their fluid as a combination power steering and automatic transmission fluid. Assembly instructions must be followed carefully to be sure that all parts are put in their proper place with the correct adjustments. After assembly, the steering gear unit should be tested before installation in the vehicle by connecting it to the power steering hoses. With fluid in the system and the engine running, the gear should be operated throughout its range to purge air from the system. Details for each gear type are given in the applicable service manual.

Power steering worm gear preload is set as part of the assembly procedure. The Pitman shaft backlash adjustment is done in the same way as it is done on standard steering gears.

REVIEW QUESTIONS

1. In what way is steering gear ratio important? [18-1]
2. Describe a recirculating ball-type steering gear. [18-1]
3. How is sector backlash adjusted? (18-1)
4. Where is the steering gear positioned to adjust sector backlash? [18-1]
5. Describe a rack and pinion steering gear. [18-1]
6. Describe the operation of an open center power steering control valve. [18-2]
7. What is the advantage of having power steering? [18-2]
8. What are the two types of spool valves used for power steering control? [18-2]
9. What is the purpose of using a torsion bar to operate a spool valve? [18-2]
10. Why is reaction control important in power steering gears? [18-2]
11. What forms the pressure surfaces in an integral recirculating ball-type power steering gear? [18-2]

12. What forms the pressure surfaces in a rack and pinion power steering gear? [18-2]

13. What type of power steering pumps are in common usage? [18-3]

14. What is a positive displacement pump? [18-3]

15. When are power assist requirements the highest? [18-3]

16. How does the operation of the flow control valve differ from the operation of a pressure relief valve? [18-3]

17. How can a problem in a power steering gear be positively identified with a pressure test unit? [18-4]

18. What types of problems require the disassembly of the power steering gear? [18-4]

19. How is air purged from the power steering system after overhaul of the steering gear? [18-4]

Glossary

A

AFTERMARKET: The sales market designed for the consumer after he has purchased a product from the dealer.

AMPLITUDE: The extent or size limit of vibrating motion.

ACKERMAN STEERING: Steering geometry that turns the inner wheel more than the outer wheel to minimize scuff during a turn by producing toe-out on turns.

AERATION: Air mixed with a liquid. Air in the brake fluid or shock absorber.

ASPECT RATIO: The ratio of the width to the length. On tires it is the fully inflated height section divided by the cross section.

AXIAL: Having the same direction or being parallel to the axis of rotation.

ARCH: The curve on a leaf spring. If the center is lower than the ends it is called *positive* arch. If the center is higher than the ends it is called *negative* arch.

ARCING: Grinding new brake linings to the same diameter (arc) as the brake drum surface.

B

BALANCE: Having equal weight on each side of a supporting point. No change in the position of the center of gravity when the part is in motion.

BALL JOINT: A suspension attachment connecting the knuckle to the control arms. Allows movement up and down as well as rotation.

BALL NUT: A casting with internal threads used in steering gear. The ball bearings rolling in the threads move the casting along the worm to rotate a sector, which in turn, moves the Pitman arm.

Glossary

BEAD: A reinforcing ridge. The wire reinforcing around a tire opening where it fits the wheel rim.

BEARING CAGE: Spacer to keep balls and rollers in proper relationship to each other in a bearing.

BELL CRANK: A moving arm that pivots near the middle. It is used to change the direction of motion.

BEARING CONE: The inner race, roller, and cage assembly of a taper roller bearing.

BEARING CUP: The outer race of a taper roller bearing.

BEARING RACE: The surface upon which a roller or balls of a bearing rotate.

BIAS: Lines running at an angle. Cord angle of tire plys. Differences in front and rear vehicle weight and brake system split.

BIAS-BELTED: Tires with the carcass ply cords at an angle, overlayed with reinforcing belts.

BIAS PLY: Tires with the carcass ply cords at an angle. No belts are used.

BINDERS: Compounds that hold the friction materials together in brake linings.

BODY ROLL: The vehicle body leaning sideways as a vehicle turns.

BOOST: An added force applied to the drivers input effort.

BRAKE BIAS: The stopping effort of the front wheels compared to that of the rear wheels.

BRAKE FADE: The loss of braking power due to heat from previous brake applications.

BRINNELLING: A bearing wear condition that results from impact loading or vibration.

BUMP STEER: The steering effect caused by the suspension moving through its travel.

BUSHING: A member that takes up space and usually allows movement at the attachment point. It may be a bearing, or a rubber or elastomer sleeve.

C

CAMBER ANGLE: The outward (positive) or inward (negative) angle of the wheel center line to absolute vertical.

CAM GRINDING: Grinding a brake lining so that it has a smaller diameter on the ends than in the center of the lining.

CASTER: The tendency of a wheel to follow the direction of the pivot point movement.

CASTER ANGLE: The rearward (positive) or forward (negative) angle of the steering axis to absolute vertical.

CASTELLATED NUT: A nut with six raised portions between which a cotter pin is placed to secure the nut (see Fig. 2-27).

CENTER OF GRAVITY: A point around which all weights balance.

CENTER OF PRESSURE: A point around which all pressures balance.

CENTRIFUGAL: A force in a direction away from the turning center of an object that is moving in a curved path.

CHAMFER: An angle cut across the corner of an otherwise sharp edge.

CHASSIS: The vehicle frame, suspension, and running gear.

CHECK VALVE: A valve that will allow flow in one direction and stops flow in the opposite direction.

CHUNKING: Pieces of rubber coming out of the tire tread.

COEFFICIENT OF DYNAMIC FRICTION: The retarding force of the brakes compared to the force used to apply them.

COEFFICIENT OF FRICTION: An expression that relates the force to move an object to the weight or load produced by the object. Coefficient of friction = force/load.

COMPLIANCE: The rearward movement of the suspension as the tire encounters road irregularities.

CONCENTRIC: Located evenly around a common center.

CONICITY: In the shape of a cone.

CONTACT PATCH: The part of a tire that is in contact with the road surface.

CONTROL ARM: A suspension member mounted horizontally with one end attached to the frame and the other end to the knuckle or axle housing.

CORD: A string or thread that makes up the fabric used in tire plys.

CREEP: Slight or slow movement.

CROCUS CLOTH: An extremely fine abrasive cloth that is used for polishing.

CROSS-SHAFT: The Pitman shaft in a steering gear. The shaft that connects the frame end of the control arm.

CUPPING: Heavy, spotty tire wear.

CURB WEIGHT: The weight of a complete vehicle with its normal load, less driver and passengers but with a full tank of fuel.

D

DAMPEN: To slow or reduce oscillations or movement.

DEFLECTION: Bending or moving to a new position as the result of an external force.

DEPENDENT SUSPENSION: Wheel connected through an axle member so that movement of one wheel moves the other wheel.

DIFFERENTIAL: The part of a drive line that allows the wheels to rotate at different speeds as the vehicle is driven through a turn.
DIVE: The front of the vehicle lowering during braking.
DYNAMIC: Moving, as in dynamic load, when the vehicle is moving.
DYNAMIC BALANCE: The center of gravity and the center of rotation remain stationary as the parts move.
DYNAMIC SEAL: A seal to prevent leakage around a moving joint.

E

ECCENTRIC: Two or more circles, one surrounding the other and each having a different center.
ELASTOMER: A rubber-like plastic or synthetic material.
END PLAY: Movement parallel to a shaft or axle.
EQUALIZER LINK: A common connector in the parking brake system that causes both rear brakes to be applied with the same cable tension.
EVASIVE MANEUVER: Rapid steering changes to avoid obstacles in the path of the vehicle.
EXTRUDED: Forced through an opening under great pressure to change the shape of the material.

F

FADE: A condition that occurs when there is little braking effect with full brake pedal force.
FAST IDLE: Engine idle speed when the carburetor fast idle cam is in position. This is normal on an engine while the choke is in operation.
FATIGUE: Natural stress buildup in the structure of the material that leads to failure.
FOOT PRINT: The area on the road that is in contact with the tire.
FRAME: The base structure upon which the rest of the structure is mounted or attached.
FRICTION FORCE: The force that resists slipping or skidding.

G

GALLING: The transfer of material from one part to the contacting part when movement is present.
GASKET: Sealing material placed in a nonmoving joint that prevents leakage.

H

HANDLING: The ease of maneuvering a vehicle without slipping or skidding.
HARSHNESS: A bumpy ride produced by a stiff suspension.
HEEL: The anchor end of a brake shoe.

HONE: Smoothing a metal surface with a fine stone.
HOTCHKISS DRIVE: A rear suspension with an open propeller shaft. Acceleration and braking torques are transferred to the frame through links, control arms, or leaf springs.
HYDRAULICS: A study of fluids in motion.
HYDROPLANING: High speed movement over water causing a tire to be lifted from the road surface and be supported on a wedge of water.
HYGROSCOPIC: Will absorb water.
HYSTERSIS: A retardation of effect, when the forces acting on a body are changed, as from viscous or internal friction.

I

IMPENDING SKID: The tire traction point at which any increase in load will produce tire skid.
INDEPENDENT SUSPENSION: A suspension that allows up and down movement of the wheel without affecting the opposite wheel.
INERTIA: The tendency of a body at rest to stay at rest or when moving to keep moving.
INCLUDED ANGLE: An angle between two intersecting lines.

J

JOUNCE: A compression load on the springs as the space between the frame and axle is reduced.

K

KNOCK BACK: Slight axial movement that pushes the caliper pistons into their bore. This causes clearance between the lining and the rotor.
KNUCKLE: The part of the suspension that connects the control arms and supports the wheel spindle.

L

LATERAL LOAD: The force on the side of the tire tread.
LOAD RANGE: The amount of weight that can be safely carried by a tire.
LEADING SHOE: A brake shoe that has the drum rotating from the toe toward the heel.
LUG: The flange stud on an axle or hub on which the drum and wheel are fastened.

M

MECHANICAL ADVANTAGE: The increased force that the operator is able to apply through a mechanical lever system.

MODULATOR: A device that changes pressures as required in the system.

MULTIVISCOSITY GRADE OIL: An oil that has its viscosity graded at both 0°F (−18°C) and 210°F (99°C).

N

NEUTRAL STEER: A vehicle that will maintain the selected turn with no driver input.

NONDIRECTIONAL SENSE: Steering that does not lead in any direction.

O

OFFSET: Having parallel centerlines on different planes. The wheel rim center plane does not match the bolt flange center plane.

ORIFICE: A carefully sized opening that controls a flow of fluid.

OVERSTEER: The tendency of a vehicle to turn sharper than the turn selected by the driver.

P

PAL NUT: A sheet metal nut placed on a stud on top of a fastening nut to keep it from accidentally loosening.

PANHARD ROD: A control rod that connects the frame on one side of the vehicle to the axle housing on the other side to keep the axle housing centered under the vehicle.

PLY: A layer of cord fabric in a tire carcass.

PLY STEER: The tendency of a tire to always turn in one direction as it rolls. This is the result of the way the tire was constructed.

PORT: An opening or passage through which liquid or gasses flow.

PRELOAD: A load caused when assembling a part before the operating load is applied.

PRESTRESS: Stress that occurs from assembly before the operating stress is applied.

PRESS FIT: Forcing a part into an opening that is slightly smaller than the part to make a solid fit.

PREVENTIVE MAINTENANCE: Performing service and making repairs before failure occurs.

PRIMARY SHOE: A brake shoe moved by a wheel cylinder.

PROPORTIONING: Maintaining the same ratio between input and output pressures throughout the operating range.

PROPORTIONING VALVE: A valve used to reduce the rear brake pressure as a proportion of the front brake pressure.

R

RACE: The part of a bearing on which the rollers or balls roll.

RADIAL: The direction that is moving straight out from the center of a circle.

RADIAL LOAD: A load applied through a bearing from the wheel toward the spindle.

RADIAL PLY: Tire cords running directly across the tire carcass from bead to bead.

RADIAL SPRING RATE: The amount of radial load required to deflect a tire one inch.

REACTION CONTROL: A feed back mechanism that gives the driver a feel of the amount of input effort being applied.

REBOUND: An expansion of a suspension spring after it has been compressed as the result of jounce.

RESONANT FREQUENCY: The natural vibrating characteristic of a component.

RIDE: The characteristic feel as one rides in a vehicle.

ROLL STEER: The steering effect as a result of body lean during a turn.

RUN OUT: The amount of wobble a shaft or disc has as it rotates.

S

SATURATES: Materials that fully saturate a lining so no openings are left that could form pockets for water to enter.

SCRUB RADIUS: The distance on the road surface under the front tire between an extension of the pivot axis and the center of weight.

SCUFF: Tire slide on the road surface during operation.

SCUFF TRAVEL: The amount of side travel of the tire as the wheel moves from maximum jounce to maximum rebound.

SECONDARY SHOE: A brake shoe that is operated by a primary shoe.

SELF-ALIGNING TORQUE: The natural tendency of the tire to return to the neutral position after being turned.

SERIES: The designation of a tire aspect ratio.

SERVO ACTION: A large output that results from a small input action.

SHIMMY: A violent front wheel shake.

SHOE: The part of a brake that supports the lining.
SHOP CLOTH: A wiping cloth especially made for use in service shops. It is usually provided by and cleaned by a laundry service.
SHOT PEEN: High-pressure blasting a part with steel shot to toughen the surface and increase the parts strength.
SIPES: Slits in the tire tread to produce more blade surface for traction.
SKID: A tire sliding on the road surface.
SLAVE CYLINDER: A cylinder that operates as a result of the action of a primary cylinder.
SLIP ANGLE: The angle between the tire heading and the actual direction of movement.
SPALLING: A fatigue-related action causing the bearing surface material to break away from the base metal.
SMEARING: Shiny polished areas caused by a bearing race slipping in its mount.
SPRING RATE: The amount of force required to deflect a spring one inch, rated in lbs/inch.
SPRUNG WEIGHT: The weight supported by a spring.
SPUR GEAR: The simplest form of toothed wheel with radial teeth parallel to its axis.
SQUIRM: The twist of the tire tread in the foot print.
STAR WHEEL: An adjustable link between the primary and secondary brake shoes.
STATIC BALANCE: Balance that causes no movement when the wheel is stopped in any position.
STATIC SEAL: A seal on a joint that has no movement.
STEER ANGLE: The angle the wheels are turned to from straight ahead.
STEERING AXIS: A line connecting the center of the ball joints and extended to the road surface.
STEERING AXIS INCLINATION (SAI): The inward tilt of the front wheel pivot or steering axis.
STEERING RATIO: The number of degrees the steering wheel is turned divided by the number of degrees the wheels are turned.
SUMMATION: Adding the effective action of the parts together to determine the effect of their combined action.
SUSPENSION COMPLIANCE: Rearward and upward movement of the suspension when the tire meets an obstacle in the road surface.
SUSPENSION GEOMETRY: The angular action of the suspension as it goes from its static position to the extremes of travel (compared to vertical lines).

T

TANG: Projection from a part used for alignment, as of a driving lug.
TIRE CONTACT PATCH: The part of a tire that contacts the road surface.
TIRE FOOTPRINT: The area on the road in contact with the tire.

TIRE FORCE VARIATION: Changes in the tire's radial spring rate as it rolls under radial loads.
TIRE SERIES: The groupings of tire sizes having the same aspect ratio.
TIRE SLIP: A slight tire slide turning a turn.
TOE: (1) The leading edge of the brake shoe. (2) The angle between the centerlines of the front tires.
TORQUE: Twisting force on a shaft.
TORSIONAL LOAD: Loads on the brakes and suspension caused by torque.
TORSIONAL VIBRATION: Back and forth motion around a turning center.
TRAILING SHOE: A brake shoe with its anchor at the toe end.
TRACTIVE FORCE: The friction force in the contact patch that causes torque on the wheel.
TRAMP: The uncontrolled vertical motion of a tire, usually caused by imbalance.
TRIM HEIGHT: Specified level vehicle height above the road surface.

U

UNDERSTEER: The tendency of the vehicle to not turn as much as the wheels are turned.

V

VACUUM SUSPENDED: A brake system with vacuum on both sides of the diaphragm when the booster is in the released position.
VECTOR SUM: The resulting effect of a number of forces on a point determined by drawing the force angles with the lengths of the lines proportional to the amount of forces.
VISCOSITY: Thickness or body of oil.
VULCANIZING: A process of bonding and curing rubber; used in tire repairs to replace damaged rubber. Involves using heat and pressure.

W

WADDLE: A sideways vehicle shake as a faulty radial tire moves slowly.
WANDER: The tendency of a vehicle to randomly drift in one direction or the other.
WHEEL ALIGNMENT: The process of setting the specified front suspension and steering geometry.

Glossary

WHEEL FIGHT: The tendency of a steering system to be easily deflected by uneven road surfaces; causes changes in toe that result in tire wear.

WHEEL OFFSET: The distance between the wheel attachment flange and the wheel rim center plane.

WHEEL SHIMMY: A violent front wheel shake caused by overcorrection by the steering geometry.

WHEEL SIDEWAYS DISPLACEMENT: Sideways movement of the wheel as the suspension goes from jounce to rebound.

WHEEL SLIP: Sideways movement of the tire tread across the foot print.

WORKING ANGLE: The angle through which the universal joint must operate.

Index

A

Ackerman layout, 230
Adjusting alignment, 239
Air leveling compressors, 205
Alignment rack, 235
American Petroleum Institute (API), 13
    gear lubricants, 14
Angular linkages, 189
Anti-roll bar, 195
Anti-skid system, 110
Anti-sway bar, 195
API, 13
API-GL, 14
Axle, 61
    full floating, 60
    housing, 62
    loads, 62

B

Backing plate service, 136
Back lash, steering gear, 279
Balance, static, 47
Balancing valve, 106
Ball joints, 5, 202
    angle, 226
    inspection, 246
    removal, 252
Bell crank steering linkage, 232
Bearing, 58
    axle, 61
    ball, 58
    cup, 61
    fluid film, 58
    needle, 58
    problem diagnosis, 65
    race, 61
    roller:
        straight, 58
        taper, 59
    service, 64
    support, 60
Bias, front weight, 218
Bleeder valve, 98

Index

Body construction, 1
  separate body and frame, 1
  stub frame, 2
  unitized body, 1
Body roll, 26
Brake:
  alternate designs, 82
  application force, 102
  bias, 86
  bleeding, 160
    general procedures, 161
    gravity bleeding, 132, 163
    pedal bleeding, 163
    pressure bleeding, 162
  design considerations, 77
  disassembly, 129
  disc, 82
  dual servo design, 80
Brake drum:
  cracking, 123
  discard dimension, 122
  hard spots, 123
  inspection, 120
  out-of-round, 122
  removal, 116
  resurfacing, 126
  taper, 122
  wear, 120
Brake dual master cylinder, 94
Brake fade, 77, 78
Brake fluid, 14, 109
  boiling point, 109
  inspection, 116
  requirements, 109
  silicone, 109
  viscosity, 109
Brake force-to-pressure ratio, 93
Brake heat loads, 77
Brake hoses, 107
  inspection, 120
  replacement, 132
Brake hydraulic system, 93
Brake lines, 107
  inspection, 120
Brake lining, 85
  coefficient of friction, 86
  contamination, 117
  edge coating, 85
  glazed, 119
  heat cracks, 119

Brake lining *(cont.)*:
  identification, 85
  inspection, 117
  materials, 86
  replacement, 133
  surface area, 86
  wear, 117
Brake master cylinder size, 93
Brake pedal fade, 116
Brake pedal reserve, 77
Brake proportioning, 82
Brake self-energizing, 80, 81
Brake shoe:
  adjustment, 140
  arcing, 136
  cam grinding, 134
  installation, 138
  radius grinding, 135
Brake split systems, 94
Brake struts, 192
Brake system:
  flushing, 160
  inspection, 115
Brake tubing, 108
Brake types, 78
Brake warning light, 104, 163
  centering the switch, 164
Brake wheel cylinders, 98
Braking requirements, 77
Braking standards, 77
Braking torque, 213
Brinelling, 67
Bubble balancer, 47
Bump steer, 216
Bushing preload, 251, 258
Bushing, suspension, 198

C

Caliper brake shoe removal, 145
Caliper design, 100
Caliper piston, 101
  force, 99
  knock back, 101
  removal, 149
  seal, 101
Caliper reassembly, 151
Caliper removal, 145
Caliper service, 149
Camber angle:
  adjustment, 240
  defined, 210, 212
  effect, 211, 224
  measurement, 236

Caster:
    adjusting, 240
    effects, 225
    measurement, 237
Center link replacement, 260
Chassis lubrication, 15, 16
Coil springs, 187
Combination valve, 107
Compliance, 198
Connecting shaft, 74
Control arm servicing, rear, 270
Control arm bushing removal, 256, 257
Curb weight, 233

D

Department of Transportation (DOT), 14
Dependent-type front suspension, 182
Differential, 4
Directional drift, 242
Disc brakes, 82
    assembly, 151
    caliper, 83, 99
    disassembly, 144
    inspection, 123
    lining, 101
    operation, 83
    rotor parallelism, 124
    rotor runout, 123
    DOT, 14
Drag link, 232
    steering linkage, 233
Drill use, 10
Drive axle service, 68
    C-lock type, 69
    retainer plate type, 70
Drive line, 3
    working angle, 270
Drum brake:
    automatic adjuster, 79
    design, 78
    names of parts, 78, 79
    operation, 80
    wheel cylinders, 98
Dynamic loading, 224

E

Engine, 2
    diesel, 2
    oil, 13
    rotating combustion chamber, 3
    turbine, 3
    turbocharger, 2

F

Fade, brake, 77, 78
Fluid:
    automatic transmission, 13
    brake, 14
    gear oils, 14
    power steering, 14
Flushing the brake system, 160
Forward weight transfer, 213
Front brake rotor removal, 146
Front suspension:
    dependent-type, 182
    independent-type, 183
    inspection, 245
    spring removal, 256
Front weight bias, 218

G

Galling, 67
Gaskets, 64
Gravity bleeding, 132
Grease, 14

H

Half shaft, 74
Handling, 209
High ratio steering, 181
Hoists, 12
Hold-off valve, 105
Hydraulic brake booster operation, 174
Hydraulic control valve, 153
Hydraulic power brake booster, 172
Hydraulic principles, 91
Hydraulic valves, 102
Hydroplaning, 26, 30

I

I-beam axle, 210
Idler arm, 231
    service, 261
Independent front suspensions, 183
Included angle, 238
Instant center, 195

J

Jack:
    floor jack, 12
    hydraulic jack, 11
    one-post bumper jack, 10

K

King pin, 182, 202
King pin inclination, 226

L

Leaf spring:
   center bolt, 185
   failure, 186
   interleaves, 185
   monoleaf, 185
   multiple leaf, 185
   service, 268
   shape, 186
Leading arms, 194
Level control systems, 204
Loaded runout, 51
Long-short arm suspension, 210
Low speed harshness, 221
Lubrication, chassis, 16
Lubrication procedures, 16
Lubricants, 13
Lug bolt replacement, 22

M

MacPherson strut, 195
   shock strut service, 249
Maintenance inspection, 16
Master cylinder, 94
   bench bleeding, 159
   compensation port, 96
   design, 95
   disassembly, 157
   inspection, 116
   installation, 160
   leakage, 156
   operation, 95
   partial system failure, 97
   pushrod adjustment, 156
   pump-up, 96
   service, 156
Metering valve, 105
   checking, 154

N

National Lubricating Grease Institute (NLGI), 14
Neutral steer, 218
NLGI, 14
Nondirectional sense, 223

O

Oversteer, 33, 218

P

Panhard rod, 189
Parking brakes, 87
   cable service, 137
   caliper, 88
   operation, 87
   system, 88
Personal safety, 8
Pitman arm replacement, 261
Ply steer, 40
Power brake:
   booster:
      installation, 179
      removal, 178
   dual power system, 176
   hydraulic assisted, 169
   on-the-vehicle testing, 177
   need for, 167
   pushrod check, 160
   reaction disc, 171
   runout, 168
   service, 177
   tandem boosters, 167
   theory, 166
   vacuum assist, 167
      multiplier type, 172
   vacuum reserve, 171
   vacuum suspended, 169, 170
Power steering, 182
   ball-nut power piston, 284
   concentric spool valve, 282
   control valve, 280
   gear, 280
   fluid, 290
   power chambers, 284
   pumps, 285
      flow control, 286
      operation 173
      pressure relief valve, 287
   problem diagnosis, 289
   reaction control, 284
   service, 288, 290
   rotary spool valve, 283
   sliding spool valve, 282
Pressure differential switch centering, 164
Pressure differential valve, 103
Proportioning valve, 106
   checking, 155

R

Radial spring rate, 32
   Rack and pinion steering linkage, 233

Radius arm, 192
  service, 251
Rear brake rotor removal, 146
Rear coil spring replacement, 266
Rear control arm service, 270
Rear stabilizer bar service, 266
Rear suspension alignment, 272
Rear wheel alignment, 273
Residual check valve, 103
Rear wheel toe measurement, 274
Ride height, 216
  adjustment, 259
Ride quality, 212, 220
Road clearance, 220
Roll axis, 195
Roll bar, 195
Roll center, 195
Roll steer, 216
Rotor removal, front, 146
Rotor resurfacing, 147
Running gear, 4

S

SAE, 13, 14
SAI, 226
  measurement, 237
Safety glasses, 8
Safety habits, 7
Safety stands, 8, 10, 12
Scrub radius, 227
Scuff travel, 211
Seals, 63
Shimmy, 243
Shock absorbers, 199
  dampening control, 201
  gas filled, 200
  level-type, 204
  mounting, 201
  operation, 200
  service, 248, 264
Sideways displacement, 210
Slop angle, 217, 230
Smearing, bearing, 67
Society of Automotive Engineers (SAE), 13
Spalling, 63, 66
Spindle, 4
Spring:
  leaf spring service, 268
  rear coil spring replacement, 226
  types, 4, 185
Spring rate:
  definition, 184
  variable, 188
Spring requirements, 184, 185
Sprung weight, 183

Stabilizer bar, 195
  front service, 251
  rear service, 266
Standard steering gear, 278
Steer angle, 218
Steering arm replacement, 262
Steering axis inclination (SAI), 266
  measurement, 237
Steering gear:
  rack and pinion, 233, 279
  recirculating ball type, 278
  sector backlash, 279
  worm and sector, 278
Steering knuckle, 4
Steering linkages, 231
  center link replacement, 260
  drag link, 233
  characteristics, 215
  Haltenberger, 232
  service, 259
Steering pivots, 202
Steering ratio, 181
Steering system geometry, 224
Strut bar service, 251
Strut bushing service, 251
Suspension:
  assist, 185
  bushings, 198
  characteristics, 210
  control devices, 188
  geometry, 210
  types, 182
Sway bar, 195

T

Tie rod, 215
  service, 259
Tire:
  aspect ratio, 35
  balance, 47
  bias belted, 27
  bias ply, 27
  characteristics, 26
  construction, 27
  design, 27
  deterioration, 34
  dismounting, 54
  dynamic balance, 50
  effect on fuel economy, 34
  failure, 37

Index

Tire *(cont.):*
  flexing, 32
  footprints, 4
  force variation, 45
  grinding, 51
  hysteresis loss, 34
  inspection, 37
  lateral runout, 43
  load range, 35
  loaded runout, 45
  loaded runout correction, 51
  loads, 25
  materials, 29
  metric sizes, 35
  mounting, 55
  performance, 30
  ply steer, 40
  radial runout, 44
  radial ply, 27
  radial spring rate, 32
  repair, 52
  replacement, 34
  requirements, 25
  retreading, 56
  ride quality, 32
  rotation, 42
  self-aligning torque, 32
  series, 35
  service, 42
  sidewall rubber, 29
  sipes, 31
  slip angle, 31, 32, 230
  squirm, 29
  specifications, 34
  steering sensitivity, 32
  tread rubber, 29
  tread wear indicators, 53
  trimming, 51
  truing, 51
  waddle, 40
  wear, 37
  wet traction, 31
Toe, 215, 229
  adjusting, 241
  measurement, 238
Toe-out on turns, 230
  measurement, 238
Tool condition, 9
Tool selection, 9
Torque tube, 189, 190, 192, 193

Torsion bar, 186
Towing the vehicle, 23
Track alignment, 275
Track bars, 189
  service, 266
Trailing arm, 189, 192
Transmission, 3
Trim height, 204
Turning radius, 230
Twin I-beam front suspension, 183

U

Understeer, 33, 217

V

Vacuum booster pushrod, 179
Vacuum power brake operation, 169
Variable spring rate, 188
Vehicle curb weight, 233
Vehicle spring requirements, 185
Vehicle steer, 217
Viscosity, 13

W

Wander, 242
Weight transfer, 213
Wheel alignment, 223, 233
Wheel bearing:
  adjustment, 20
  assembly, 68
  disassembly, 66
  maintenance, 17
  packing, 19
  preload, 20
  service, 66
Wheel balance, 47
Wheel cylinder assembly, 132
Wheel cylinder bleeder valve, 98
Wheel cylinder service, 130
Wheel drop center, 54
Wheel dynamic balance, 50
Wheel failure, 46
Wheel fight, 216
Wheel hub removal, 17
Wheel installation, 21
Wheel lateral run out, 44
Wheel lug nuts, 21
Wheel pull, 40
  during braking, 242
Wheel radial runout, 44
Wheel rim size, 34
Wheels, 36
Wheel slip control, 110
  brake modulators, 112

Wheel slip control *(cont.)*:
   components, 111
   speed sensor, 112
Wheel shimmy, 50
Wheel service, 46
Wheel spindle, 60
Weight transfer, 213